WIDEBAND BEAMFORMING

Wiley Series on Wireless Communications and Mobile Computing

Series Editors: Dr Xuemin (Sherman) Shen, *University of Waterloo, Canada*
Dr Yi Pan, *Georgia State University, USA*

The "Wiley Series on Wireless Communications and Mobile Computing" is a series of comprehensive, practical and timely books on wireless communication and network systems. The series focuses on topics ranging from wireless communication and coding theory to wireless applications and pervasive computing. The books provide engineers and other technical professionals, researchers, educators, and advanced students in these fields with invaluable insights into the latest developments and cutting-edge research.

Other titles in the series:

Misic and Misic: *Wireless Personal Area Networks: Performance, Interconnection, and Security with IEEE 802.15.4*, January 2008, 978-0-470-51847-2

Takagi and Walke: *Spectrum Requirement Planning in Wireless Communications: Model and Methodology for IMT-Advanced*, April 2008, 978-0-470-98647-9

Pérez-Fontán and Espiñeira: *Modeling the Wireless Propagation Channel: A simulation approach with MATLAB®*, August 2008, 978-0-470-72785-0

Ippolito: *Satellite Communications Systems Engineering: Atmospheric Effects, Satellite Link Design and System Performance*, August 2008, 978-0-470-72527-6

Lin and Sou: *Charging for Mobile All-IP Telecommunications*, September 2008, 978-0-470-77565-3

Myung and Goodman: *Single Carrier FDMA: A New Air Interface for Long Term Evolution*, October 2008, 978-0-470-72449-1

Wang, Kondi, Luthra and Ci: *4G Wireless Video Communications*, April 2009, 978-0-470-77307-9

Cai, Shen and Mark: *Multimedia Services in Wireless Internet: Modeling and Analysis*, June 2009, 978-0-470-77065-8

Stojmenovic: *Wireless Sensor and Actuator Networks: Algorithms and Protocols for Scalable Coordination and Data Communication*, February 2010, 978-0-470-17082-3

Liu and Weiss, *Wideband Beamforming – Concepts and Techniques*, March 2010, 978-0-470-71392-1

Riccharia and Westbrook, *Satellite Systems for Personal Applications: Concepts and Technology*, July 2010, 978-0-470-71428-7

Hart, Tao and Zhou: *Mobile Multi-hop WiMAX: From Protocol to Performance*, October 2010, 978-0-470-99399-6

Qian, Muller and Chen: *Security in Wireless Networks and Systems*, November 2010, 978-0-470-512128

WIDEBAND BEAMFORMING CONCEPTS AND TECHNIQUES

Wei Liu
University of Sheffield, UK

Stephan Weiss
University of Strathclyde, UK

WILEY
A John Wiley and Sons, Ltd., Publication

This edition first published 2010

Registered office
John Wiley & Sons Ltd, The Atrium, Southern Gate, Chichester, West Sussex, PO19 8SQ, United Kingdom

For details of our global editorial offices, for customer services and for information about how to apply for permission to reuse the copyright material in this book please see our website at www.wiley.com.

Library of Congress Cataloging-in-Publication Data

Liu, Wei, 1974-
Wideband beamforming : concepts and techniques / Wei Liu, Stephan Weiss.
p. cm.
Includes bibliographical references and index.
ISBN 978-0-470-71392-1 (cloth)
1. Beamforming. 2. Antenna radiation patterns. 3. Adaptive antennas. 4. Adaptive signal processing.
5. Adaptive filters. 6. Broadband communication systems. I. Weiss, Stephan, 1968- II. Title.
TK7871.67.A33L58 2010
621.382′2 – dc22

2009052109

A catalogue record for this book is available from the British Library.

ISBN 978-0-470-71392-1 (H/B)

Typeset in 10/12 Times by Laserwords Private Limited, Chennai, India.
Printed and bound in the United Kingdom by CPI Antony Rowe, Chippenham, Wiltshire.

In memory of my mother

Wei Liu

About the Series Editors

Xuemin (Sherman) Shen (M'97–SM'02) received the B.Sc degree in electrical engineering from Dalian Maritime University, China in 1982, and the M.Sc. and Ph.D. degrees (both in electrical engineering) from Rutgers University, New Jersey, USA, in 1987 and 1990 respectively. He is a Professor and University Research Chair, and the Associate Chair for Graduate Studies, Department of Electrical and Computer Engineering, University of Waterloo, Canada. His research focuses on mobility and resource management in interconnected wireless/wired networks, UWB wireless communications systems, wireless security, and ad hoc and sensor networks. He is a co-author of three books, and has published more than 300 papers and book chapters in wireless communications and networks, control and filtering. Dr Shen serves as a Founding Area Editor for IEEE Transactions on Wireless Communications; Editor-in-Chief for Peer-to-Peer Networking and Application; Associate Editor for IEEE Transactions on Vehicular Technology; KICS/IEEE Journal of Communications and Networks, *Computer Networks*; ACM/Wireless Networks; and Wireless Communications and Mobile Computing (Wiley), etc. He has also served as Guest Editor for IEEE JSAC, IEEE Wireless Communications, and IEEE Communications Magazine. Dr Shen received the Excellent Graduate Supervision Award in 2006, and the Outstanding Performance Award in 2004 from the University of Waterloo, the Premier's Research Excellence Award (PREA) in 2003 from the Province of Ontario, Canada, and the Distinguished Performance Award in 2002 from the Faculty of Engineering, University of Waterloo. Dr Shen is a registered Professional Engineer of Ontario, Canada.

Dr Yi Pan is the Chair and a Professor in the Department of Computer Science at Georgia State University, USA. Dr Pan received his B.Eng. and M.Eng. degrees in computer engineering from Tsinghua University, China, in 1982 and 1984, respectively, and his Ph.D. degree in computer science from the University of Pittsburgh, USA, in 1991. Dr Pan's research interests include parallel and distributed computing, optical networks, wireless networks, and bioinformatics. Dr Pan has published more than 100 journal papers with over 30 papers published in various IEEE journals. In addition, he has published over 130 papers in refereed conferences (including

IPDPS, ICPP, ICDCS, INFOCOM, and GLOBECOM). He has also co-edited over 30 books. Dr Pan has served as an Editor-in-Chief or an editorial board member for 15 journals including five IEEE Transactions and has organized many international conferences and workshops. Dr Pan has delivered over 10 keynote speeches at many international conferences. Dr Pan is an IEEE Distinguished Speaker (2000–2002), a Yamacraw Distinguished Speaker (2002), and a Shell Oil Colloquium Speaker (2002). He is listed in Men of Achievement, Who's Who in America, Who's Who in American Education, Who's Who in Computational Science and Engineering, and Who's Who of Asian Americans.

Contents

Preface

Beamforming is a spatial filtering technique for receiving signals illuminating an array of sensors from some specific directions, whilst attenuating signals from other directions. Depending on the signal bandwidth, it can be divided into two categories: narrowband beamforming and wideband beamforming. For narrowband beamforming, it is achieved by an instantaneous linear combination of the received array signals. However, when the involved signals are wideband, we have to employ an additional processing dimension for effective operation, such as tapped delay-lines (or FIR/IIR filters), or the recently proposed sensor delay-lines, which lead to a wideband beamforming system.

Wideband beamforming has been studied extensively in the past due to its applications in various areas ranging from radar, sonar, microphone arrays, radio astronomy, seismology, medical diagnosis and treatment, to communications. In particular, since speech/sound is a natural source of wideband signals, much of the research and development in wideband beamforming has been focused on the area of microphone arrays.

Traditionally, beamforming is considered as part of the wider area of array signal processing and chapters relating to beamforming can be found in many books on array signal processing. Recently, due to its importance in the wireless communications area, there have been some books dedicated to beamforming in the form of smart antenna techniques.

However, since in many current wireless communication applications the signal bandwidth is still relatively narrow, almost all of the books within the smart antenna literature are focused on narrowband beamforming and the topic of wideband beamforming is by and large ignored. With the introduction of ultra-wideband systems, one or two chapters on wideband beamforming have recently appeared in books about ultra-wideband communications and wideband radar, etc. With the increasing importance of wideband beamforming and recent advances in this area, it appears timely to have a book dedicated to this topic for the benefit of the wireless communications community. However, the concepts and techniques presented in this book for wideband beamforming are general and not limited to the wireless communications area, or any other specific applications.

There has been a huge amount of work going on in the past half a century in the area of wideband beamfoming and it is impossible to cover all of them in the first attempt of producing a single book dedicated to this area. Although we have tried our best to give an extensive review of this topic in Chapters 1, 2 and 4 about both fixed and adaptive beamforming techniques, the remaining part of the book is mostly based on our own research over the past ten years in this area. Our primary goal is to give a systematic introduction to the various concepts and techniques in wideband beamforming in the form of a self-contained monograph and also present some of the most recent research and development in this area.

The contents of the book are organized into eight chapters.

Chapter 1 is a brief introduction to the general area of array signal processing, including both narrowband and wideband beamforming, with a detailed analysis for the beam steering process for both cases. It will be shown that unlike the narrowband case, where the steered beam response is a circularly shifted version of the original one given a half wavelength spacing, a more complicated relationship exists for a wideband beamformer.

In Chapter 2, we will study a range of basic approaches to adaptive wideband beamforming. The latter can be achieved by a standard adaptive filtering structure when a reference signal is available. When we know the direction of arrival (DOA) angle of the signal of interest, a linearly constrained minimum variance (LCMV) beamformer can be constructed and realized by either a constrained adaptive algorithm or an unconstrained one through the structure of a generalized sidelobe canceller (GSC). In addition to the standard LCMV beamformer, two other minimum variance beamformers will also be studied, including the soft-constrained beamformer and the correlation constrained beamformer. To improve the robustness of the beamformer in the presence of steering vector errors, the topic of robust adaptive beamforming is addressed at the end of this chapter.

Chapter 3 is focused on various subband techniques and structures for adaptive wideband beamforming, which can normally achieve a higher convergence rate and a lower computational complexity. Since the discrete Fourier transform (DFT) and inverse DFT (IDFT) pair can be considered as a simple maximally decimated filter banks system, frequency-domain adaptation techniques are also studied in this chapter.

Chapters 4 and 5 are devoted to the fixed wideband beamformer design problem, with Chapter 4 for a general design using the iterative optimisation method, the least squares method and the eigenfilter method, and Chapter 5 for a special class of fixed wideband beamformers – the frequency invariant beamformer. The design of a frequency invariant beamformer can be achieved by many different methods and at the end of Chapter 5, an application of the frequency invariant beamforming technique to the adaptive wideband beamforming problem is also studied, which leads to a beamspace adaptive wideband beamformer.

Chapter 6 is focused on a different class of adaptive beamformers: the blind adaptive wideband beamformer, which is based on the concept of blind source separation. For this class of beamformers, neither a reference signal nor the DOA information of the desired signal is needed and only some assumptions on the statistical properties of the source signals are required.

In Chapter 7, we will introduce a totally different approach to wideband beamforming based on the recently proposed sensor delay-line system. A special property of the resultant wideband beamforming structure is that there is not any form of temporal processing required, such as tapped delay-lines or FIR/IIR filters. Therefore it can be considered as a wideband beamforming structure with spatial-only information. Most of the techniques developed for the traditional wideband beamformers can be applied to this new structure directly. However, further studies are needed in the future to fully exploit its potential for wideband beamforming in various signal environments and applications.

In the last chapter, Chapter 8, we will study the wideband beamforming problem in a multipath environment. For the case with a small number of multhpath signals, two solutions will be provided employing the wideband beamspace adaptive beamforming structure studied in previous chapters. When a large number of multipath signals is present to the beamformer, we will have a generalized signal mixing problem independent of the array geometry and the original array system can be considered as a general multiple input multiple output (MIMO) system. A brief introduction to the MIMO system is then provided from the viewpoint of beamforming at the end.

Acknowledgements

I would like to thank Professor S. C. Chan and Dr K. L. Ho for introducing me to the research area of signal processing and in particular filter banks and wavelets when my background was still mainly in space physics. I am very grateful to my Ph.D. supervisors Dr S. Weiss and Professor L. Hanzo. With their support I moved to the area of wideband beamforming which has defined a major part of my research life today. Many thanks are due to Professor S. Chen who has been supporting me in various ways since the very start of my studies at the University of Southampton, UK and Dr D. P. Mandic who led me into the area of blind source separation during my postdoctoral research work, which opened up a new horizon for my work in wideband beamforming. I would also like to thank Professors P. A. Houston, B. Chambers and R. J. Langley in the Department of Electronic and Electrical Engineering at the University of Sheffield, UK for their support and help, especially during my 'starting years' of my academic career at the University of Sheffield. Special thanks must go to Professors I. K. Proudler, J. G. McWhirter, A. Cichocki and R. Wu, Dr D. C. McLernon and Professor M. Ghogho. Much of the research presented here was conducted in collaboration with them. Thanks also must go to Dr Y. Zhang and Professors R. W. Stewart and J. Li for their valuable comments and suggestions during the preparation of this book.

Wei Liu
University of Sheffield, UK

1

Introduction

1.1 Array Signal Processing

Array signal processing is one of the major areas of signal processing and has been studied extensively in the past due to its wide applications in various areas ranging from radar, sonar, microphone arrays, radio astronomy, seismology, medical diagnosis and treatment, to communications (Allen and Ghavami, 2005; Brandstein and Ward, 2001; Fourikis, 2000; Haykin, 1985; Hudson, 1981; Johnson and Dudgeon, 1993; Monzingo and Miller, 2004; Van Trees, 2002). It involves multiple sensors (microphones, antennas, etc.) placed at different positions in space to process the received signals arriving from different directions. An example for a simple array system consisting of four sensors with two impinging signals is shown in Figure 1.1 for illustrative purposes, where the direction of arrival (DOA) of the signals is characterized by two parameters: an elevation angle θ and an azimuth angle ϕ.

We normally assume the array sensors have the same characteristics and they are omnidirectional (or isotropic), i.e. their responses to an impinging signal are independent of their DOA angles. According to the relative locations of the sensors, arrays can be divided into three classes (Van Trees, 2002):

- one-dimensional (1-D) arrays or linear arrays;
- two-dimensional (2-D) arrays or planar arrays;
- three-dimensional (3-D) arrays or volumetric arrays.

Each of them can be further divided into two categories:

- regular spacing, including uniform and nonuniform spacings;
- irregular or random spacing.

Our study in this book will be based on arrays with regular spacings.

For the impinging signals, we always assume that they are plane waves, i.e. the array is located in the far field of the sources generating the waves and the received signals have a planar wavefront.

Now consider a plane wave with a frequency f propagating in the direction of the z-axis of the Cartesian coordinate system as shown in Figure 1.2. At the plane defined

Wideband Beamforming Wei Liu and Stephan Weiss

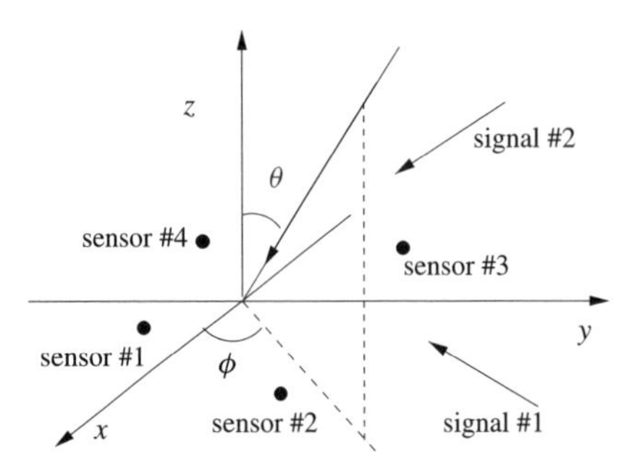

Figure 1.1 An illustrative array example with four sensors and two impinging signals

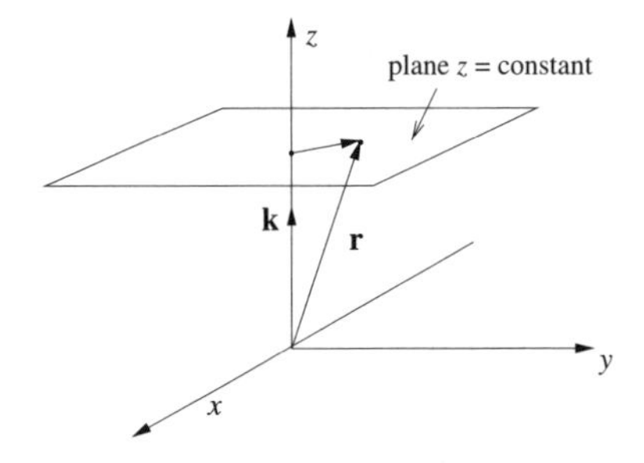

Figure 1.2 A plane wave propagating in the direction of the z-axis of the Cartesian coordinate system

by $z =$ constant, the phase of the signal can be expressed as:

$$\phi(t, z) = 2\pi f t - kz \tag{1.1}$$

where t is time and the parameter k is referred to as the wavenumber and defined as (Crawford, 1968):

$$k = \frac{\omega}{c} = \frac{2\pi}{\lambda} \tag{1.2}$$

where ω is the (temporal) angular frequency, c denotes the speed of propagation in the specific medium and λ is the wavelength. Similar to ω, which means that in a temporal interval t the phase of the signal accumulates to the value ωt, the interpretation of k is that over a distance z, measured in the propagation direction, the phase of the signal accumulates to kz radians. As a result, k can be referred to as the spatial frequency of a signal.

Different from the temporal frequency ω, which is one-dimensional, the spatial frequency k is three-dimensional and its direction is opposite to the propagating direction of the signal. In a Cartesian coordinate system, it can be denoted by a three-element vector:

$$\mathbf{k} = [k_x, k_y, k_z]^T \tag{1.3}$$

with a length of:

$$k = \sqrt{k_x^2 + k_y^2 + k_z^2} \tag{1.4}$$

This vector is referred to as the wavenumber vector. In the case shown in Figure 1.2, we have $k_x = k_y = 0$ and $k_z = -k$. Let $\hat{\mathbf{z}} = [0, 0, 1]^T$ denote the unit vector along the z-axis direction, then we have $\mathbf{k} = -k\hat{\mathbf{z}}$.

These two quantities are not independent of each other and as shown in Equation (1.2), they are related by the following equation:

$$k = \frac{2\pi f}{c} \tag{1.5}$$

Any point in a 3-D space can be represented by a vector $\mathbf{r} = [r_x, r_y, r_z]^T$, where r_x, r_y and r_z are the coordinates of this point in the Cartesian coordinate system. With the definition of the wavenumber vector $\mathbf{k}$, the phase function $\phi(t, \mathbf{r})$ of a plane wave can be expressed in a general form:

$$\phi(t, \mathbf{r}) = 2\pi f t + \mathbf{k}^T \mathbf{r} \tag{1.6}$$

For the case in Figure 1.2, we have:

$$\mathbf{k}^T \mathbf{r} = -k(\hat{\mathbf{z}}^T \mathbf{r}) = -k r_z \tag{1.7}$$

Therefore, as long as the points have the same coordinate r_z in the z-axis direction, they have the same phase value at a fixed time instant t.

For the general case, where the signal impinges upon the array from an elevation angle θ and an azimuth angle ϕ, as shown in Figure 1.1, the wavenumber vector $\mathbf{k}$ is given by:

$$\mathbf{k} = \begin{bmatrix} k_x \\ k_y \\ k_z \end{bmatrix} = k \begin{bmatrix} \sin\theta\cos\phi \\ \sin\theta\sin\phi \\ \cos\theta \end{bmatrix} \tag{1.8}$$

Then the time independent phase term $\mathbf{k}^T \mathbf{r}$ changes to:

$$\mathbf{k}^T \mathbf{r} = k(r_x \sin\theta\cos\phi + r_y \sin\theta\sin\phi + r_z \cos\theta) \tag{1.9}$$

The wavefront of the signal is still represented by the plane perpendicular to its propagation direction.

There are three major research areas for array signal processing:

1. Detecting the presence of an impinging signal and determine the signal numbers.
2. Finding the DOA angles of the impinging signals.
3. Enhancing the signal of interest coming from some known/unknown directions and suppress the interfering signals (if present) at the same time.

The third research area is the task of beamforming, which can be divided into narrowband beamforming and wideband beamforming depending on the bandwidth of the impinging signals, and wideband beamforming will be the focus of this book. In the next sections, we will first introduce the idea of narrowband beamforming and then extend it to the wideband case.

1.2 Narrowband Beamforming

In beamforming, we estimate the signal of interest arriving from some specific directions in the presence of noise and interfering signals with the aid of an array of sensors. These sensors are located at different spatial positions and sample the propagating waves in space. The collected spatial samples are then processed to attenuate/null out the interfering signals and spatially extract the desired signal. As a result, a specific spatial response of the array system is achieved with 'beams' pointing to the desired signals and 'nulls' towards the interfering ones.

Figure 1.3 shows a simple beamforming structure based on a linear array, where M sensors sample the wave field spatially and the output $y(t)$ at time t is given by an instantaneous linear combination of these spatial samples $x_m(t)$, $m = 0, 1, \ldots, M-1$, as:

$$y(t) = \sum_{m=0}^{M-1} x_m(t) w_m^* \tag{1.10}$$

where * denotes the complex conjugate.

The beamformer associated with this structure is only useful for sinusoidal or narrowband signals, where the term 'narrowband' means that the bandwidth of the impinging signal should be narrow enough to make sure that the signals received by the opposite ends of the array are still correlated with each other (Compton, 1988b), and hence it is termed a narrowband beamformer.

We now analyse the array's response to an impinging complex plane wave $e^{j\omega t}$ with an angular frequency ω and a DOA angle θ, where $\theta \in [-\pi/2\ \pi/2]$ is measured with respect to the broadside of the linear array, as shown in Figure 1.3. For convenience, we assume the phase of the signal is zero at the first sensor. Then the signal received by the first sensor is $x_0(t) = e^{j\omega t}$ and by the mth sensor is $x_m(t) = e^{j\omega(t-\tau_m)}$, $m = 1, 2, \ldots, M-1$, where τ_m is the propagation delay for the signal from sensor 0 to sensor m and is a function of θ. Then the beamformer output is:

$$y(t) = e^{j\omega t} \sum_{m=0}^{M-1} e^{-\omega\tau_m} w_m^* \tag{1.11}$$

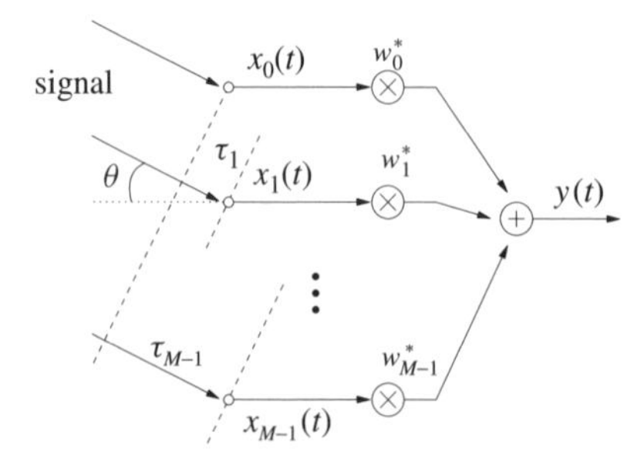

Figure 1.3 A general structure for narrowband beamforming

with $\tau_0 = 0$. The response of this beamformer is given by:

$$P(\omega, \theta) = \sum_{m=0}^{M-1} \mathrm{e}^{-j\omega\tau_m} w_m^* = \mathbf{w}^H \mathbf{d}(\omega, \theta) \tag{1.12}$$

where the weight vector $\mathbf{w}$ holds the M *complex conjugate* coefficients of the sensors, given by:

$$\mathbf{w} = [w_0 \; w_1 \; \ldots \; w_{M-1}]^T \tag{1.13}$$

and the vector $\mathbf{d}(\omega, \theta)$ is given by:

$$\mathbf{d}(\omega, \theta) = \left[1 \;\; \mathrm{e}^{-j\omega\tau_1} \;\; \ldots \;\; \mathrm{e}^{-j\omega\tau_{M-1}}\right]^T \tag{1.14}$$

We refer to $\mathbf{d}(\theta, \omega)$ as the array response vector, which is also known as the steering vector or direction vector (Van Veen and Buckley, 1988). We will use the term 'steering vector' to avoid confusion with the response vector used in linearly constrained minimum variance beamforming introduced later in Section 2.2 of Chapter 2.

In our notation, we generally use lowercase bold letters for vector valued quantities, while uppercase bold letters denote matrices. The operators $\{\cdot\}^T$ and $\{\cdot\}^H$ represent transpose and Hermitian transpose operations, respectively.

Based on the steering vector, we briefly discuss the spatial aliasing problem encountered in array processing. In analogue to digital conversion, we sample the continuous-time signal temporally and convert it into a discrete-time sequence. In this temporal sampling process, aliasing is referred to as the phenomenon that signals with different frequencies have the same discrete sample series, which occurs when the signal is sampled at a rate lower than the Nyquist sampling rate, i.e. twice the highest frequency of the signal (Oppenheim and Schafer, 1975). With temporal aliasing, we will not be able to recover the original continuous-time signal from their samples. In array processing, the sensors sample the impinging signals spatially and if the signals from different spatial locations are not sampled by the array sensors densely enough, i.e. the inter-element spacing of the array is too large, then sources at different locations will have the same array steering vector and we cannot uniquely determine their locations based on the received array signals. Similar to the temporal sampling case, now we have a spatial aliasing problem, due to the ambiguity in the directions of arrival of source signals.

For signals having the same angular frequency ω and the corresponding wavelength λ, but different DOAs θ_1 and θ_2 satisfying the condition $(\theta_1, \theta_2) \in [-\pi/2 \; \pi/2]$, aliasing implies that we have $\mathbf{d}(\theta_1, \omega) = \mathbf{d}(\theta_2, \omega)$, namely:

$$\mathrm{e}^{-j\omega\tau_m(\theta_1)} = \mathrm{e}^{-j\omega\tau_m(\theta_2)} \tag{1.15}$$

For a uniformly spaced linear array with an inter-element spacing d, we have $\tau_m = m\tau_1 = m(d \sin\theta)/c$ and $\omega\tau_m = m(2\pi d \sin\theta)/\lambda$. Then Equation (1.15) changes to:

$$\mathrm{e}^{-jm(2\pi d \sin\theta_1)/\lambda} = \mathrm{e}^{-jm(2\pi d \sin\theta_2)/\lambda} \tag{1.16}$$

In order to avoid aliasing, the condition $|2\pi(\sin\theta)d/\lambda|_{\theta=\theta_1,\theta_2} < \pi$ has to be satisfied. Then we have $|d/\lambda \sin\theta| < 1/2$. Since $|\sin\theta| \leq 1$, this requires that the array distance d should be less than $\lambda/2$.

In the following, we will always set $d = \lambda/2$, unless otherwise specified, then $\omega\tau_m = m\pi \sin\theta$ and the response of the uniformly spaced narrowband beamformer is given by:

$$P(\omega,\theta) = \sum_{m=0}^{M-1} \mathrm{e}^{-jm\pi\sin\theta} w_m^* \tag{1.17}$$

Note for an FIR (finite impulse response) filter with the same set of coefficients (Oppenheim and Schafer, 1975), its frequency response is given by:

$$P(\Omega) = \sum_{m=0}^{M-1} \mathrm{e}^{-jm\Omega} w_m^* \tag{1.18}$$

with $\Omega \in [-\pi\ \pi]$ being the normalized frequency. For the response of the beamformer given by Equation (1.17), when θ changes from $-\pi/2(-90^\circ)$ to $\pi/2(90^\circ)$, $\pi\sin\theta$ changes from $-\pi$ to π accordingly, which is in the same range as Ω in Equation (1.18). With this correspondence, the design of uniformly spaced linear arrays can be achieved by the existing FIR filter design approaches directly.

As a simple example, if we want to form a flat beam response pointing to the directions $\theta \in [-\pi/6\ \pi/6]([-30^\circ\ 30^\circ])$, while suppressing signals from directions $\theta \in [-\pi/2\ -\pi/4]$ and $[\pi/4\ \pi/2]$, then it is equivalent to designing an FIR filter with a passband of $\Omega \in [-0.5\pi\ 0.5\pi]$ and a stopband of $\Omega \in [-\pi\ \ -0.71\pi]$ and $[0.71\pi\ \pi]$ ($\sin\pi/6 = 0.5$ and $\sin\pi/4 = 0.71$). We can use the MATLAB© function *remez* to design such a filter (Mat, 2001), and then use the result directly as the coefficients of the desired beamformer. One of the design results is given by ($M = 10$):

$$\begin{aligned}\mathbf{w}^H = [&0.0422\ \ 0.0402\ \ -0.1212\ \ 0.0640\ \ 0.5132\\ &0.5132\ \ 0.0640\ \ -0.1212\ \ 0.0402\ \ 0.0422]\end{aligned} \tag{1.19}$$

Substituting this result into Equation (1.17), we can draw the resultant amplitude response $|P(\theta,\omega)|$ of the beamformer with respect to the DOA angle θ. $|P(\theta,\omega)|$ is called the beam pattern of the beamformer to describe the sensitivity of the beamformer with respect to signals arriving from different directions and with different frequencies. Figure 1.4 shows the beam pattern (BP) in dB, which is defined as follows:

$$\mathrm{BP} = 20\log_{10}\frac{|P(\theta,\omega)|}{\max|P(\theta,\omega)|} \tag{1.20}$$

For the general case of $d = \alpha\lambda/2$, $\alpha \le 1$, the response of the beamformer given by Equation (1.17), will change to:

$$P(\omega,\theta) = \sum_{m=0}^{M-1} \mathrm{e}^{-jm\alpha\pi\sin\theta} w_m^* \tag{1.21}$$

Its design can be obtained in a similar way as above and the only difference is that the FIR filter can have an arbitrary response over the regions $\Omega \in [-\pi\ \ -\alpha\pi]$ and $[\alpha\pi\ \pi]$ without affecting that of the narrowband beamformer.

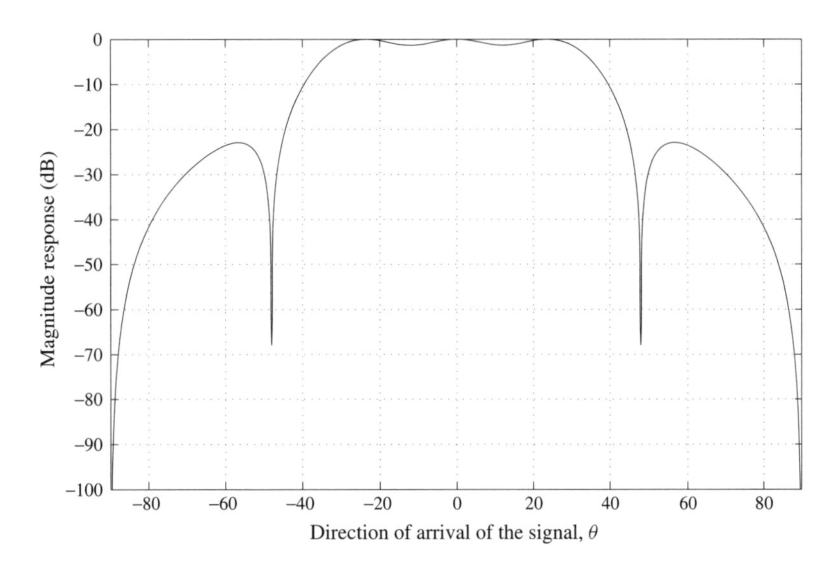

Figure 1.4 The beam pattern of the resultant narrowband beamformer with $M = 10$ sensors

1.3 Wideband Beamforming

The beamforming structure introduced in the last section works effectively only for narrowband signals. When the signal bandwidth increases, its performance will degrade significantly. This can be explained as follows.

Suppose there are in total M impinging signals $s_m(t)$, $m = 0, 1, \ldots, M-1$, from directions of θ_m, $m = 0, 1, \ldots, M-1$, respectively. The first one $s_0(t)$ is the signal of interest and the others are interferences. Then the array's steering vector $\mathbf{d}_m$ for these signals is given by:

$$\mathbf{d}_m(\omega, \theta) = \begin{bmatrix} 1 & \mathrm{e}^{-j\omega\tau_1(\theta_m)} & \ldots & \mathrm{e}^{-j\omega\tau_{M-1}(\theta_m)} \end{bmatrix}^T \tag{1.22}$$

Ideally, for beamforming, we aim to form a fixed response to the signal of interest and zero response to the interfering signals. Note for simplicity, we do not consider the effect of noise here. This requirement can be expressed as the following matrix equation:

$$\begin{pmatrix} 1 & \mathrm{e}^{-j\omega\tau_1(\theta_0)} & \ldots & \mathrm{e}^{-j\omega\tau_{M-1}(\theta_0)} \\ 1 & \mathrm{e}^{-j\omega\tau_1(\theta_1)} & \ldots & \mathrm{e}^{-j\omega\tau_{M-1}(\theta_1)} \\ \vdots & \vdots & \ddots & \vdots \\ 1 & \mathrm{e}^{-j\omega\tau_1(\theta_{M-1})} & \ldots & \mathrm{e}^{-j\omega\tau_{M-1}(\theta_{M-1})} \end{pmatrix} \begin{pmatrix} w_0^* \\ w_1^* \\ \vdots \\ w_{M-1}^* \end{pmatrix} = \begin{pmatrix} \text{constant} \\ 0 \\ \vdots \\ 0 \end{pmatrix} \tag{1.23}$$

Obviously, as long as the matrix on the left has full rank, we can always find a set of array weights to cancel the $M-1$ interfering signals and the exact value of the weights for complete cancellation of the interfering signals is dependent on the signal frequency (certainly also on their directions of arrival).

For wideband signals, since each of them consists of infinite number of different frequency components, the value of the weights should be different for different frequencies and we can write the weight vector in the following form:

$$\mathbf{w}(\omega) = [w_0(\omega)\ w_1(\omega)\ \ldots,\ w_{M-1}(\omega)]^T \tag{1.24}$$

This is why the narrowband beamforming structure with a single constant coefficient for each received sensor signal will not work effectively in a wideband environment.

The frequency dependent weights can be achieved by sensor delay-lines (SDLs), which were proposed only recently and will be studied in Chapter 7. Traditionally, an easy way to form such a set of frequency dependent weights is to use a series of tapped delay-lines (TDLs), or FIR/IIR filters in its discrete form (Compton, 1988a; Frost, 1972; Mayhan et al., 1981; Monzingo and Miller, 2004; Rodgers and Compton, 1979; Van Veen and Buckley, 1988; Vook and Compton, 1992).

Both TDLs and FIR/IIR filters perform a temporal filtering process to form a frequency dependent response for each of the received wideband sensor signals to compensate the phase difference for different frequency components. Such a structure is shown in Figure 1.5. The beamformer obeying this architecture samples the propagating wave field in both space and time. The output of such a wideband beamformer can be expressed as:

$$y(t) = \sum_{m=0}^{M-1} \sum_{i=0}^{J-1} x_m(t - iT_s) \times w_{m,i}^* \tag{1.25}$$

where $J - 1$ is the number of delay elements associated with each of the M sensor channels in Figure 1.5 and T_s is the delay between adjacent taps of the TDLs.

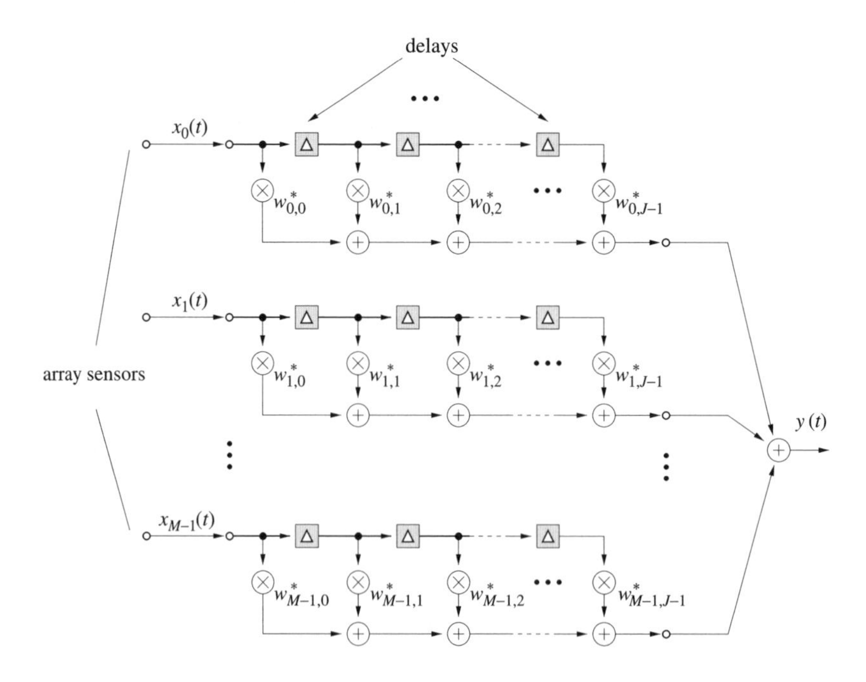

Figure 1.5 A general structure for wideband beamforming

In vector form, Equation (1.25) can be rewritten as:

$$y(t) = \mathbf{w}^H \mathbf{x}(t) \tag{1.26}$$

The weight vector $\mathbf{w}$ holds all MJ sensor coefficients with:

$$\mathbf{w} = \begin{bmatrix} \mathbf{w}_0 \\ \mathbf{w}_1 \\ \vdots \\ \mathbf{w}_{J-1} \end{bmatrix} \tag{1.27}$$

where each vector $\mathbf{w}_i$, $i = 0, 1, \cdots, J-1$, contains the M *complex conjugate* coefficients found at the ith tap position of the M TDLs, and is expressed as:

$$\mathbf{w}_i = [w_{0,i} \; w_{1,i} \; \cdots \; w_{M-1,i}]^T \tag{1.28}$$

Similarly, the input data are also accumulated in a vector form $\mathbf{x}$ as follows:

$$\mathbf{x} = \begin{bmatrix} \mathbf{x}_0(t) \\ \mathbf{x}_1(t - T_s) \\ \vdots \\ \mathbf{x}_{J-1}(t - (J-1)T_s] \end{bmatrix} \tag{1.29}$$

where $\mathbf{x}_i(t - iT_s)$, $i = 0, 1, \ldots, J-1$, holds the ith data slice corresponding to the ith coefficient vector $\mathbf{w}_i$:

$$\mathbf{x}(t - iT_s) = \left[x_0(t - iT_s) \;\; x_1(t - iT_s) \;\; \cdots \;\; x_{M-1}(t - iT_s)\right]^T \tag{1.30}$$

Note that this notation incorporates the narrowband beamformer with the special case of $J = 1$.

Now, for an impinging complex plane wave signal $\mathrm{e}^{j\omega t}$, assume $x_0(t) = \mathrm{e}^{j\omega t}$. Then we have:

$$x_m(t - iT_s) = \mathrm{e}^{j\omega(t-(\tau_m + iT_s))} \tag{1.31}$$

with $m = 0, 1, \ldots, M-1$, $i = 0, \ldots, J-1$. The array output is given by:

$$\begin{aligned} y(t) &= \mathrm{e}^{j\omega t} \sum_{m=0}^{M-1} \sum_{i=0}^{J-1} \mathrm{e}^{-j\omega(\tau_m + iT_s)} \cdot w_{m,i}^* \\ &= \mathrm{e}^{j\omega t} \times P(\theta, \omega) \end{aligned} \tag{1.32}$$

where $P(\theta, \omega)$ is the beamformer's angle and frequency dependent response. It can be expressed in vector form as:

$$P(\theta, \omega) = \mathbf{w}^H \mathbf{d}(\theta, \omega) \tag{1.33}$$

where $\mathbf{d}(\theta, \omega)$ is the steering vector for this new wideband beamformer and its elements correspond to the complex exponentials $e^{-j\omega(\tau_m + iT_s)}$:

$$\mathbf{d}(\theta, \omega) = [e^{-j\omega\tau_0} \ \dots \ e^{-j\omega\tau_{M-1}} \ e^{-j\omega(\tau_0+T_s)} \ \dots \ e^{-j\omega(\tau_{M-1}+T_s)} \\ \dots \ e^{-j\omega(\tau_0+(J-1)T_s)} \ \dots \ e^{-j\omega(\tau_{M-1}+(J-1)T_s)}]^T \tag{1.34}$$

For $J = 1$, it is reduced to the steering vector introduced for the narrowband beamformer in Equation (1.14).

For an equally spaced linear array with an inter-element spacing d, we have $\tau_m = m\tau_1$ and $\omega\tau_m = m(2\pi d \sin\theta)/\lambda$ for $m = 0, 1, \dots, M-1$. To avoid aliasing, $d < \lambda_{\min}/2$, where $\lambda_{\min}$ is the wavelength of the signal component with the highest frequency $\omega_{\max}$. Assume the operating frequency of the array is $\omega \in [\omega_{min} \ \omega_{max}]$ and $d = \alpha\lambda_{\min}/2$ with $\alpha \leq 1$. In its discrete form, T_s is the temporal sampling period of the system and should be no more than half the period $T_{\min}$ of the signal component with the highest frequency according to the Nyquist sampling theorem (Oppenheim and Schafer, 1975), i.e. $T_s \leq T_{\min}/2$.

With the normalized frequency $\Omega = \omega T_s$, $\omega(m\tau_1 + iT_s)$ changes to $m\mu\Omega\sin\theta + i\Omega$ with $\mu = d/(cT_s)$, then the steering vector $\mathbf{d}(\theta, \omega)$ changes to:

$$\mathbf{d}(\theta, \omega) = [1 \ \dots \ e^{-j(M-1)\mu\Omega\sin\theta} \ e^{-j\Omega} \ \dots \ e^{-j\Omega(\mu\sin\theta(M-1)+1)} \\ \dots \ e^{-j(J-1)\Omega} \ \dots \ e^{-j\Omega(\mu\sin\theta(M-1)+J-1)}]^T \tag{1.35}$$

and we have:

$$\begin{aligned} P(\theta, \omega) &= \sum_{m=0}^{M-1}\sum_{i=0}^{J-1} e^{-j\Omega(m\mu\sin\theta+i)} \times w_{m,i}^* \\ &= \sum_{m=0}^{M-1} e^{-jm\mu\Omega\sin\theta} \sum_{i=0}^{J-1} e^{-ji\Omega} \times w_{m,i}^* \\ &= \sum_{m=0}^{M-1} e^{-jm\mu\Omega\sin\theta} \times W_m(e^{j\Omega}) \end{aligned} \tag{1.36}$$

where $W_m(e^{j\Omega}) = \sum_{i=0}^{J-1} e^{-ji\Omega} \times w_{m,i}^*$ is the Fourier transform of the TDL coefficients attached to the mth sensor. For the case where $\alpha = 1$ and $T_s = T_{\min}/2$, we have $\mu = 1$.

Now given the coefficients of the wideband beamformer, we can draw its 3-D beam pattern $|P(\theta, \omega)|$ with respect to frequency and DOA angle, according to Equation (1.36). To calculate the beam pattern for N_θ number of discrete DOA values and N_Ω number of discrete temporal frequencies, an $N_\theta \times N_\Omega$ matrix is obtained holding the response samples on the defined DOA/frequency grid.

As an example, consider an array with $M = 5$ sensors and a TDL length $J = 3$. Suppose the weight vector is given as:

$$\mathbf{W} = [0 \ 0 \ 0 \ 0 \ 0 \ 0.2 \ 0.2 \ 0.2 \ 0.2 \ 0.2 \ 0 \ 0 \ 0 \ 0 \ 0]^T \tag{1.37}$$

The beam pattern of such an array is shown in Figure 1.6 for $N_\Omega = 50$ and $N_\theta = 60$, where the gain is displayed in dB as defined in Equation (1.20).

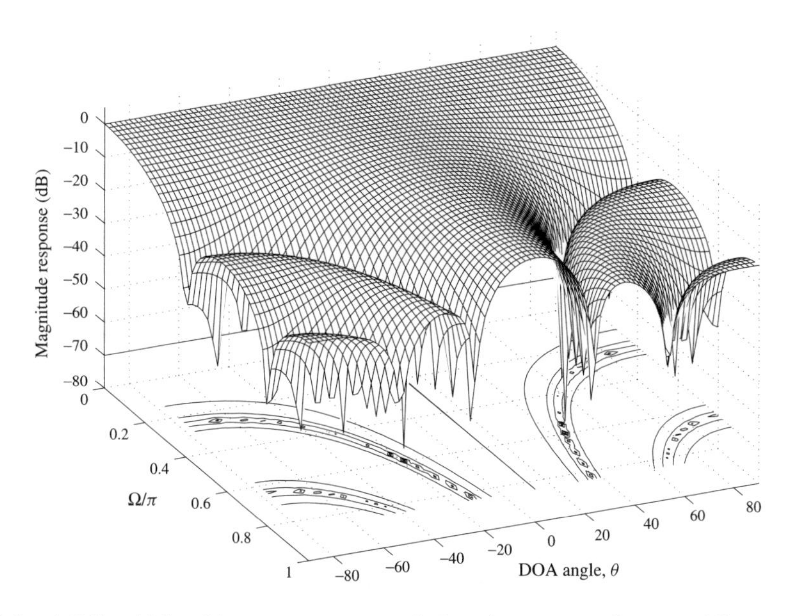

Figure 1.6 A 3-D wideband beam pattern example based on an equally spaced linear array with $M = 5$, $J = 3$ and $\mu = 1$

In the above example, the values of the weight coefficients are fixed and the resultant beamformer will maintain a fixed response independent of the signal/interference scenarios. In statistically optimum beamforming, the weight coefficients need to be updated based on the statistics of the array data. When the data statistics are unknown or time varying, adaptive optimization is required (Haykin, 1996), where according to different signal environments and application requirements, different beamforming techniques may be employed. Both kinds of beamformers will be studied later in this book.

1.4 Wideband Beam Steering

For a narrowband beamformer, we can steer its main beam to a desired direction by adding appropriate steering delays or phase shifts (Johnson and Dudgeon, 1993; Van Trees, 2002). The relationship between the steered response and the original one is simple for a half wavelength spaced linear array: the former one is a circularly shifted version of the latter one, i.e. the sidelobe shifted out from one side is simply shifted back from the other side.

Intuitively, we may think that adding steering delays for wideband beamformers has the same effect as in the narrowband case. However, this is not true and in general there is not a one-to-one correspondence between the original beam response and the steered one (Liu and Weiss, 2008c, 2009a).

In this section we will give a detailed analysis about this relationship. We will see that after adding steering delays to the originally received wideband array signals, the main beam will be shifted to the desired direction; however for the sidelobe region, for one side, it is shifted out of the visible area and for the other side, it is not a simple shifted-back of those shifted out, but exhibits a very complicated pattern.

1.4.1 Beam Steering for Narrowband Arrays

For a uniformly spaced narrowband linear array, its response can be expressed as:

$$P(\sin\theta) = \sum_{m=0}^{M-1} w_m^* \mathrm{e}^{-jm\mu\Omega\sin\theta} \tag{1.38}$$

which is a special case ($J = 1$) of Equation (1.36).

Suppose the set of coefficients w_m^*, $m = 0, 1, \ldots, M-1$, forms a main beam pointing to the broadside of the array ($\theta = 0$). In order to steer the beam to the direction θ_0, we can add a delay of $(M-1)(d\sin\theta_0)/c$ to the first received array signal, a delay of $(M-2)(d\sin\theta_0)/c$ to the second received array signal, and so on. Then the new response with a main beam pointing to θ_0 is given by:

$$P(\sin\theta - \sin\theta_0) = \mathrm{e}^{-j(M-1)\mu\Omega\sin\theta_0} \sum_{m=0}^{M-1} w_m^* \mathrm{e}^{-jm\mu\Omega(\sin\theta - \sin\theta_0)} \tag{1.39}$$

where the term $\mathrm{e}^{-j(M-1)\mu\Omega\sin\theta_0}$ represents a constant delay for all signals and will be ignored in the following equations and discussions.

To avoid spatial aliasing, $d = \lambda/2$, where λ is the signal wavelength. We also assume the sampling frequency is twice that of the signal frequency. Then, we have $\mu = 1$ and $\Omega = \pi$. As a result, Equation (1.39) changes to:

$$P(\sin\theta - \sin\theta_0) = \sum_{m=0}^{M-1} w_m^* \mathrm{e}^{-jm\pi(\sin\theta - \sin\theta_0)} \tag{1.40}$$

Since the function $\mathrm{e}^{-jm\pi x}$ is periodic with a period of 2, compared to Equation (1.38), the response given by Equation (1.40) is simply a circularly shifted version of the response in Equation (1.38) for one period $\sin\theta \in [-1\ 1]$. As an example, suppose we have a broadside main beam response $P(\sin\theta)$ with a maximum response at $\sin\theta = 0$ for $\sin\theta \in [-1\ 1]$, as shown in Figure 1.7. Then after shifting it by $\sin\theta_0 < 0$, the new response will be given by Figure 1.8.

Now we consider the effect as a function of θ. For the remaining part of Section 1.4, without loss of generality, we always assume $\theta_0 < 0$. For $-1 < \sin\theta \leq (1 + \sin\theta_0)$, we have:

$$-1 < -1 - \sin\theta_0 < \sin\theta - \sin\theta_0 \leq 1 \tag{1.41}$$

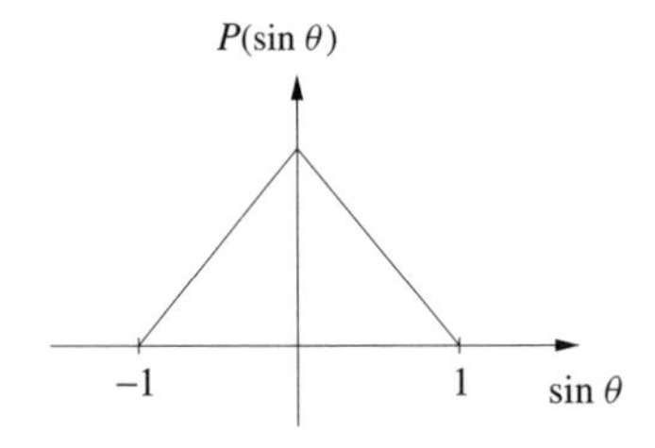

Figure 1.7 A broadside main beam example for an equally spaced narrowband linear array

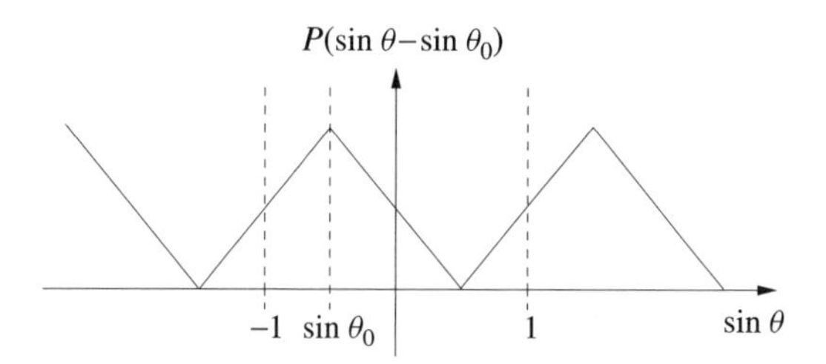

Figure 1.8 The broadside main beam is shifted to the direction θ_0 for the example in Figure 1.7

Then, the response at θ after steering for $-1 < \sin\theta \leq (1 + \sin\theta_0)$ will be the same as the response of the original broadside one at:

$$\hat{\theta} = \arcsin(\sin\theta - \sin\theta_0) \tag{1.42}$$

For $(1 + \sin\theta_0) < \sin\theta \leq 1$, we have:

$$2 \geq (1 - \sin\theta_0) \geq (\sin\theta - \sin\theta_0) > 1 \tag{1.43}$$

Then:

$$0 \geq (\sin\theta - \sin\theta_0 - 2) > -1 \tag{1.44}$$

Therefore, for this range of θ, the steered response at θ will be the same as the response of the original one at:

$$\breve{\theta} = \arcsin(\sin\theta - \sin\theta_0 - 2) \tag{1.45}$$

1.4.2 Beam Steering for Wideband Arrays

1.4.2.1 Wideband Arrays with TDLs or FIR/IIR Filters

As discussed in Section 1.3, for wideband beamforming, we will need the structure shown in Figure 1.5. Recall that its beam response has been given in Equation (1.36) as follows:

$$P(\Omega, \sin\theta) = \sum_{m=0}^{M-1}\sum_{i=0}^{J-1} w_{m,i}^* \times e^{-jm\mu\Omega\sin\theta} \times e^{-ji\Omega} \tag{1.46}$$

Suppose the set of coefficients $w_{m,i}^*$, $m = 0, 1, \ldots, M-1$, $i = 0, 1, \ldots, J-1$ forms a broadside main beam ($\theta = 0$), with an example shown in Figure 1.10. In order to steer the beam to the direction θ_0, we add delays in the same way as in the narrowband case and the new response is given by:

$$P(\Omega, \sin\theta - \sin\theta_0) = \sum_{m=0}^{M-1}\sum_{i=0}^{J-1} w_{m,i}^* \times e^{-jm\mu\Omega(\sin\theta - \sin\theta_0)} \times e^{-ji\Omega} \tag{1.47}$$

To avoid aliasing, $d = \lambda_{\min}/2$ and $T_s = \pi/\omega_{\max}$. Then we have $\mu = 1$ and:

$$P(\Omega, \sin\theta - \sin\theta_0) = \sum_{m=0}^{M-1}\sum_{i=0}^{J-1} w_{m,i}^* \times \mathrm{e}^{-jm\Omega(\sin\theta - \sin\theta_0)} \times \mathrm{e}^{-ji\Omega} \tag{1.48}$$

Note the required steering delays can be implemented by some analogue devices and digital interpolation methods (Pridham and Mucci, 1978, 1979; Schafer and Rabiner, 1973), or FIR/IIR filters with fractional delays (Lu and Morris, 1999). A special case is a delay over the whole normalized frequency range $[0\ \pi]$, which can be realized by a series of truncated sinc functions.

As an example, we steer the main beam in Figure 1.9 in this way to the direction $\theta_0 = -30^\circ$ and the result is shown in Fig. 1.10. Although the main beam is indeed steered to the desired direction, there are some problems. The first one is the distorted response at around the frequency $\Omega = \pi$, which is due to the fact that the delay cannot be approximated well by the sinc function at $\Omega = \pi$. More importantly, the relationship between Figure 1.9 and Figure 1.10 is clearly not a simple shift between each other because we can see the irregularity in the steered response at the sidelobe region between about 40° and 90° and we cannot find it in the original broadside main beam in Figure 1.9. This difference indicates that beam steering with delays for wideband beamformers has quite a different effect. In the next section, we will give a detailed analysis about this relationship.

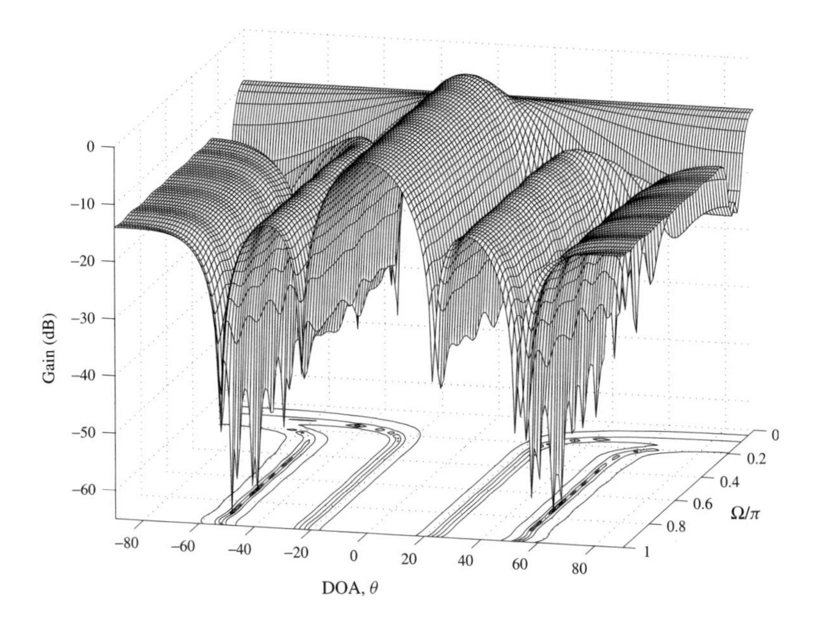

Figure 1.9 A broadside main beam for a linear wideband array with $M = 21$ sensors and $J = 25$ coefficients for each of the attached FIR filters ($\mu = 1$)

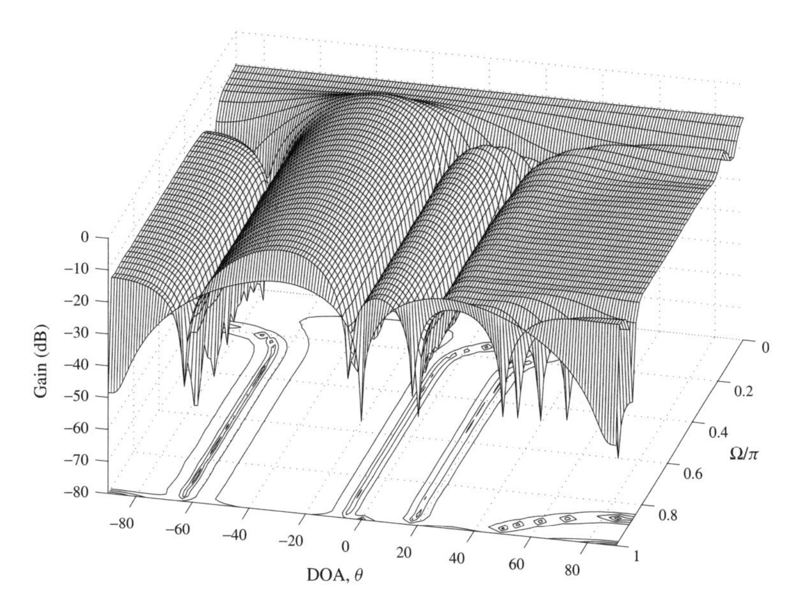

Figure 1.10 The result after the broadside main beam in Figure 1.9 is steered to an off-broadside direction (−30°)

Since $\theta_0 < 0$, for $-1 < \sin\theta \le (1 + \sin\theta_0)$, we have:

$$-1 < -1 - \sin\theta_0 < \sin\theta - \sin\theta_0 \le 1 \tag{1.49}$$

Then the steered response at θ for $-1 < \sin\theta \le (1 + \sin\theta_0)$ will be the same as the response of the original broadside main beam design at:

$$\hat{\theta} = \arcsin(\sin\theta - \sin\theta_0) \tag{1.50}$$

However, for $(1 + \sin\theta_0) < \sin\theta \le 1$, we have $(\sin\theta - \sin\theta_0) > 1$. Then the shift relationship cannot be expressed as Equation (1.50) any more and we need to further consider the following two cases bearing in mind the periodicity of the function $e^{-jm\Omega}$:

1. For $\Omega \le \pi/(\sin\theta - \sin\theta_0)$, we have $\Omega(\sin\theta - \sin\theta_0) \le \pi$, since $(\sin\theta - \sin\theta_0) > 1$, it seems that we cannot find any correspondence between the steered response and the original one for this case.

2. For $\Omega > \pi/(\sin\theta - \sin\theta_0)$, we have:

$$\Omega(\sin\theta - \sin\theta_0) > \pi \tag{1.51}$$

Then we have

$$e^{-jm\Omega(\sin\theta - \sin\theta_0)} = e^{-jm\Omega(\sin\theta - \sin\theta_0 - 2\pi/\Omega)} \tag{1.52}$$

If $\sin\theta - \sin\theta_0 - 2\pi/\Omega < -1$, namely:

$$\Omega < \frac{2\pi}{1 + \sin\theta - \sin\theta_0} \tag{1.53}$$

then we come to the same conclusion as in the first case. Otherwise, we have:

$$\Omega \geq \frac{2\pi}{1 + \sin\theta - \sin\theta_0} \tag{1.54}$$

Then we can assume:

$$\sin\tilde{\theta} = \sin\theta - \sin\theta_0 - 2\pi/\Omega \tag{1.55}$$

Then the steered response for this case will be the same as the response of the original one at frequency Ω and DOA angle $\tilde{\theta} = \arcsin(\sin\theta - \sin\theta_0 - 2\pi/\Omega)$.

Note since $(\sin\theta - \sin\theta_0) > 1$, we have:

$$\frac{2\pi}{1 + \sin\theta - \sin\theta_0} > \frac{\pi}{\sin\theta - \sin\theta_0} \tag{1.56}$$

Then the above two cases for $(1 + \sin\theta_0) < \sin\theta \leq 1$ can be simplified as:

1. For $\Omega < 2\pi/(1 + \sin\theta - \sin\theta_0)$, there is no correspondence between the two beam responses.
2. For $\Omega \geq 2\pi/(1 + \sin\theta - \sin\theta_0)$, the steered response will be the same as the response of the original one at frequency Ω and DOA angle $\tilde{\theta}$.

1.4.2.2 Wideband Arrays with a Narrowband Structure

Sometimes we also use the narrowband beamforming structure for wideband signals and then Equation (1.46) changes back to the one given in Equation (1.38). In this case, the steered response for $\mu = 1$ is given by:

$$P(\sin\theta - \sin\theta_0) = \sum_{m=0}^{M-1} w_m^* \mathrm{e}^{-jm\Omega(\sin\theta - \sin\theta_0)} \tag{1.57}$$

Now for $-1 < \sin\theta \leq (1 + \sin\theta_0)(\theta_0 < 0)$, the relationship between the steered response and the original one is the same as the one given in Equation (1.50). For $(1 + \sin\theta_0) < \sin\theta \leq 1$, we have $(\sin\theta - \sin\theta_0) > 1$ and we again need to consider two different cases:

1. For $\Omega \leq \pi/(\sin\theta - \sin\theta_0)$, we have:

$$\mathrm{e}^{-jm\Omega(\sin\theta - \sin\theta_0)} = \mathrm{e}^{-jm\pi[\Omega(\sin\theta - \sin\theta_0)]/\pi} \tag{1.58}$$

and:

$$\frac{\Omega(\sin\theta - \sin\theta_0)}{\pi} \leq 1 \tag{1.59}$$

Then the steered beam response for this case will be the same as the response of the original broadside one at frequency $\Omega = \pi$ and DOA angle:

$$\bar{\theta} = \arcsin(\frac{\Omega(\sin\theta - \sin\theta_0)}{\pi}) \tag{1.60}$$

2. For $\Omega > \pi/(\sin\theta - \sin\theta_0)$, we have:

$$e^{-jm\Omega(\sin\theta - \sin\theta_0)} = e^{-jm\pi(\Omega(\sin\theta - \sin\theta_0) - 2\pi)/\pi} \tag{1.61}$$

and:

$$\frac{\Omega(\sin\theta - \sin\theta_0) - 2\pi}{\pi} > -1 \tag{1.62}$$

Then the steered beam response for this case will be the same as the response of the original one at frequency $\Omega = \pi$ and a DOA angle of:

$$\breve{\theta} = \arcsin\left(\frac{\Omega(\sin\theta - \sin\theta_0) - 2\pi}{\pi}\right) \tag{1.63}$$

1.4.3 A Unified Interpretation

In summary, the relationship between the steered beam response and the original one, given in Sections 1.4.2 and 1.4.2, respectively, is complicated and not as straightforward as in the narrowband one. However, there is another way to understand the relationship between the steered beam response and the original one.

Since $\theta_0 < 0$, we have $|\sin\theta - \sin\theta_0| \leq (1 - \sin\theta_0)$ and $1 - \sin\theta_0 = \hat{\mu} > 1$, then Equation (1.48) can be rewritten as:

$$P(\Omega, \sin\theta) = \sum_{m=0}^{M-1}\sum_{i=0}^{J-1} w_{m,i}^{*} e^{-jm\hat{\mu}\Omega(\sin\theta - \sin\theta_0)/\hat{\mu}} e^{-ji\Omega} \tag{1.64}$$

Since $|(\sin\theta - \sin\theta_0)/\hat{\mu}| \leq 1$, we can assume:

$$\sin\ddot{\theta} = \frac{\sin\theta - \sin\theta_0}{\hat{\mu}} \tag{1.65}$$

Then Equation (1.64) changes to:

$$P(\Omega, \theta) = \sum_{m=0}^{M-1}\sum_{i=0}^{J-1} w_{m,i}^{*} e^{-jm\hat{\mu}\Omega\sin\ddot{\theta}} e^{-ji\Omega} \tag{1.66}$$

When $d_x = \lambda_{\min}/2$, we have $\mu = 1$ in Equation (1.46). Then when $\mu = \hat{\mu}$, it is equivalent to $d_x = \hat{\mu}\lambda_{\min}/2$, i.e. the inter-element spacing is increased by $\hat{\mu} > 1$. As a result, the steered beam response at $\theta \in [-\pi/2\ \pi/2]$ will be the same as the response of the original broadside design at $\ddot{\theta} \in [\arcsin((-1 - \sin\theta_0)/\hat{\mu})\ \pi/2]$ with the inter-element spacing increased by $\hat{\mu}$ and subject to a nonlinear mapping between θ and $\ddot{\theta}$ given in Equation (1.65).

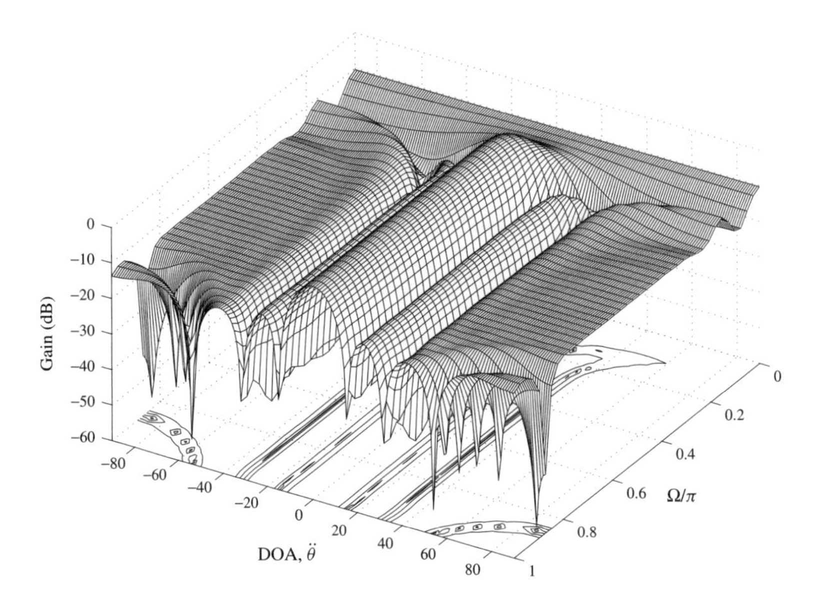

Figure 1.11 The resultant beam pattern with the inter-element spacing increased by 1.5, given the same set of coefficients for obtaining Figure 1.9

Now for the example of $\theta_0 = -30°$, we have $\hat{\mu} = 1.5$ and $\arcsin((-1 - \sin\theta_0)/\hat{\mu}) \approx -20°$. We can draw the response of Equation (1.66) given the same set of coefficients for the example of Figure 1.9. The result is shown in Figure 1.11. Compared to the beam pattern of Figure 1.10, we can see a clear match between Figure 1.10 and that of Figure 1.11 for $\ddot{\theta} \in [-20° \; 90°]$, taking into consideration the nonlinear mapping effect of the sinusoidal function.

1.5 Summary

In this chapter, we have given a brief introduction to array signal processing and in particular narrowband beamforming, including how to calculate its beam pattern and obtain a desired beamformer using existing FIR filter design techniques. We then extend the narrowband beamforming structure to the wideband case by considering the need of forming a series of frequency dependent weight coefficients, which can be realized by tapped delay-lines or FIR/IIR filters. Another possibility is to employ sensor delay-lines for wideband beamforming, which will be the topic of Chapter 7.

A detailed analysis is provided at the end for the beam steering process in both narrowband and wideband beamforming. It is shown that unlike the narrowband case, where the steered beam response is a circularly shifted version of the original one given a half wavelength spacing, a more complicated relationship exists for wideband beamformers and a unified interpretation is provided for an easy understanding by considering a wideband beamformer with an increased inter-element spacing.

2

Adaptive Wideband Beamforming

In many of the beamforming applications, we often need to adjust the beamformer's coefficients according to the received array data to achieve an optimum solution to the specified scenario. When the environment keeps changing, the coefficients need change too. This class of beamformers often employ all kinds of adaptive algorithms to update their coefficients either block by block or continuously. For block-based adaptation, we normally assume that the signals are relatively stationary for the duration of the received block of array data, so that the statistics of the signals can be estimated accurately for the block period to calculate the optimum solution. For a more time-varying environment or where the number of coefficients is too large, we may prefer to update the coefficients continuously, i.e. they are adjusted for each new set of signal samples.

In this chapter, we will discuss two commonly used adaptive wideband beamforming structures: the reference signal based adaptive beamformer and the linearly constrained minimum variance (LCMV) beamformer. Although the assumption for the reference signal based beamformer may look unrealistic in practice, it provides an opportunity to review the standard adaptive algorithms employed in many adaptive wideband beamformers discussed in this chapter and the following chapters.

The LCMV beamformer will be our real focus and we will also discuss the constraint design problem and its alternative implementation–the generalized sidelobe canceller (GSC). Then two other minimum variance beamformers will be introduced and at the end we will discuss some robust adaptive beamforming techniques to deal with the problem of steering vector errors.

2.1 Reference Signal-Based Beamformer

If there is a reference signal available, then beamforming can be performed using a standard multi-channel adaptive filter (MCAF), as shown in Figure 2.1, where the M received sensor signals $x_0[n]$, $x_1[n]$, ..., $x_{M-1}[n]$ are fed into the MCAF and its coefficients are adjusted by minimizing a cost function based on the error signal $e[n]$ between the reference signal $r[n]$ and the MCAF output $y[n]$. The adaptive filter length for each channel is J and there are in total M channels and MJ adaptive coefficients. Note for convenience, we have written all of the signals in the discrete form. We have:

$$\begin{aligned} e[n] &= r[n] - y[n] \\ &= r[n] - \mathbf{w}^H \mathbf{x}[n] \end{aligned} \tag{2.1}$$

Wideband Beamforming Wei Liu and Stephan Weiss

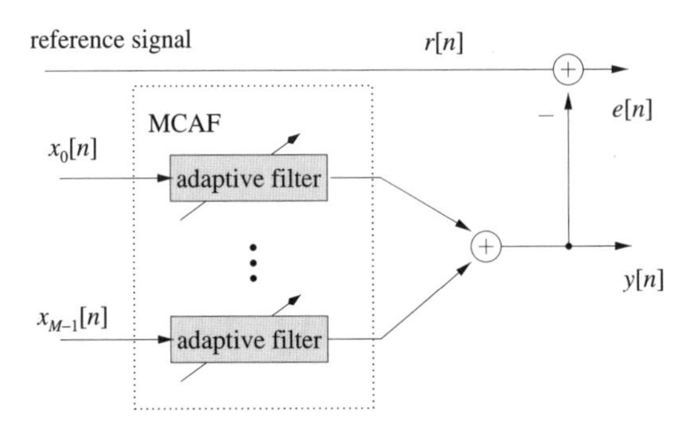

Figure 2.1 The reference signal based wideband beamforming structure, where 'MCAF' represents 'multi-channel adaptive filter'

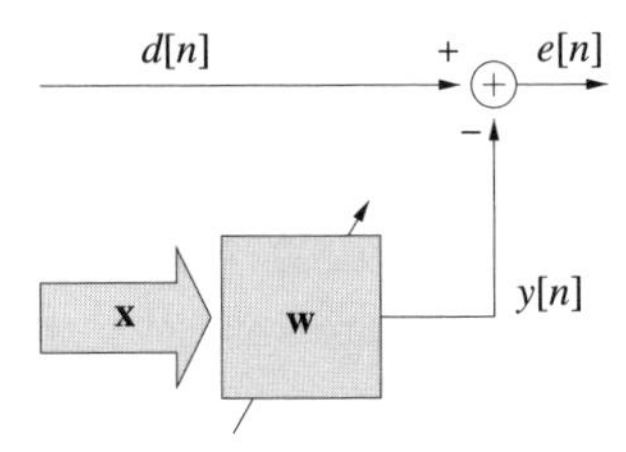

Figure 2.2 A standard adaptive filter setup

where the adaptive weight vector $\mathbf{w}$ consists of the array coefficients as defined in Equation (1.27) and $\mathbf{x}[n]$ is defined in the same way as in Equation (1.29) and simply its discrete form.

This is a standard adaptive filtering problem and has the same setup as a general adaptive filter (Haykin, 1996; Widrow and Stearns, 1985), as shown in Figure 2.2, where the only difference from Figure 2.1 is that we have renamed the reference signal $r[n]$ to a more commonly used name $d[n]$, the desired signal. The error signal $e[n] = d[n] - y[n]$ is used for adjusting the weights $\mathbf{w}$ according to some criterion and usually this criterion is to minimize the error in a mean square or weighted sum of squares sense (Haykin, 1996).

In the following, based on the standard adaptive filter setup in Figure 2.2, a short review of the least mean square (LMS), the normalized LMS (NLMS) as well as the recursive least squares (RLS) algorithms will be provided. We will also mention the frequency-domain and subband implementations very briefly at the end.

Note that here $\mathbf{x}$ is a general input signal vector and $\mathbf{w}$ a general weight vector and we will include the time index n and use $\mathbf{w}[n]$ to indicate its value at the time instant n in the following discussions.

2.1.1 Least Mean Square Algorithm

The LMS algorithm is a stochastic gradient technique based on the particular shape of the cost function employed. This cost function ξ, which is constructed by the mean square error (MSE), can be formulated as:

$$\begin{aligned}\xi &= E\{e[n]e^*[n]\} \\ &= E\{(d[n] - \mathbf{w}[n]^H\mathbf{x}[n])(d[n] - \mathbf{w}[n]^H\mathbf{x}[n])^*\} \\ &= \sigma_{dd}^2 - \mathbf{w}[n]^H\mathbf{p} - \mathbf{p}^H\mathbf{w}[n] + \mathbf{w}[n]^H\boldsymbol{R}_{xx}\mathbf{w}[n]\end{aligned}$$

where $\sigma_{dd}^2 = E\{|d[n]|^2\}$, $\mathbf{p} = E\{\mathbf{x}[n]d[n]^*\}$ and $\boldsymbol{R}_{xx} = E\{\mathbf{x}[n]\mathbf{x}^H[n]\}$.

The MSE $\xi(\mathbf{w})$ is dependent on the elements of the weight vector and has the shape of a hyperparabola with a unique global minimum ξ_{min} for a full-rank covariance matrix $\boldsymbol{R}_{xx}$, as illustrated for the 2-D case in Figure 2.3.

The minimum point of the cost function can be found by an update rule involving successive corrections of the weight vector $\mathbf{w}[n]$ from an initial vector in the direction of the negative gradient of the MSE, which can be expressed as:

$$\mathbf{w}[n+1] = \mathbf{w}[n] - \mu\nabla\xi[n] \tag{2.2}$$

The factor μ is a positive real-valued constant weighting the amount of innovation applied at each step and referred to as the step-size parameter. The variable $\nabla\xi[n]$ denotes the value of the gradient vector at time n, which is formulated as:

$$\nabla\xi[n] = \frac{\partial\xi[n]}{\partial\mathbf{w}} \tag{2.3}$$

Using Wirtinger's calculus (Weiss and Stewart, 1998), the gradient vector can be evaluated as:

$$\nabla\xi[n] = -\mathbf{p} + \boldsymbol{R}_{xx}\mathbf{w}[n] \tag{2.4}$$

Thus, the update equation for the standard adaptive filter configuration can be computed as:

$$\mathbf{w}[n+1] = \mathbf{w}[n] + \mu(\mathbf{p} - \boldsymbol{R}_{xx}\mathbf{w}[n]) \tag{2.5}$$

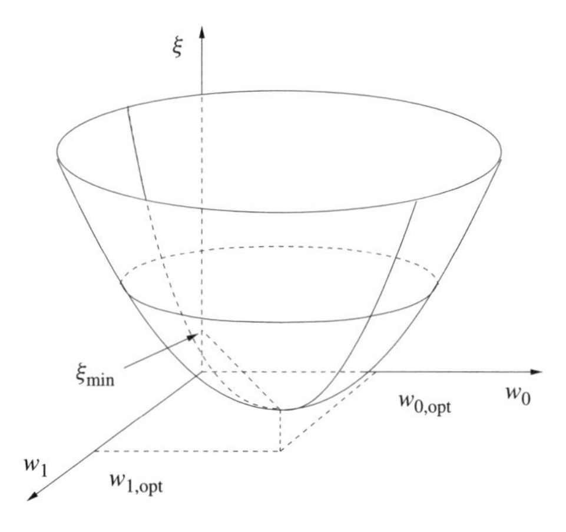

Figure 2.3 The cost function ξ for a special case of two adaptive coefficients

The above update equation is based on the method of steepest descent (Haykin, 1996; Widrow and Stearns, 1985), and we need to have the exact second-order statistics information of the received signals, i.e. the cross-correlation vector $\mathbf{p}$ and the covariance matrix $\boldsymbol{R}_{xx}$. In reality, this is impossible and we have to estimate them from the available data. The simplest choice to is to replace each of the expectation values ($\boldsymbol{R}_{xx}$ and $\mathbf{p}$) by an instantaneous single sample estimate based on the input vector $\mathbf{x}[n]$ and the desired signal $d[n]$, given by:

$$\hat{\boldsymbol{R}}_{xx}[n] = \mathbf{x}[n]\mathbf{x}^H[n] \qquad \text{and} \qquad \hat{\mathbf{p}} = \mathbf{x}[n]d^*[n] \tag{2.6}$$

Then the steepest descent algorithm of Equation (2.5) is changed to:

$$\begin{aligned}\mathbf{w}[n+1] &= \mathbf{w}[n] + \mu(d^*[n] - \mathbf{x}^H[n]\mathbf{w}[n])\mathbf{x}[n] \\ &= \mathbf{w}[n] + \mu e^*[n]\mathbf{x}[n]\end{aligned} \tag{2.7}$$

which leads directly to the well-known least mean square algorithm.

The convergence and stability of the LMS algorithm depend on the correct choice of the step size μ (Haykin, 1996; Widrow and Stearns, 1985). A large step size results in a fast convergence speed but also a large excess mean square error after adaptation, i.e. the algorithm is not very precise in reaching and staying at the exact minimum of the cost function ξ. If μ is small, the adaptation is slow but the excess mean square error after adaptation is small. Thus a trade-off exists between the convergence speed and the steady state mean square error and usually a compromise has to be made.

For the stability of the algorithm, μ should satisfy (Haykin, 1996):

$$0 < \mu < \frac{2}{\lambda_{\max}} \tag{2.8}$$

where $\lambda_{\max}$ is the maximum eigenvalue of the covariance matrix $\boldsymbol{R}_{xx}$. Since $\boldsymbol{R}_{xx}$ is positive semidefinite, we have:

$$\lambda_{\max} \leq \sum_{i=0}^{l_a-1} \lambda_i = \mathrm{tr}\{\boldsymbol{R}_{xx}\} = l_a\sigma_{xx}^2 \tag{2.9}$$

where l_a is the dimension of $\mathbf{x}$, λ_i the eigenvalues of $\boldsymbol{R}_{xx}$, $\mathrm{tr}\{\cdot\}$ the trace of the matrix argument and σ_{xx}^2 is the variance of input signal. Then we have:

$$\mu < \frac{2}{l_a\sigma_{xx}^2} \tag{2.10}$$

as an upper limit for μ. When exceeding this limit, μ is likely to cause the LMS algorithm unstable.

2.1.2 Normalized Least Mean Square Algorithm

In the LMS algorithm, the step size μ should never exceeds its upper bound in Equation (2.10). In a non-stationary environment or where σ_{xx}^2 is not known a priori, the worst case has to be assumed and we have to choose a very small value for μ, which leads to a rather slow convergence rate for the algorithm.

According to Equation (2.9), it is possible to *normalize* the step size to ensure an approximately constant rate of adaptation. Equation (2.9) can be approximated by an instantaneous estimate as:

$$l_a \sigma_{xx}^2 \approx \mathbf{x}[n]^H \mathbf{x}[n] \tag{2.11}$$

Then a normalization of the step size is given by:

$$\mu = \frac{\mu_0}{\mathbf{x}[n]^H \cdot \mathbf{x}[n]} \tag{2.12}$$

where μ_0 is the new step size. Substituting Equation (2.12) into the LMS update equation yields a constant convergence rate independent of the power of the input signal $\mathbf{x}$.

2.1.3 Recursive Least Squares Algorithm

The recursive least squares algorithm is another class of adaptive algorithms and the cost function to minimize is a sum of squared errors as given below:

$$\xi_{\mathrm{LS}}[n] = \sum_{\nu=0}^{n} \beta^{\nu} |e[n-\nu]|^2 = \sum_{\nu=0}^{n} \beta^{\nu} |d[n-\nu] - \mathbf{w}^H[n-\nu]\mathbf{x}[n-\nu]|^2 \tag{2.13}$$

The parameter β, $(0 < \beta \leq 1)$ is the forgetting factor, which ensures that the recent error signal is given more weighting in the cost function and previous errors tend to be 'forgotten' in an exponential rate.

The minimization of the cost function is performed by solving:

$$\nabla \xi_{\mathrm{LS}}[n] = \mathbf{0} \tag{2.14}$$

In a similar way to Equation (2.4) we obtain:

$$\boldsymbol{R}_{xx}[n]\mathbf{w}[n] = \mathbf{p}[n] \tag{2.15}$$

with:

$$\begin{aligned} \boldsymbol{R}_{xx}[n] &= \sum_{\nu=0}^{n} \beta^{\nu} \mathbf{x}[n-\nu]\mathbf{x}^H[n-\nu] \\ \mathbf{p}[n] &= \sum_{\nu=0}^{n} \beta^{\nu} d^*[n-\nu]\mathbf{x}[n-\nu] \end{aligned} \tag{2.16}$$

$\boldsymbol{R}_{xx}[n]$ and $\mathbf{p}[n]$ can be calculated iteratively by:

$$\begin{aligned} \boldsymbol{R}_{xx}[n] &= \beta \boldsymbol{R}_{xx}[n-1] + \mathbf{x}[n]\mathbf{x}^H[n] \\ \mathbf{p}[n] &= \beta \mathbf{p}[n-1] + d^*[n]\mathbf{x}[n] \end{aligned} \tag{2.17}$$

Then, the weight vector $\mathbf{w}$ could be calculated with a matrix inversion of $\boldsymbol{R}_{xx}[n]$ by solving Equation (2.15) for each time instant n, which can be very time consuming, especially for a large value of l_a.

To avoid this matrix inversion operation, we can exploit the matrix inversion lemma (Haykin, 1996)

$$(\boldsymbol{A}+\boldsymbol{BCD})^{-1}=\boldsymbol{A}^{-1}-\boldsymbol{A}^{-1}\boldsymbol{B}(\boldsymbol{C}^{-1}+\boldsymbol{DA}^{-1}\boldsymbol{B})^{-1}\boldsymbol{DA}^{-1} \tag{2.18}$$

with $\boldsymbol{A}=\beta\boldsymbol{R}_{xx}[n-1]$, $\boldsymbol{B}=\mathbf{x}[n]$, $\boldsymbol{C}=1$ and $\boldsymbol{D}=\mathbf{x}^H[n]$.

Denoting $\boldsymbol{S}[n]=\boldsymbol{R}_{xx}^{-1}[n]$, we have:

$$\boldsymbol{S}[n]=\frac{1}{\beta}\big(\boldsymbol{S}[n-1]-\boldsymbol{G}[n]\mathbf{x}[n]\mathbf{x}^H[n]\boldsymbol{S}[n-1]\big) \tag{2.19}$$

with:

$$\boldsymbol{G}[n]=\frac{\boldsymbol{S}[n-1]}{\beta+\mathbf{x}^H[n]\boldsymbol{S}[n-1]\mathbf{x}[n]} \tag{2.20}$$

Note that the initial conditions should be chosen to ensure that $\boldsymbol{R}_{xx}[0]$ is not singular. By rearranging Equation (2.20), we can find that $\boldsymbol{G}[n]=\boldsymbol{S}[n]$. Inserting Equation (2.17) and Equation (2.19) into $\mathbf{w}[n+1]=\boldsymbol{R}_{xx}^{-1}[n]\mathbf{p}[n]$ leads to the following update equation:

$$\begin{aligned}\mathbf{w}[n+1]&=\mathbf{w}[n]+\boldsymbol{S}[n]\mathbf{x}[n]\big[d[n]-\mathbf{w}^H[n]\mathbf{x}[n]\big]^*\\&=\mathbf{w}[n]+\boldsymbol{S}[n]\mathbf{x}[n]e^*[n]\end{aligned} \tag{2.21}$$

From Equation (2.21), we can see that the main difference between the LMS and RLS algorithms is in replacing μ by $s[n]$ in the RLS case. By this modification, the convergence speed of the RLS algorithm is independent of the eigenvalue distribution of the correlation matrix and typically an order of magnitude faster than that of the LMS algorithm. For a detailed discussion of the RLS's properties and its recent developments, please refer to the following (Haykin, 1996; Skidmore and Proudler, 2001; Weiss and Stewart, 1998).

2.1.4 Comparison of Computational Complexities

The computational complexity of an algorithm is not only determined by itself, but is also dependent on the processors we use, e.g. fix-point processors or floating-point processors. We here only consider the number of real multiplications in each step as an indication of the computational complexity of an adaptive algorithm. Since the computational complexities for real-valued and complex-valued signals are different, we will discuss them separately. Moreover, for single-channel and multi-channel adaptive filters, as in the case of beamformers, the input signal vector $\mathbf{x}$ has different signal structures, which can also affect the computational complexity.

In the following discussion, l_a is the total number of adaptive weights for both the single and multi-channel cases. For the multi-channel reference-signal based beamformer, $l_a=MJ$, where M is the sensor number or channel number, and J is the TDL/FIR filter length.

2.1.4.1 Real-Valued Input Signal

For the LMS algorithm, l_a multiplications are required to calculate the output $e[n]$, one multiplication for the product of $e[n]$ and μ and l_a multiplications for the final multiplication with $\mathbf{x}[n]$, totaling to $2l_a + 1$ multiplications.

The NLMS algorithm needs 2 additional multiplications to update the value in Equation (2.12) for a single-channel system and $M + 1$ multiplications for an $M-$channel system. In total this yields $2l_a + 3$ and $2l_a + M + 2$ multiplications for single and $M-$channel realizations, respectively.

For the RLS algorithm, l_a multiplications are required to calculate the output $e[n]$, $2l_a^2 + l_a$ multiplications to calculate $S[n]$ according to Equation (2.20), and $l_a^2 + l_a$ multiplications for the weight update in Equation (2.21). In total, its computational complexity will be $3l_a^2 + 3l_a$ multiplications.

For convenience, we summarize these results in Table 2.1.

2.1.4.2 Complex-Valued Input Signal

For the LMS algorithm, we need $4l_a$ real multiplications to calculate the output $e[n]$, two real multiplications for the product between $e[n]$ and μ and $4l_a$ real multiplications for the final multiplication with $\mathbf{x}[n]$. It totals to $8l_a + 2$ real multiplications.

The NLMS algorithm needs an additional three real multiplications in the single-channel case and $2M + 1$ in the M-channel case to update μ in Equation (2.12) and the total computational complexity will be $8l_a + 5$ or $8l_a + 2M + 3$ real multiplications.

For the RLS algorithm, it requires $4l_a$ real multiplications to calculate the output $e[n]$, $4l_a^2 + 4l_a$ real multiplications to calculate the complex result of $\beta + \mathbf{x}^H[n]S[n-1]\mathbf{x}[n]$ in Equation (2.20), additional $4l_a^2 + 4$ real multiplications to have $S[n]$ and $4l_a^2 + 4l_a$ real multiplications for the weight update in Equation (2.21). In total, it is $12l_a^2 + 12l_a + 4$ real multiplications.

These results are summarized in Table 2.2.

As shown in Tables 2.1 and 2.2, the RLS algorithm has a much higher computational complexity than the LMS-type algorithms. Although recently a fast stable RLS algorithm has been introduced with a computational complexity of order $\mathcal{O}(l_a)$ (Skidmore and Proudler, 2001), it still has a much higher complexity than the LMS-type algorithms. In our simulations throughout this book, we will only use the LMS-type algorithms, especially the normalized LMS algorithm.

Table 2.1 Computational complexities for real-valued input signals

Adaptive algorithms	Real multiplications (single-channel)	Real multiplications (M-channel)
LMS	$2l_a + 1$	$2l_a + 1$
NLMS	$2l_a + 3$	$2l_a + M + 2$
RLS	$3l_a^2 + 3l_a$	$3l_a^2 + 3l_a$

Table 2.2 Computational complexities for complex-valued input signal

Adaptive algorithms	Real multiplications (single-channel)	Real multiplications (M-channel)
LMS	$8l_a + 2$	$8l_a + 2$
NLMS	$8l_a + 5$	$8l_a + 2M + 3$
RLS	$12l_a^2 + 12l_a + 4$	$12l_a^2 + 12l_a + 4$

2.1.5 Frequency-Domain and Subband Adaptive Algorithms

Tables 2.1 and 2.2 show that the computational complexities of an adaptive algorithm are in general of the order $\mathcal{O}(l_a)$ for LMS-type or $\mathcal{O}(l_a^2)$ for RLS-type algorithms, which increase dramatically with the length of the adaptive filter.

To reduce the large computational complexity imposed by using long adaptive filters, we can use frequency-domain adaptive filtering (FDAF) algorithms (Ferrara, 1985; Shynk, 1992), which are based on block by block updating strategies whereby the filter convolution and the gradient correlation can be performed efficiently using fast Fourier transform (FFT) algorithms (Burrus and Parks, 1985). Since the output and weight update are computed only after a large block of data has been accumulated, the computational complexity can be significantly reduced. Moreover, due to the approximately uncorrelated output signals after the transform, different step sizes can be used for different frequency bin outputs. As a result, the convergence speed of the algorithm may be improved significantly.

The discrete Fourier transform (DFT), as employed in frequency-domain adaptive algorithms, can be viewed as a filter bank with maximal decimation (Akansu and Haddad, 1992; Vaidyanathan, 1993). Because of its relatively poor frequency resolution, the DFT filter bank has a large degree of spectral overlap between the adjacent frequency bands, which can lead to severe aliasing distortion and may cause problems when the input data do not exactly lie on a frequency bin (Weiss and Proudler, 2002). As a solution and also an alternative, subband adaptive filtering (SAF) algorithms have been developed (Gilloire and Vetterli, 1988, 1992; Shynk, 1992; Weiss and Stewart, 1998; Yamada *et al.*, 1994; Yang *et al.*, 1995).

We will discuss the details of both the frequency-domain and subband adaptive algorithms and their implementations in adaptive wideband beamforming in Chapter 3.

2.1.6 Simulations

In this part, we provide a simulation result using the NLMS algorithm. There are $M = 10$ sensors and the FIR filter length J for each sensor is set to be $J = 96$. Our aim is to receive a signal of interest from the broadside ($\theta = 0°$) and adaptively suppress two wideband interfering signals with a normalized frequency bandwidth $\Omega \in [0.25\pi\ \pi]$ arriving from DOA angles $\theta = 20°$ and $-30°$, respectively. The signal-to-interference ratio (SIR) for each interfering signal is 0 dB. For this reference signal-based beamformer, a copy of the signal of interest is used as the reference signal $r[n]$. Additionally, the sensor signals are corrupted by independent white Gaussian noise with a signal-to-noise ratio (SNR) of 20 dB. A normalized LMS algorithm with a step size of 0.08 is used.

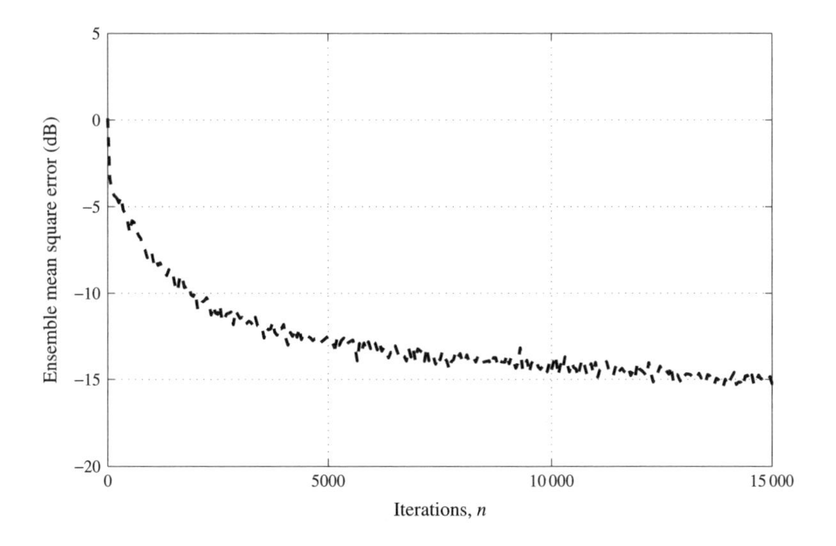

Figure 2.4 An average leaning curve for the reference signal based beamformer

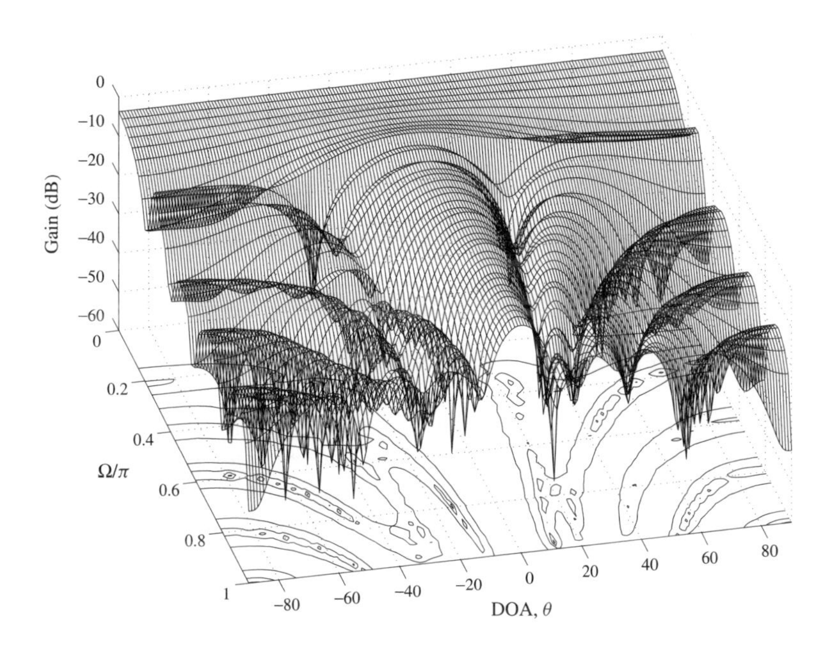

Figure 2.5 The resultant beam pattern for the reference signal-based wideband beamformer

The ensemble mean square value of $e[n]$ obtained by 500 independent runs is shown in Figure 2.4 and the resultant beam pattern for this scenario is shown in Figure 2.5, where the two nulls at $\theta = 20^\circ$ and -30° are formed to suppress the two interfering signals.

2.2 Linearly Constrained Minimum Variance Beamforming

If a reference signal is not available, but we know the DOA angle of the signal of interest and their bandwidth range, then we can impose some constraints on the array coefficients and adaptively minimize the variance or power $E\{y(t)^*y(t)\}$ of the beamformer output subject to the imposed constraints. This leads to the well-known linearly constrained minimum variance (LCMV) beamformer (Frost, 1972), where the response of the beamformer is constrained such that the desired signals impinging on the array from some specific directions are preserved subject to a specified gain and phase response, while the contributions in the beamformer output due to interfering signals arriving from other directions are minimized.

We have seen that the beamformer's response to a signal having a frequency ω and DOA angle θ can be expressed as in Equation ((1.33)). In order to ensure that any signal having a frequency ω_0 and DOA angle θ_0 passes the beamformer with a specified response G_0, where G_0 is a complex constant, we can set this constraint to:

$$\mathbf{w}^H \mathbf{d}(\theta_0, \omega_0) = G_0 \tag{2.22}$$

Note that the value of the output power or variance is given by (Frost, 1972):

$$E\{|y[n]|^2\} = \mathbf{w}^H \boldsymbol{R}_{xx} \mathbf{w} \tag{2.23}$$

where $\boldsymbol{R}_{xx}$ is the observed array data's correlation matrix, and assumed to be positive definite and given in the form of:

$$\boldsymbol{R}_{xx} = E\{\mathbf{x}\mathbf{x}^H\} \tag{2.24}$$

Then the LCMV beamforming problem can be formulated as:

$$\mathbf{w} = \arg\min_{\mathbf{w}} \mathbf{w}^H \boldsymbol{R}_{xx} \mathbf{w} \qquad \text{subject to} \qquad \mathbf{d}^H(\theta_0, \omega_0)\mathbf{w} = G_0^* \tag{2.25}$$

This single constraint formulation can be generalized to multiple linear constraints for an enhanced control of the beamformer's response, for example, by specifying more DOA angles and frequencies (Ahmed and Evans, 1984; Steyskal, 1983; Takao and Komiyama, 1980). If there are $r < MJ$ number of linearly independent constraints imposed on $\mathbf{w}$, we can formulate the constraints in matrix form as:

$$\boldsymbol{C}^H \mathbf{w} = \mathbf{f} \tag{2.26}$$

where the $MJ \times r$ dimensional matrix $\boldsymbol{C}$ is termed as the constraint matrix, while the r dimensional vector $\mathbf{f}$ is the response vector. In the next section we will discuss a special class of constraints and more details about the constraints design problem in LCMV beamforming will be provided in Section 2.3.

2.2.1 A Simple Formulation of Constraints

The constraints imposed on the LCMV beamformer ensure that the beamformer has the required response to signals arriving from the specified angles and at given frequencies, no matter what values are assigned to the weights. For different applications, there are different constraints, one of which is that for a prescribed direction, the response of the array is maintained constant. The resultant beamformer is referred to as the minimum variance distortionless response (MVDR) beamformer (Capon, 1969; Owsley, 1985). A formulation of such constraints is based on a simple relation between the response in the look direction and the weights in the array. Based on Figure 1.5, in the following we will briefly introduce this approach.

Assume that the signal of interest arrives from the broadside of a linear array, i.e. $\theta = 0$. If this is not the case, the array can be steered either mechanically or electrically by imposing appropriate time delays (or phase shifts in the narrowband beamforming scenario) immediately after each sensor output, such that the signals incident on the array from the directions of interest other than the broadside appear as identical replicas of one another at the outputs of the steering delay elements.

With this pre-steering, the signal of interest can be treated as if it had arrived from the broadside. Thus, identical signal components appear at the sensors simultaneously and pass in parallel through the tapped delay-lines following the sensors. Hence, the FIR filters seen in Figure 1.5 appear to have a common input. As far as the signal of interest is concerned, the array processor is equivalent to a single FIR filter in which each weight is equal to the sum of weights in the corresponding vertical column, as indicated in Figure 2.6 (Frost, 1972), where we have:

$$f^*[j] = \sum_{m=0}^{M-1} w^*_{m,j} \tag{2.27}$$

with $j = 0, 1, \ldots, J-1$.

These summed weights in the equivalent tapped delay-line form one single temporal FIR filter, specifying the frequency response of the beamformer to the signal incident from the broadside and must be selected appropriately to give the desired response characteristic in the look direction. In MVDR beamforming, this response could be a pure integer delay, i.e. one of the taps $f^*[j]$, $j = 0, 1, \ldots, J-1$ will be 1 and all the others are zero.

Thus, the broadside constraint can be formulated in the following way:

$$\mathbf{C}^H \mathbf{w} = \mathbf{f} \tag{2.28}$$

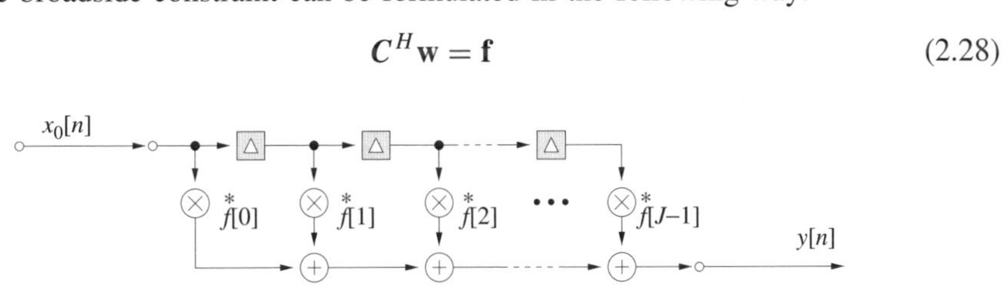

Figure 2.6 The equivalent processor based on the wideband beamforming structure in Figure 1.5 for a signal arriving from the broadside

where:

$$\mathbf{f} = \begin{bmatrix} f[0] \\ f[1] \\ \vdots \\ f[J-1] \end{bmatrix} \tag{2.29}$$

and:

$$\boldsymbol{C} = \begin{bmatrix} \mathbf{c}_0 & & \mathbf{0} \\ & \ddots & \\ \mathbf{0} & & \mathbf{c}_0 \end{bmatrix} \in \mathbf{C}^{MJ \times J} \tag{2.30}$$

with:

$$\mathbf{c}_0 = [1 \;\; 1 \;\; \cdots \;\; 1]^T \in \mathbf{C}^{M \times 1} \tag{2.31}$$

Note that the response vector $\mathbf{f}$ is defined as containing the complex conjugate of a desired gain.

2.2.2 *Optimum Solution to the LCMV Problem*

The solution to the general LCMV problem in Equations (2.25) and (2.26) can be obtained by the method of Lagrange multipliers (Haykin, 1996; Johnson and Dudgeon, 1993), which is outlined here.

The Lagrangian is formed by the objective function $E\{|y[n]|^2\} = \mathbf{w}^H \boldsymbol{R}_{xx}\mathbf{w}$, plus the real part of the constraint function of $\boldsymbol{C}^H\mathbf{w} - \mathbf{f}$, weighted element wise by the r-dimensional vector of undetermined Lagrange multipliers $\boldsymbol{\lambda}$, which is given by:

$$\mathbf{w}^H \boldsymbol{R}_{xx}\mathbf{w} + \boldsymbol{\lambda}^H(\boldsymbol{C}^H\mathbf{w} - \mathbf{f}) + \boldsymbol{\lambda}^T(\boldsymbol{C}^T\mathbf{w}^* - \mathbf{f}^*) \tag{2.32}$$

Note that the gradient of the constraint function constituted by the second and third terms of Equation (2.32) must be linearly independent of each other for the Lagrange multipliers to hold, i.e. the columns of $\boldsymbol{C}$ must have full rank. Differentiating the function in Equation (2.32) with respect to $\mathbf{w}^*$, we have:

$$\boldsymbol{R}_{xx}\mathbf{w} + \boldsymbol{C}\boldsymbol{\lambda} \tag{2.33}$$

Setting this result equal to zero, we obtain the optimal weight vector $\mathbf{w}_{opt}$ in term of the Lagrange multipliers as follows:

$$\mathbf{w}_{opt} = -\boldsymbol{R}_{xx}^{-1}\boldsymbol{C}\boldsymbol{\lambda} \tag{2.34}$$

Since the optimal weight vector must satisfy Equation (2.26), we have:

$$-\boldsymbol{C}^H\boldsymbol{R}_{xx}^{-1}\boldsymbol{C}\boldsymbol{\lambda} = \mathbf{f} \tag{2.35}$$

Solving this equation for $\boldsymbol{\lambda}$ and finally substituting $\boldsymbol{\lambda}$ into Equation (2.34) yields:

$$\mathbf{w}_{opt} = \boldsymbol{R}_{xx}^{-1}\boldsymbol{C}(\boldsymbol{C}^H\boldsymbol{R}_{xx}^{-1}\boldsymbol{C})^{-1}\mathbf{f} \tag{2.36}$$

which represents the solution to the LCMV optimization problem in Equations (2.25) and (2.26) (Johnson and Dudgeon, 1993).

2.2.3 Frost's Algorithm for LCMV Beamforming

From Equation (2.36), we know that for the LCMV beamformer the optimum solution $\mathbf{w}_{opt}$ is based on the statistics of the array data. However, in many applications the second order statistics of the array data required in the correlation matrix in Equation (2.24) are unknown or may change over time. In this case, constrained adaptive algorithms can be employed for determining the coefficients of $\mathbf{w}$. One such approach is given by the Frost's algorithm as proposed in Frost (1972), which will be introduced in the following.

At the beginning, we set the weight vector to $\mathbf{w}[0] = \boldsymbol{C}(\boldsymbol{C}^H\boldsymbol{C})^{-1}\mathbf{f}$ for initialization, which satisfies the constraint in Equation (2.26). At each iteration, the vector $\mathbf{w}$ is updated in the direction of the negative gradient expressed in Equation (2.33) by a step proportional to a scaling factor μ according to:

$$\mathbf{w}[n+1] = \mathbf{w}[n] - \mu\big(\boldsymbol{R}_{xx}\mathbf{w}[n] + \boldsymbol{C}\boldsymbol{\lambda}[n]\big) \tag{2.37}$$

Since $\mathbf{w}[n+1]$ must satisfy the constraint in Equation (2.26), we can substitute Equation (2.37) into Equation (2.26) and solve for the Lagrange multipliers $\boldsymbol{\lambda}[n]$. Then we substitute $\boldsymbol{\lambda}[n]$ into the iteration equation in (2.37) and arrive at:

$$\begin{aligned}\mathbf{w}[n+1] = \mathbf{w}[n] - &\\ &\mu\big(\boldsymbol{I} - \boldsymbol{C}(\boldsymbol{C}^H\boldsymbol{C})^{-1}\boldsymbol{C}^H\big)\boldsymbol{R}_{xx}\mathbf{w}[n] + \boldsymbol{C}(\boldsymbol{C}^H\boldsymbol{C})^{-1}\big(\mathbf{f} - \boldsymbol{C}^H\mathbf{w}[n]\big)\end{aligned} \tag{2.38}$$

Upon defining the short-hand of $\boldsymbol{P} = \boldsymbol{I} - \boldsymbol{C}(\boldsymbol{C}^H\boldsymbol{C})^{-1}\boldsymbol{C}^H$, the algorithm in Equation (2.38) can be rewritten as:

$$\mathbf{w}[n+1] = \boldsymbol{C}(\boldsymbol{C}^H\boldsymbol{C})^{-1}\mathbf{f} + \boldsymbol{P}\big(\mathbf{w}[n] - \mu\boldsymbol{R}_{xx}\mathbf{w}[n]\big) \tag{2.39}$$

Not knowing the true second order statistics $\boldsymbol{R}_{xx}$, the correlation matrix can be replaced by its simple approximation $\tilde{\boldsymbol{R}}_{xx} = \mathbf{x}\mathbf{x}^H$. This results in the minimization of the instantaneous square error rather than the mean square error, and leads to the following stochastic constrained algorithm:

$$\mathbf{w}[n+1] = \boldsymbol{C}(\boldsymbol{C}^H\boldsymbol{C})^{-1}\mathbf{f} + \boldsymbol{P}\big(\mathbf{w}[n] - \mu e^*[n]\mathbf{x}[n]\big) \tag{2.40}$$

which is also known as the Frost's algorithm.

2.2.4 Simulations

In this part, we provide a simulation result using the Frost's algorithm. There are $M = 10$ sensors and the FIR filter length J for each sensor is set to $J = 48$. The aim is to receive a signal of interest from the broadside ($\theta = 0^\circ$) and adaptively suppress two wideband interfering signals with a normalized frequency bandwidth $\Omega \in [0.4\pi\ \pi]$ arriving from DOA angles $\theta = -30^\circ$ and 40°, respectively. The SIR for each interfering signal is -20 dB and the SNR is 20 dB. A step size of 8×10^{-7} is used in the adaptation.

To assess the convergence performance of the LCMV beamformer, we consider the ensemble mean square value of the residual error, which is defined as the difference between the array output $y[n]$ and the appropriately delayed desired signal from the

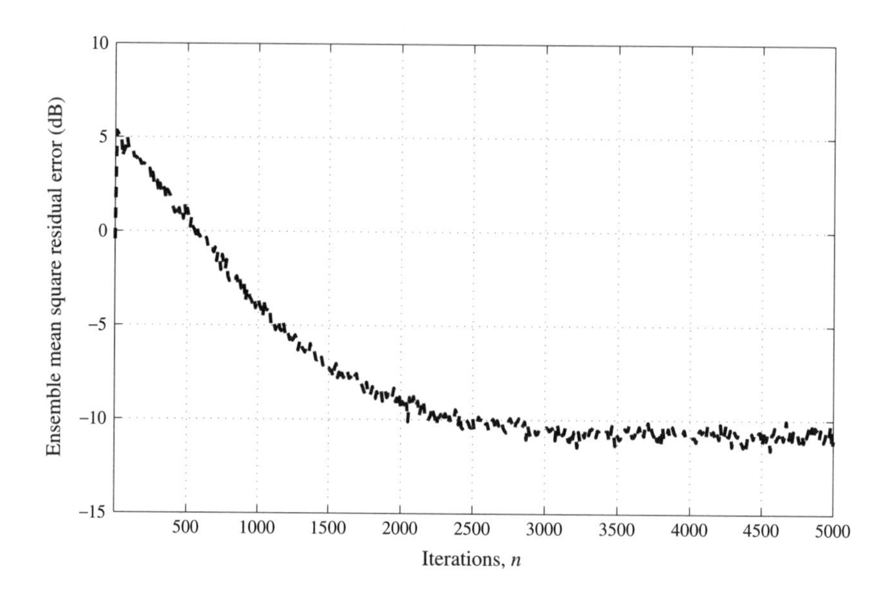

Figure 2.7 An averaged leaning curve for the LCMV beamformer

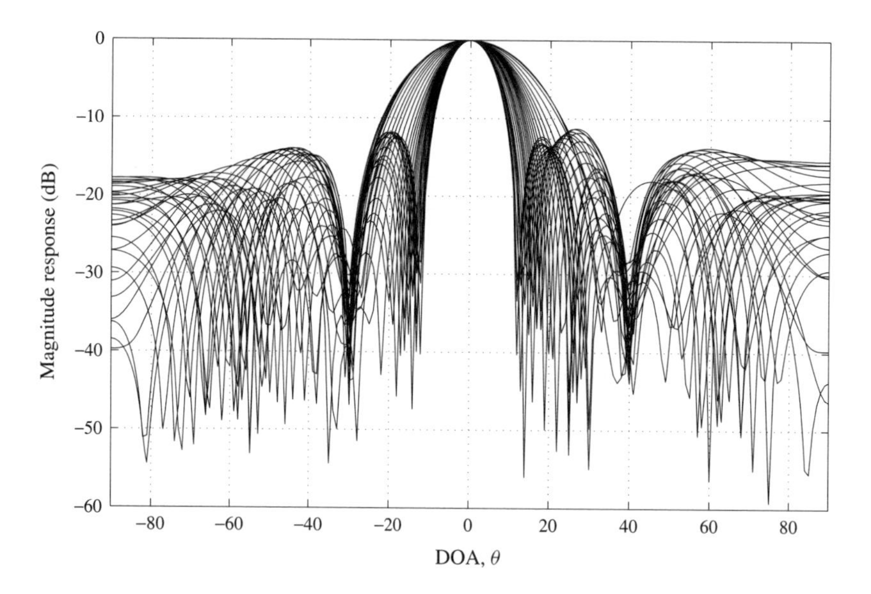

Figure 2.8 A resultant beam pattern for the LCMV beamformer

broadside. This ensemble mean square residual error $e[n]$ obtained by averaging 500 independent runs is shown in Figure 2.7 and the resultant beam pattern for this scenario is shown in Figure 2.8 over the bandwidth of $\Omega \in [0.4\pi \ \pi]$, where the two nulls at $\theta = -30^\circ$ and 40° are formed to suppress the two interfering signals.

2.3 Constraints Design for LCMV Beamforming

In Section 2.2.1, we have provided a simple formulation of the constraint matrix when the signal of interest comes from the the broadside of a linear array. For the more general case with a non-broadside arrival, we can formulate the constraint matrix by sampling the frequency band of interest of the signal and constrain the response of the beamformer to those frequency points to be the desired ones, which are usually some pure delays or zeros if we want to null out this signal.

This is a multiple linear point constraints approach and can be easily extended to multiple source directions with different bandwidths (Ahmed and Evans, 1984; Steyskal, 1983; Takao and Komiyama, 1980). The more sampling points we adopt in the constraint formulation, the better control we can achieve over the response of the beamformer for those specific directions and frequencies. However, a constraint matrix directly formed by sampling the frequency range of interest for specific directions is not the most efficient way to utilize the degrees of freedom of the array and a modified version called the eigenvector constraint approach was developed based on the low-rank representation of wideband source signals (Buckley, 1987), which is an extension of the design approach proposed for efficiently blocking wideband interfering signals (Er and Cantoni, 1986a). In this section, we will describe the eigenvector constraint approach in detail.

2.3.1 Eigenvector Constraint Design

Suppose that we want to form r linearly independent constraints for the direction θ, i.e. the rank of the constraint matrix $\boldsymbol{C}$ is r, and the frequency band of the signal is $\omega \in [\omega_{\min}\ \omega_{\max}]$. In the first step of the eigenvector constraint approach, we uniformly sample the frequency band of interest with $\hat{r} \gg r$ frequency points $\omega_i, i = 0, 1, \ldots, \hat{r} - 1$. The corresponding desired response of the beamformer for the frequency point ω_i with the direction θ is given by $G(\theta, \omega_i)$. Recall that the steering vector of a wideband array is given by Equation (1.34) as follows:

$$\mathbf{d}(\theta, \omega) = [\mathrm{e}^{-j\omega\tau_0} \ \ldots \ \mathrm{e}^{-j\omega\tau_{M-1}} \ \mathrm{e}^{-j\omega(\tau_0+T_s)} \ \ldots \ \mathrm{e}^{-j\omega(\tau_{M-1}+T_s)} \\ \ldots \ \mathrm{e}^{-j\omega(\tau_0+(J-1)T_s)} \ \ldots \ \mathrm{e}^{-j\omega(\tau_{M-1}+(J-1)T_s)}]^T \tag{2.41}$$

Then the constraint for the $\hat{r}$ frequency points can be formulated as:

$$\left[\mathbf{d}(\theta, \omega_0)\ \mathbf{d}(\theta, \omega_1)\ \ldots\ \mathbf{d}(\theta, \omega_{\hat{r}-1})\right]^H \mathbf{w} = \begin{bmatrix} G^*(\theta, \omega_0) \\ G^*(\theta, \omega_1) \\ \vdots \\ G^*(\theta, \omega_{\hat{r}-1}) \end{bmatrix} \in \mathbf{C}^{\hat{r}\times 1} \tag{2.42}$$

This is the multiple linear point constraint formulation and the constraint matrix $\boldsymbol{C}$ and the response vector $\mathbf{f}$ are given by:

$$\boldsymbol{C} = \left[\mathbf{d}(\theta, \omega_0)\ \mathbf{d}(\theta, \omega_1)\ \ldots\ \mathbf{d}(\theta, \omega_{\hat{r}-1})\right] \tag{2.43}$$

and:

$$\mathbf{f} = [G^*(\theta, \omega_0)\ G^*(\theta, \omega_1)\ \ldots\ G^*(\theta, \omega_{\hat{r}-1})]^T \tag{2.44}$$

In the second step of the eigenvector constraint approach, the matrix $\boldsymbol{C}$ is decomposed into the product of three matrices with a singular value decomposition (SVD) operation (Golub and Van Loan, 1996; Haykin, 1996; Strang, 1980), given by:

$$\boldsymbol{C} = \boldsymbol{U}\boldsymbol{\Sigma}\boldsymbol{V}^H \tag{2.45}$$

where $\boldsymbol{\Sigma}$ is an $MJ \times \hat{r}$ diagonal matrix containing the singular values of $\boldsymbol{C}$ in a descending order, $\boldsymbol{U}$ is an $MJ \times MJ$ unitary matrix and $\boldsymbol{V}$ an $\hat{r} \times \hat{r}$ unitary matrix.

To find a rank r approximation matrix $\boldsymbol{C}_r$ to the matrix $\boldsymbol{C}$, we separate matrix $\boldsymbol{U}$ into two parts as follows:

$$\boldsymbol{U} = \left[\boldsymbol{U}_r \tilde{\boldsymbol{U}}_r\right] \tag{2.46}$$

where $\boldsymbol{U}_r$ holds the first r columns of the matrix $\boldsymbol{U}$, and $\tilde{\boldsymbol{U}}_r$ holds the remaining columns of $\boldsymbol{U}$. The matrix $\boldsymbol{V}$ is split in the same way with the first sub-matrix $\boldsymbol{V}_r$ holding its first r column vectors and the second sub-matrix $\tilde{\boldsymbol{V}}_r$ holding the remaining ones, given by:

$$\boldsymbol{V} = \left[\boldsymbol{V}_r \tilde{\boldsymbol{V}}_r\right] \tag{2.47}$$

Then the approximation matrix $\boldsymbol{C}_r$ can be expressed as:

$$\boldsymbol{C}_r = \boldsymbol{U}_r \boldsymbol{\Sigma}_r \boldsymbol{V}_r^H \tag{2.48}$$

where $\boldsymbol{\Sigma}_r$ is the diagonal matrix holding the first r largest singular values in $\boldsymbol{\Sigma}$. The matrix $\boldsymbol{C}_r$ thus obtained is the best rank r approximation to $\boldsymbol{C}$ based on minimization of the error matrix' Frobenius norm $||\boldsymbol{C} - \boldsymbol{C}_r||$, where the Frobenius norm function $||\cdot||$ is defined as the square root of the sum of the squares of all the elements in the matrix concerned (Stewart, 1973, 1993). For a proof of this rank r approximation, please see Appendix A.

Then the original multiple linear point constraint formulation is changed to:

$$\boldsymbol{C}_r^H \mathbf{w} = \boldsymbol{V}_r \boldsymbol{\Sigma}_r \boldsymbol{U}_r^H \mathbf{w} = \mathbf{f} \tag{2.49}$$

which is further simplified to:

$$\boldsymbol{U}_r^H \mathbf{w} = \mathbf{f}_r \tag{2.50}$$

with $\mathbf{f}_r = \boldsymbol{\Sigma}_r^{-1} \boldsymbol{V}_r^H \mathbf{f}$.

We can now use $\boldsymbol{U}_r$ as the final constraint matrix in the LCMV problem and $\mathbf{f}_r$ as the corresponding response vector. Although the above discussion is focused on a single signal direction θ, it can be easily extended to the multiple directions case by including the steering vector $\mathbf{d}(\theta, \omega)$ for those directions in the constraint matrix $\boldsymbol{C}$ in the first step.

One key question about this eigenvector constraint approach is how to choose r, given the signal bandwidth B_θ and DOA angle θ. It has been found that the performance of a beamformer is limited by the 'time-bandwidth' product $B_\theta T_\theta$ of the signal (Ahmed and Evans, 1983; Fenn, 1985; Gabriel, 1976; Mayhan *et al.*, 1981), where T_θ is the temporal duration for the source to propagate through the beamformer to the output from the time it first reaches the array, which is given by $T_\theta = \tau_{M-1} + (J-1)T_s$ as indicated in the steering vector Equation (1.34).

A detailed study has shown that such a wideband source can be represented accurately by $D_\theta \geq \lceil B_\theta T_\theta/\pi + 1\rceil$ orthogonal basis functions, where $\lceil\cdot\rceil$ is the ceiling function rounding its element to the next integer towards infinity (Buckley, 1987). Then as a guideline, r is chosen to be D_θ to span the constraint space effectively. Note that for real-valued lowpass signals, B_θ is defined as the highest frequency component of the signal, for real-valued bandpass signals, $B_\theta = \omega_{\max} - \omega_{\min}$ and for complex-valued signals, $B_\theta = (\omega_{\max} - \omega_{\min})/2$ (Buckley, 1987).

2.3.2 Design Example

In this section, we give a design example based on a uniformly spaced wideband linear array for a real-valued signal with a normalized frequency band $\Omega \in [0.25\pi\ \pi]$.

The steering vector in Equation (1.34) is based on a complex signal $e^{j\omega t}$. To form the constraint matrix using this equation for a real-valued signal, we need to uniformly sample both the negative and the positive frequency bands. Another way is to consider a real-valued sinusoidal signal $\cos(\omega_i t)$ directly in the formulation and then extend it to multiple frequency points ω_i, $i = 0, 1, \ldots, \hat{r} - 1$.

The signal vector $\mathbf{x}$ for such a signal is given by:

$$\begin{aligned}\mathbf{x}(\theta, \omega_i) = [&\cos(\omega_i(t - \tau_0)) \ \ldots \ \cos(\omega_i(t - \tau_{M-1})) \\ &\cos(\omega_i(t - (\tau_0 + T_s))) \ \ldots \ \cos(\omega_i(t - (\tau_{M-1} + T_s))) \\ &\ldots \cos(\omega_i(t - (\tau_0 + (J-1)T_s))) \\ &\ldots \ \cos(\omega_i(t - (\tau_{M-1} + (J-1)T_s)))]^T\end{aligned} \tag{2.51}$$

For a uniformly spaced linear array ($\tau_i = i\tau_1$), according to Equation (1.36), we have:

$$\begin{aligned}\mathbf{x}(\theta, \omega_i) &= [\cos(\omega_i t) \ \ldots \ \cos(\omega_i(t - (M-1)\tau_1)) \\ &\quad \cos(\omega_i(t - T_s)) \ \ldots \ \cos(\omega_i(t - ((M-1)\tau_1 + T_s))) \\ &\quad \ldots \ \cos(\omega_i(t - (J-1)T_s)) \ \ldots \ \cos(\omega_i(t - ((M-1)\tau_1 + (J-1)T_s)))]^T \\ &= [\cos(\omega_i t) \ \ldots \ \cos(\omega_i(t - (M-1)\mu\sin\theta T_s)) \ \cos(\omega_i(t - T_s)) \\ &\quad \ldots \ \cos(\omega_i(t - ((M-1)\mu\sin\theta + 1)T_s)) \ \ldots \\ &\quad \cos(\omega_i(t - (J-1)T_s)) \ \ldots \ \cos(\omega_i(t - ((M-1)\mu\sin\theta + (J-1))T_s))]^T \\ &= \cos(\omega_i t)\mathbf{x}_c - \sin(\omega_i t)\mathbf{x}_s\end{aligned} \tag{2.52}$$

where $\mu = d/(cT_s)$:

$$\begin{aligned}\mathbf{x}_c = [1 \ &\ldots \ \cos(\Omega_i(M-1)\mu\sin\theta) \ \cos(\Omega_i) \ \ldots \ \cos(\Omega_i((M-1)\mu\sin\theta + 1)) \\ &\ldots \ \cos((J-1)\Omega_i) \ \ldots \ \cos(\Omega_i((M-1)\mu\sin\theta + (J-1)))]^T\end{aligned} \tag{2.53}$$

and:

$$\begin{aligned}\mathbf{x}_s = [1 \ &\ldots \ \sin(\Omega_i(M-1)\mu\sin\theta) \ \sin(\Omega_i) \ \ldots \ \sin(\Omega_i((M-1)\mu\sin\theta + 1)) \\ &\ldots \ \sin((J-1)\Omega_i) \ \ldots \ \sin(\Omega_i((M-1)\mu\sin\theta + (J-1)))]^T\end{aligned} \tag{2.54}$$

with $\Omega_i = \omega_i T_s$. The decomposition of $\mathbf{x}(\theta, \omega_i)$ in the third step of Equation (2.52) is obtained by employing the trigonometric identity.

Let the desired response of the beamformer to this sinusoidal signal be a pure delay T_0, i.e. the output should be:

$$\begin{aligned}\mathbf{w}^H\mathbf{x}(\theta, \omega_i) &= \cos(\omega_i(t - T_0)) \\ &= \cos(\omega_i t)\cos(\omega_i T_0) - \sin(\omega_i t)\sin(\omega_i T_0) \\ &= \cos(\omega_i t)\cos(\Omega_i \frac{T_0}{T_s}) - \sin(\omega_i t)\sin(\Omega_i \frac{T_0}{T_s})\end{aligned} \tag{2.55}$$

Comparing Equations (2.52) and (2.55), the following constraint can be imposed on the weight vector $\mathbf{w}$:

$$\begin{bmatrix} \mathbf{x}_c^T \\ \mathbf{x}_s^T \end{bmatrix} \mathbf{w} = \begin{bmatrix} \cos(\Omega_i T_0/T_s) \\ \sin(\Omega_i T_0/T_s) \end{bmatrix} \tag{2.56}$$

For the standard setup, where the inter-element spacing d is half the shortest wavelength of the signal and T_s half the corresponding period, we have $\mu = 1$, as mentioned at the end of Section 1.3. For the design example, the sensor number is $M = 10$, the FIR filter length for each sensor is $J = 96$, $\theta = -30^\circ$ and $\mu = 1$. In this case, we have:

$$D_\theta \geq \left\{ \frac{(\pi - 0.25\pi)}{T_s} \frac{[(10-1)|\sin(-30^\circ)| + (96-1)]T_s}{\pi} \right\} = 76 \tag{2.57}$$

Therefore, we choose $r = 76$. We sample the band $[0.25\pi\ \pi]$ with 50 points. Since the dimension of the resultant blocking matrix $\boldsymbol{U}_r$ is very large, instead of giving the result for the constraint matrix directly, we will show the beam response of $\mathbf{w} = \boldsymbol{U}_r(\boldsymbol{U}_r^H\boldsymbol{U}_r)^{-1}\mathbf{f}_r$, which is the least square solution to Equation (2.50) and the response should have a magnitude response of 1 over the frequency band $[0.25\pi\ \pi]$ to satisfy the constraint equation.

The design result is shown in Figure 2.9 by drawing the beam response over the frequency band $[0.25\pi\ \pi]$ with respect to the DOA angle. An almost constant response with a magnitude of 0 dB at $\theta = -30^\circ$ can be seen clearly.

2.3.3 Application to Wideband DOA Estimation

Direction of arrival (DOA) estimation is one of the major research areas in array signal processing (Krim and Viberg, 1996). For narrowband signals, there have been many approaches available and two well-known ones are the MUSIC algorithm and the ESPRIT algorithm (Roy and Kailath, 1989; Schmidt, 1986). For wideband signals, the most commonly used approach is to decompose the wideband signal into different frequency bins and then apply the narrowband DOA methods to each frequency bin or to a final coherently combined composite covariance matrix (Friedlander and Weiss, 1993; Hung and Mao, 1994; Krolik and Swingler, 1990; Lee, 1994; Wang and Kaveh, 1985; Wax *et al.*, 1984).

Nonetheless, beamforming also provides a viable approach to the DOA estimation problem and the most simple one is to employ a series of fixed beamformers pointing to different directions and their output powers are then estimated and compared to decide the right direction of the source signals (Johnson, 1982).

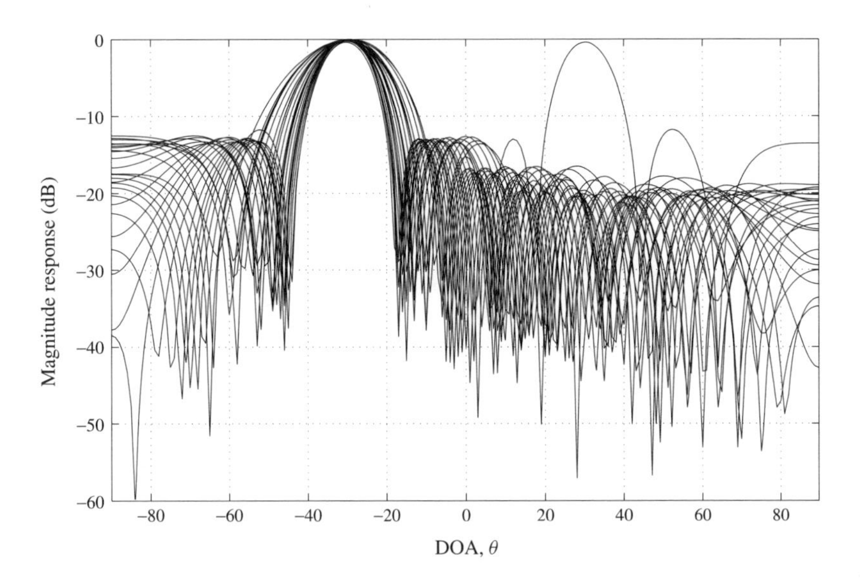

Figure 2.9 The beam response of the resultant weight vector for the frequency band $[0.25\pi\ \pi]$

A further improvement to the fixed beamformer idea is to employ the LCMV beamformer (Gabriel, 1980; Johnson, 1982; Walach, 1984). For wideband signals, at each presumed direction, we can design a set of linear constraints over the frequency band of interest using the eigenvector constraint approach and then perform the LCMV beamforming. The output power of the beamformer with respect to the DOA angle θ can then be obtained with its peaks corresponding to the real DOA angles. This process can be simplified by projecting the signal subspace of the observed data covariance matrix onto the source representation space for each DOA angle θ directly (Buckley and Griffiths, 1986). A discussion of the spatial resolution of the LCMV beamformer for wideband signals can be found in Nordholm *et al.* (1992).

For this application, we give an example based on a uniformly spaced linear array with $M = 10$ sensors and $J = 48$ taps for each sensor. A real-valued signal impinges on the array with a normalized frequency band $\Omega \in [0.4\pi\ \pi]$ and from the direction $\theta = 10^\circ$. Additionally, the sensor signals are corrupted by independent white Gaussian noise with an SNR of 20 dB.

To estimate the direction of the impinging signal, we uniformly sample the direction range from $\theta = -90^\circ$ to 90° with 181 points θ_i, $i = 0, 1, \ldots, 180$. For each θ_i, we formulate a set of constraints $\boldsymbol{C}_r$ with an appropriate choice of r according to Equation (2.56) and then calculate the corresponding optimum weight vector $\mathbf{w}_{opt}$ according to Equation (2.36). Given $\mathbf{w}_{opt}$, the output power p_θ are obtained by $p_\theta = \mathbf{w}_{opt}^H \boldsymbol{R}_{xx} \mathbf{w}_{opt}$ for θ_i. With $\boldsymbol{R}_{xx} = \boldsymbol{R}_{xx}^H$, it can be simplified as follows:

$$\begin{aligned} p_\theta &= \mathbf{f}^H (\boldsymbol{C}^H \boldsymbol{R}_{xx}^{-1} \boldsymbol{C})^{-1} \boldsymbol{C}^H \boldsymbol{R}_{xx}^{-1} \boldsymbol{R}_{xx} \boldsymbol{R}_{xx}^{-1} \boldsymbol{C} (\boldsymbol{C}^H \boldsymbol{R}_{xx}^{-1} \boldsymbol{C})^{-1} \mathbf{f} \\ &= \mathbf{f}^H (\boldsymbol{C}^H \boldsymbol{R}_{xx}^{-1} \boldsymbol{C})^{-1} \boldsymbol{C}^H \boldsymbol{R}_{xx}^{-1} \boldsymbol{C} (\boldsymbol{C}^H \boldsymbol{R}_{xx}^{-1} \boldsymbol{C})^{-1} \mathbf{f} \\ &= \mathbf{f}^H (\boldsymbol{C}^H \boldsymbol{R}_{xx}^{-1} \boldsymbol{C})^{-1} \mathbf{f} \end{aligned} \tag{2.58}$$

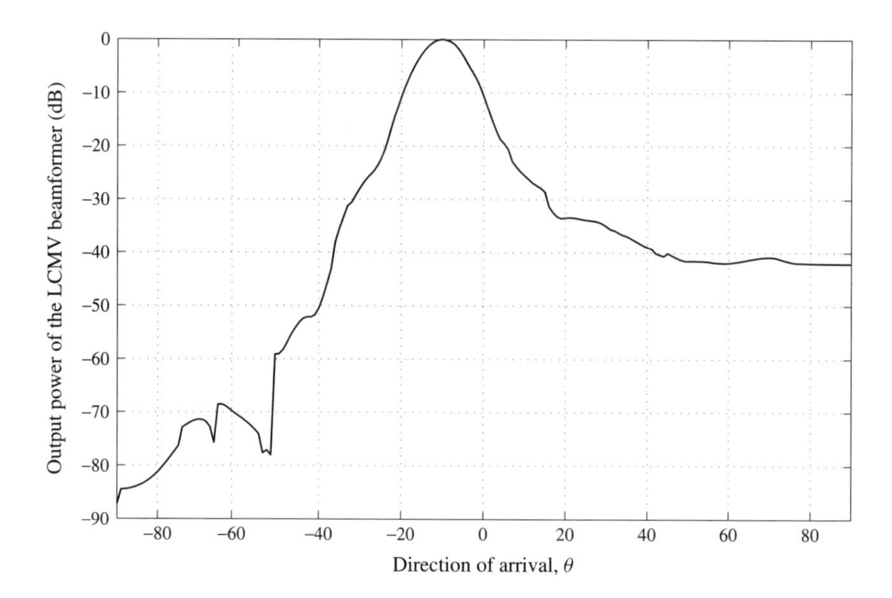

Figure 2.10 The curve showing the output power as a function of θ, where the peak at $\theta = 10°$ indicates a successful DOA estimation

In the eigenvector constraint approach, $\boldsymbol{U}_r$ is the final constraint matrix and $\mathbf{f}_r$ is the corresponding response vector. Since $\boldsymbol{U}_r$ is unitary, we have:

$$p_\theta = \mathbf{f}_r^H(\boldsymbol{U}_r^H \boldsymbol{R}_{xx} \boldsymbol{U}_r)\mathbf{f}_r \tag{2.59}$$

Then a curve of the output power p_θ with respect to different DOA angles θ_i can be drawn and the result for this example is shown in Figure 2.10, where a peak is reached at $\theta = 10°$, indicating a successful DOA estimation.

2.4 Generalized Sidelobe Canceller

Instead of using the constrained adaptive algorithm in Equation (2.40), Griffith *et al.* proposed an alternative, but efficient implementation of the LCMV beamformer, which is referred to as the generalized sidelobe canceller (GSC) (Applebaum and Chapman, 1976; Bitzer *et al.*, 1999; Breed and Strauss, 2002; Buckley, 1986; Buckley and Griffith, 1986; Cohen, 2003; Griffiths and Jim, 1982; Jim, 1977; Werner *et al.*, 2003). The GSC can be considered as a scheme for transforming the constrained minimization problem of Equation (2.25) into an unconstrained one so that the well-known standard unconstrained adaptive algorithms discussed in Section 2.1 can be employed in this new structure directly.

2.4.1 GSC Structure

The evolution of the constrained LCMV beamformer into the unconstrained GSC structure is shown in Figure 2.11(a)–(c) (Van Veen and Buckley, 1988; Weiss *et al.*, 1999b).

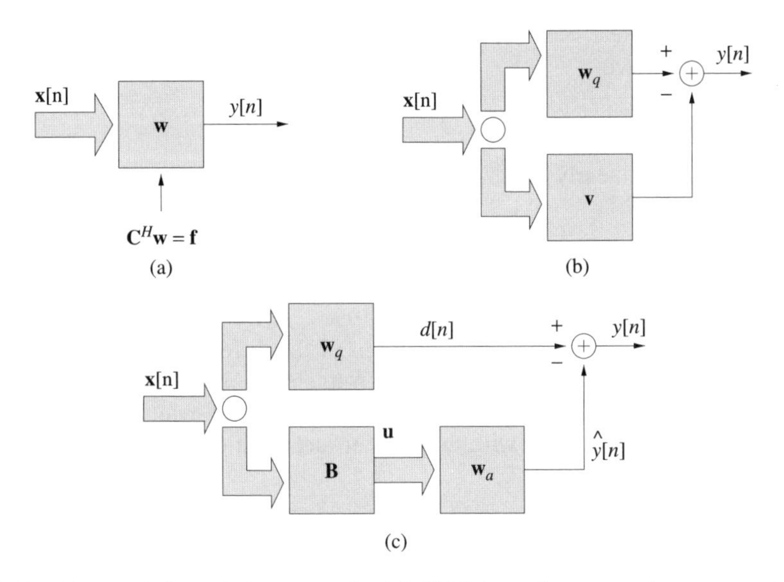

Figure 2.11 The evolution of the constrained LCMV beamformer to the unconstrained GSC structure: (a) the constrained LCMV beamformer; (b) decomposition of the weight vector into two orthogonal components, one in the range of $\boldsymbol{C}$ and one in its null space; (c) further decomposition of the null space vector into an adaptive part and a fixed part, which leads to an unconstrained adaptation problem

The LCMV beamformer has two conditions on the array weight vector $\mathbf{w}$. The first one is the constraint in Equation (2.26); the second is to minimize the output variance. In order to satisfy these two conditions simultaneously, one way is to decompose $\mathbf{w}$ into two orthogonal components $\mathbf{w}_q$ and $-\mathbf{v}$ defined in the context of $\mathbf{w} = \mathbf{w}_q - \mathbf{v}$, as seen in Figure 2.11(b). The vector $\mathbf{w}_q$ lies in the range of the matrix $\boldsymbol{C}$, while the component $\mathbf{v}$ is in the null space of $\boldsymbol{C}$, i.e. the space of all $\mathbf{v}$ fulfilling $\boldsymbol{C}^H\mathbf{v} = 0$. Since the range and null space of a matrix span the entire space (Strang, 1980), this is a general decomposition and can be used to represent any vector $\mathbf{w}$. Then, to meet the constraint in Equation (2.26), we must have:

$$\boldsymbol{C}^H\mathbf{w}_q = \mathbf{f} \tag{2.60}$$

Solving the above constraint equation using the pseudo-inverse of $\boldsymbol{C}^H$ (Golub and Van Loan, 1996), we have:

$$\mathbf{w}_q = (\boldsymbol{C}^H)^{\dagger}\mathbf{f} = \boldsymbol{C}(\boldsymbol{C}^H\boldsymbol{C})^{-1}\mathbf{f} \tag{2.61}$$

where $\{\cdot\}^{\dagger}$ indicates the pseudo-inverse. In a quiet environment (quiescent condition), where the received signal consists of white noise only, the vector $\mathbf{w}_q$ will be the optimum solution to the LCMV problem as given in Equation (2.36), since the correlation matrix $\boldsymbol{R}_{xx} = \sigma^2\boldsymbol{I}$, where σ^2 is the noise variance and $\boldsymbol{I}$ is the identity matrix. For this reason $\mathbf{w}_q$ is called the 'quiescent vector'.

Since the vector $\mathbf{v}$ lies in the null space, it can be constructed by a linear combination of the basis vectors of that null space. If the columns of a $MJ \times (MJ - r)$ matrix $\boldsymbol{B}$

form such a basis, where r is the rank of $\boldsymbol{C}$, i.e. the number of linearly independent constraints, we first have:

$$\boldsymbol{C}^H \boldsymbol{B} = \mathbf{0} \tag{2.62}$$

Using a vector $\mathbf{w}_a$ to linearly combine the basis vectors in $\boldsymbol{B}$ to form $\mathbf{v}$, we have:

$$\mathbf{v} = \boldsymbol{B}\mathbf{w}_a \tag{2.63}$$

The matrix $\boldsymbol{B}$ can be obtained from $\boldsymbol{C}$ using some orthogonalization methods such as the QR decomposition (Golub and Van Loan, 1996). Two of the widely used methods in the context of the GSC are the cascaded columns of the difference (CCD) method and the singular value decomposition (SVD) method (Applebaum and Chapman, 1976; Buckley and Griffith, 1986; Goldstein and Reed, 1997; Jablon, 1986), which will be reviewed in Section 2.4.3. The new structure with this further factorization is given in Figure 2.11(c), where:

$$\hat{y}[n] = \mathbf{w}_a^H \mathbf{u} \tag{2.64}$$

Given the choice for $\mathbf{w}_q$ and $\boldsymbol{B}$, the whole factorization $\mathbf{w} = \mathbf{w}_q - \boldsymbol{B}\mathbf{w}_a$ satisfies the constraint Equation (2.26) automatically for any choice of $\mathbf{w}_a$ and the LCMV problem is reduced to that of finding the weights $\mathbf{w}_a$, which is unconstrained and not subject to the constraints any more. Then a modified LCMV formulation is obtained as:

$$\mathbf{w}_a = \arg\min_{\mathbf{w}_a} [\mathbf{w}_q - \boldsymbol{B}\mathbf{w}_a]^H \boldsymbol{R}_{xx} [\mathbf{w}_q - \boldsymbol{B}\mathbf{w}_a] \tag{2.65}$$

The solution to the problem in Equation (2.65) can be obtained by Equation (2.36). As $\mathbf{w}_{opt} = \mathbf{w}_q - \boldsymbol{B}\mathbf{w}_{a,opt}$, we have:

$$\boldsymbol{B}\mathbf{w}_{a,opt} = \mathbf{w}_q - \boldsymbol{R}_{xx}^{-1}\boldsymbol{C}(\boldsymbol{C}^H \boldsymbol{R}_{xx}^{-1}\boldsymbol{C})^{-1}\mathbf{f} \tag{2.66}$$

Multiplying the two sides of Equation (2.66) by $\boldsymbol{B}^H \boldsymbol{R}_{xx}$, and also noting that $\boldsymbol{B}^H \boldsymbol{C} = \mathbf{0}$, we have:

$$\boldsymbol{B}^H \boldsymbol{R}_{xx} \boldsymbol{B}\mathbf{w}_{a,opt} = \boldsymbol{B}^H \boldsymbol{R}_{xx}\mathbf{w}_q - \mathbf{0} \tag{2.67}$$

Further multiplying the two sides of Equation (2.67) by the inverse of $\boldsymbol{B}^H \boldsymbol{R}_{xx} \boldsymbol{B}$, the final solution to Equation (2.65) is given by (Johnson and Dudgeon, 1993; Van Veen and Buckley, 1988):

$$\mathbf{w}_{a,opt} = (\boldsymbol{B}^H \boldsymbol{R}_{xx} \boldsymbol{B})^{-1} \boldsymbol{B}^H \boldsymbol{R}_{xx} \mathbf{w}_q \tag{2.68}$$

We can consider both the vector $\mathbf{w}_q$ and the columns of $\boldsymbol{B}$ as a set of fixed weights applied to the array data. If the constraints in Equation (2.26) are designed to present a specified gain and phase response to signals impinging on the array from a set of directions and at different frequencies, then we also have $\boldsymbol{C}^H \mathbf{w}_q = \mathbf{f}$, i.e. the upper path of the GSC in Figure 2.11(c) will have those signals pass exactly as required. The design example shown in Figure 2.9 in Section 2.3.2 can be considered as the response of such a quiescent vector $\mathbf{w}_q$ for maintaining a unit response at $\theta = -30^\circ$ over the frequency band $[0.25\pi \ \pi]$.

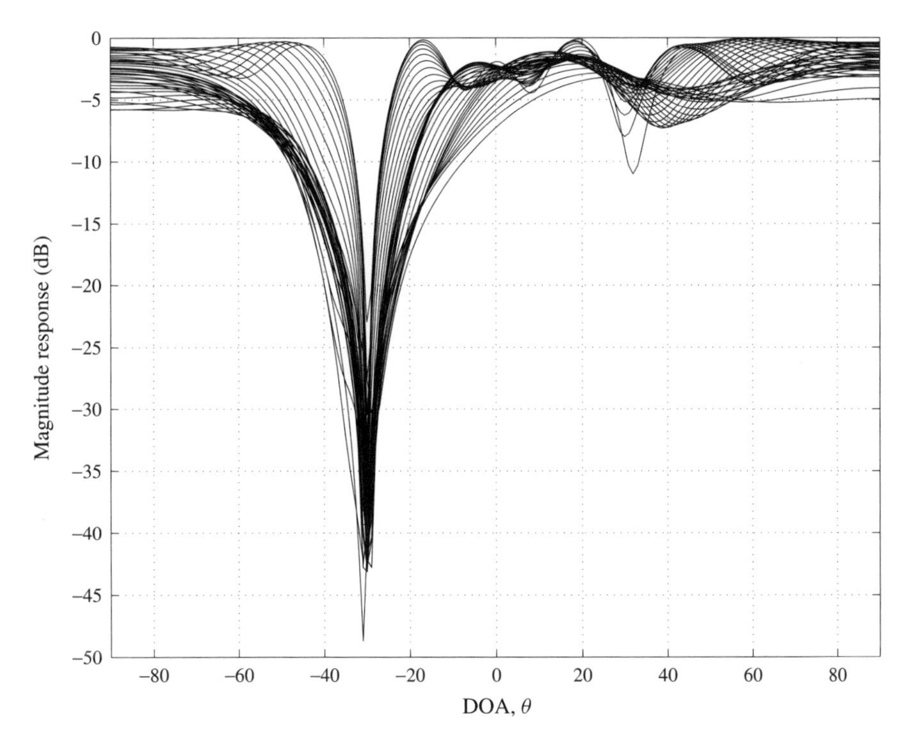

Figure 2.12 The beam response of a column of the blocking matrix corresponding to the quiescent vector example in Figure 2.9 over the frequency band $[0.25\pi\ \pi]$

On the other hand, we have $\boldsymbol{C}^H\boldsymbol{B} = \mathbf{0}$, where $\mathbf{0}$ is an $r \times (MJ - r)$ matrix with all of its elements being zero. Then the lower path of the GSC will have a zero response to those signals and equivalently we can consider the columns of $\boldsymbol{B}$ have blocked those directions and frequencies. This is why we often refer to $\boldsymbol{B}$ as the 'blocking matrix'.

Figure 2.12 shows the response of a column of the blocking matrix corresponding to the quiescent vector example of Figure 2.9, where the zero response at $\theta = -30^\circ$ indicates a total blocking of the signal for that direction.

In addition to the specified signals, there are also interfering signals and noise in the upper path, with a response also determined by $\mathbf{w}_q$. For the lower branch, since the desired signals are blocked, only the interfering signals and the noise can pass. When adapting $\mathbf{w}_a$ to minimize the variance or power of the output signal $y[n]$, the scheme will tend to cancel the interference and noise component in the upper path only.

In Figure 2.11(c), the upper branch output $d[n]$ is obtained by $d[n] = \mathbf{w}_q^H\mathbf{x}$. For the lower branch, $\mathbf{u} = \boldsymbol{B}^H\mathbf{x}$ and $\hat{y}[n] = \mathbf{w}_a^H\mathbf{u}$. To calculate the blocking matrix output $\mathbf{u}$, $MJ(MJ - r)$ multiplications are required for every sampling period. For larger values of M and J, the resultant computational complexity will increase significantly. Based on the broadside constraint formulations in Section 2.2.1, we can avoid this potential problem by a simplification of the blocking matrix and introducing a GSC with tapped delay-lines in Section 2.4.2.

2.4.2 GSC with Tapped Delay-Lines

Note that both the constraint matrix and the corresponding response vector for the broadside constraints given in Section 2.2.1 fall into the following more general form:

$$\boldsymbol{C} = \left[\hat{\boldsymbol{C}}_0 \ldots \hat{\boldsymbol{C}}_{S-1}\right] \in \mathbf{C}^{MJ\times SJ} \qquad \text{with} \quad \hat{\boldsymbol{C}}_i = \begin{bmatrix} \mathbf{c}_i & & \mathbf{0} \\ & \ddots & \\ \mathbf{0} & & \mathbf{c}_i \end{bmatrix} \in \mathbf{C}^{MJ\times J} \tag{2.69}$$

and:

$$\mathbf{f} = [a_0\mathbf{f}_0^T \;\; a_1\mathbf{f}_0^T \;\; \ldots \;\; a_{S-1}\mathbf{f}_0^T]^T \in \mathbf{C}^{SJ\times 1} \tag{2.70}$$

with:

$$\mathbf{f}_0 = [f[0] \;\; f[1] \;\; \ldots \;\; f[J-1]]^T \in \mathbf{C}^{J\times 1} \tag{2.71}$$

and $a_0 = 1$. For the broadside constraints given in Section 2.2.1, we have $S = 1$. This is actually a form obtained when imposing derivative constraints of zero up to an order of $S - 1$ on the beamformer for the broadside signal of interest, where $a_1 = \cdots = a_{S-1} = 0$ (Applebaum and Chapman, 1976; Buckley and Griffith, 1986; Er and Cantoni, 1983; Huarng and Yeh, 1992). We will discuss the derivative constraints in detail in the topic of robust adaptive wideband beamforming in Section 2.6.1.

As the blocking matrix $\boldsymbol{B}$ is composed of the basis vectors of the null space of $\boldsymbol{C}$, we have $\boldsymbol{B} \in \mathbf{C}^{MJ\times(M-S)J}$. Assume that $\boldsymbol{B}$ has the following block diagonal form:

$$\boldsymbol{B} = \begin{bmatrix} \tilde{\boldsymbol{B}} & \mathbf{0} & \ldots & \mathbf{0} \\ \mathbf{0} & \tilde{\boldsymbol{B}} & \ldots & \\ \vdots & \vdots & \ddots & \vdots \\ \mathbf{0} & \mathbf{0} & \ldots & \tilde{\boldsymbol{B}} \end{bmatrix} \tag{2.72}$$

where $\tilde{\boldsymbol{B}}$ is an $M \times (M - S)$ dimensional matrix. The condition for the blocking matrix $\boldsymbol{B}$ in Equation (2.62) can then be expressed as:

$$\boldsymbol{C}^H\boldsymbol{B} = \begin{bmatrix} \hat{\boldsymbol{C}}_0^H\boldsymbol{B} \\ \hat{\boldsymbol{C}}_1^H\boldsymbol{B} \\ \vdots \\ \hat{\boldsymbol{C}}_{S-1}^H\boldsymbol{B} \end{bmatrix} = \mathbf{0} \tag{2.73}$$

where the rows $\hat{\boldsymbol{C}}_i^H\boldsymbol{B}$, $i = 0, 1, \ldots, S-1$, can be expressed as:

$$\hat{\boldsymbol{C}}_i^H\boldsymbol{B} = \begin{bmatrix} \mathbf{c}_i^H\tilde{\boldsymbol{B}} & \mathbf{0} & \ldots & \mathbf{0} \\ \mathbf{0} & \mathbf{c}_i^H\tilde{\boldsymbol{B}} & \ldots & \\ \vdots & \vdots & \ddots & \vdots \\ \mathbf{0} & \mathbf{0} & \ldots & \mathbf{c}_i^H\tilde{\boldsymbol{B}} \end{bmatrix} \tag{2.74}$$

Then, as long as $\tilde{\boldsymbol{B}}$ fulfils:

$$\tilde{\boldsymbol{C}}^H \tilde{\boldsymbol{B}} = \mathbf{0} \qquad \text{where} \quad \tilde{\boldsymbol{C}} = \left[\mathbf{c}_0 \ \ldots \mathbf{c}_{S-1}\right] \tag{2.75}$$

the original blocking matrix $\boldsymbol{B}$ will automatically satisfy Equation (2.62).

With this construction, the blocking matrix output $\mathbf{u} = \boldsymbol{B}^H \mathbf{x}$ becomes:

$$\mathbf{u} = \begin{bmatrix} \tilde{\boldsymbol{B}}^H \mathbf{x}[n] \\ \tilde{\boldsymbol{B}}^H \mathbf{x}[n-1] \\ \vdots \\ \tilde{\boldsymbol{B}}^H \mathbf{x}[n-J+1] \end{bmatrix} \tag{2.76}$$

Assume that we have $\mathbf{u}[n] = [u_0[n], \ldots, u_{M-S-1}[n]]^T = \tilde{\boldsymbol{B}}^H \mathbf{x}[n]$. Then we see that the input signal $\mathbf{u}$ to the adaptation block $\mathbf{w}_a$ is a series of delayed versions of the signal vector $\mathbf{u}[n]$, namely:

$$\mathbf{u} = \begin{bmatrix} \mathbf{u}[n] \\ \mathbf{u}[n-1] \\ \vdots \\ \mathbf{u}[n-J+1] \end{bmatrix} \tag{2.77}$$

Thus, it is possible to apply a smaller blocking matrix $\tilde{\boldsymbol{B}}$ straight to the sensor signal vector $\mathbf{x}[n]$, rather than applying the fullsize matrix $\boldsymbol{B}$. As a result, the output vector $\mathbf{u}[n]$ from $\tilde{\boldsymbol{B}}$ is fed into the $M - S$ tapped delay-lines of length J, as shown in Figure 2.13.

Considering the quiescent vector, the sparse nature of $\boldsymbol{C}$ allows us to rearrange Equation (2.61) and obtain a simplified form for $\mathbf{w}_q$ in the following (Buckley and Griffith, 1986):

$$\mathbf{w}_q = \hat{\mathbf{f}} \bigotimes \tilde{\mathbf{w}}_q \tag{2.78}$$

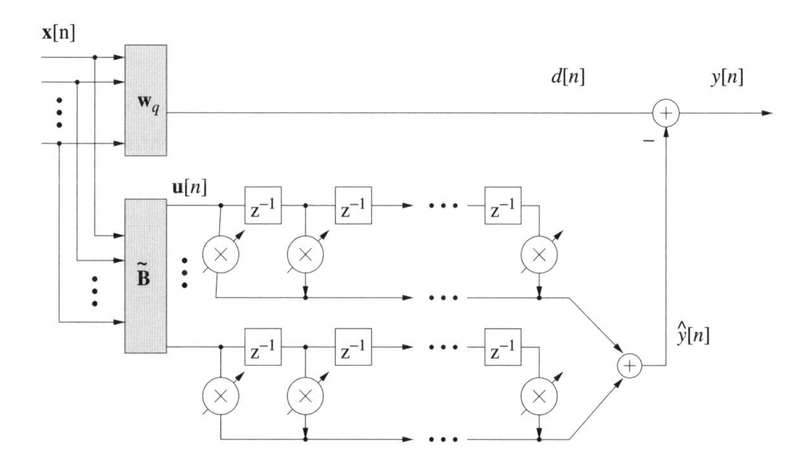

Figure 2.13 A simplified GSC structure with tapped delay-lines at the lower branch

where the operator $\bigotimes$ denotes the Kronecker product operator (Horn and Johnson, 1985), and:

$$\hat{\mathbf{f}} = \begin{bmatrix} f[0] \\ f[1] \\ \vdots \\ f[J-1] \end{bmatrix} \tag{2.79}$$

$$\tilde{\mathbf{w}}_q = \tilde{\boldsymbol{C}}(\tilde{\boldsymbol{C}}^H \tilde{\boldsymbol{C}})^{-1}\mathbf{e} \tag{2.80}$$

with $\mathbf{e} = [a_0, a_1, \ldots, a_{S-1}]^T$ being an $S \times 1$ vector.

We derive the formulation for the quiescent vector $\mathbf{w}_q$ in Equation (2.78) by first permutating $\boldsymbol{C}$ to obtain a new constraint matrix:

$$\bar{\boldsymbol{C}} = \begin{bmatrix} \tilde{\boldsymbol{C}} & \mathbf{0} & \ldots & \mathbf{0} \\ \mathbf{0} & \tilde{\boldsymbol{C}} & \ldots & \\ \vdots & \vdots & \ddots & \vdots \\ \mathbf{0} & \mathbf{0} & \ldots & \tilde{\boldsymbol{C}} \end{bmatrix}_{MJ \times SJ} \tag{2.81}$$

Correspondingly, the original constraint equation becomes:

$$\bar{\boldsymbol{C}}^H \mathbf{w} = \hat{\mathbf{f}} \bigotimes \mathbf{e} \tag{2.82}$$

From Equation (2.61), we have:

$$\begin{aligned}
\mathbf{w}_q &= \bar{\boldsymbol{C}}(\bar{\boldsymbol{C}}^H \bar{\boldsymbol{C}})^{-1}(\hat{\mathbf{f}} \bigotimes \mathbf{e}) \\
&= \bar{\boldsymbol{C}} \cdot \begin{bmatrix} \tilde{\boldsymbol{C}}^H\tilde{\boldsymbol{C}} & \mathbf{0} & \ldots & \mathbf{0} \\ \mathbf{0} & \tilde{\boldsymbol{C}}^H\tilde{\boldsymbol{C}} & \ldots & \\ \vdots & \vdots & \ddots & \vdots \\ \mathbf{0} & \mathbf{0} & \ldots & \tilde{\boldsymbol{C}}^H\tilde{\boldsymbol{C}} \end{bmatrix}^{-1} (\hat{\mathbf{f}} \bigotimes \mathbf{e}) \\
&= \begin{bmatrix} \tilde{\boldsymbol{C}}(\tilde{\boldsymbol{C}}^H\tilde{\boldsymbol{C}})^{-1} & \mathbf{0} & \ldots & \mathbf{0} \\ \mathbf{0} & \tilde{\boldsymbol{C}}(\tilde{\boldsymbol{C}}^H\tilde{\boldsymbol{C}})^{-1} & \ldots & \\ \vdots & \vdots & \ddots & \vdots \\ \mathbf{0} & \mathbf{0} & \ldots & \tilde{\boldsymbol{C}}(\tilde{\boldsymbol{C}}^H\tilde{\boldsymbol{C}})^{-1} \end{bmatrix} (\hat{\mathbf{f}} \bigotimes \mathbf{e})
\end{aligned} \tag{2.83}$$

Now we arrive at:

$$\mathbf{w}_q = \begin{bmatrix} f[0]\tilde{\boldsymbol{C}}(\tilde{\boldsymbol{C}}^H\tilde{\boldsymbol{C}})^{-1}\mathbf{e} \\ f[1]\tilde{\boldsymbol{C}}(\tilde{\boldsymbol{C}}^H\tilde{\boldsymbol{C}})^{-1}\mathbf{e} \\ \vdots \\ f[J-1]\tilde{\boldsymbol{C}}(\tilde{\boldsymbol{C}}^H\tilde{\boldsymbol{C}})^{-1}\mathbf{e} \end{bmatrix} = \begin{bmatrix} f[0]\tilde{\mathbf{w}}_q \\ f[1]\tilde{\mathbf{w}}_q \\ \vdots \\ f[J-1]\tilde{\mathbf{w}}_q \end{bmatrix} \tag{2.84}$$

which leads to Equation (2.78). As we have $d[n] = \mathbf{w}_q^H \mathbf{x}$, with the simplification of $\mathbf{w}_q$, we can express $d[n]$ as:

$$d[n] = \hat{\mathbf{f}}^H \begin{bmatrix} \tilde{\mathbf{w}}_q^H \mathbf{x}[n] \\ \tilde{\mathbf{w}}_q^H \mathbf{x}[n-1] \\ \vdots \\ \tilde{\mathbf{w}}_q^H \mathbf{x}[n-J+1] \end{bmatrix} \tag{2.85}$$

With the notation $\tilde{d}[n] = \tilde{\mathbf{w}}_q^H \mathbf{x}[n]$, Equation (2.85) can be further simplified as:

$$d[n] = \hat{\mathbf{f}}^H \begin{bmatrix} \tilde{d}[n] \\ \tilde{d}[n-1] \\ \vdots \\ \tilde{d}[n-J+1] \end{bmatrix}$$

$$= f^*[0] \cdot \tilde{d}[n] + f^*[1] \cdot \tilde{d}[n-1] + \cdots + f^*[J-1] \cdot \tilde{d}[n-J+1] \tag{2.86}$$

Now we can introduce the fully simplified GSC structure with tapped delay-lines as shown in Figure 2.14. The output of the vector $\tilde{\mathbf{w}}_q$ is processed by an FIR filter with coefficients held in $\hat{\mathbf{f}}$, while the adaptive part of the GSC after $\tilde{\boldsymbol{B}}$ becomes a multi-channel adaptive filtering system, which will form the basis for the subband adaptive GSC proposed in Chapter 3. In this simplified structure, we still refer to $\tilde{\mathbf{w}}_q$ and $\tilde{\boldsymbol{B}}$ as the quiescent vector and the blocking matrix, respectively.

For MVDR beamformers, only one coefficient of $\hat{\mathbf{f}}$ is one and all the others are zero (see Section 2.2.1). Thus the FIR filter in the upper branch of Figure 2.14 becomes a pure delay as shown in Figure 2.15. If we set $\mathbf{f} = [1 \ 0 \ \ldots \ 0]^T$, then the delay in Figure 2.15 will be zero.

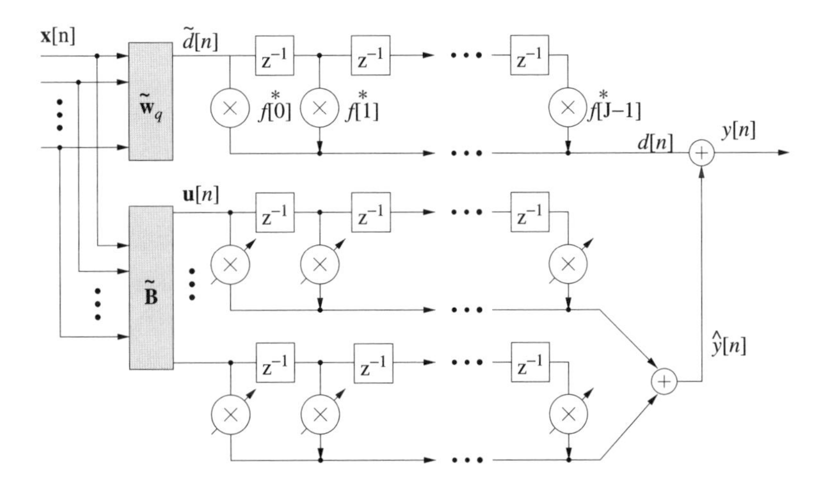

Figure 2.14 A fully simplified GSC structure with tapped delay-lines

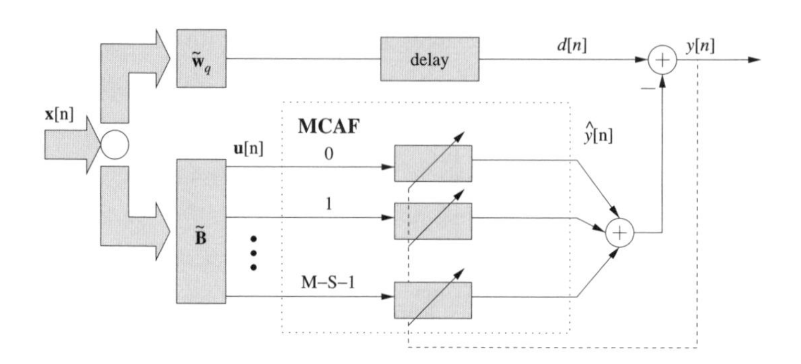

Figure 2.15 The fully simplified GSC structure with a pure delay response to the desired signal

2.4.3 Blocking Matrix Design

In the GSC structure, we have to find the proper blocking matrix $\boldsymbol{B}$, which fulfils the requirement formulated in Equation (2.62). Through the rearrangement outlined in Section 2.4.2, this problem is reduced to finding a suitable blocking matrix $\tilde{\boldsymbol{B}}$. As mentioned in Section 2.4.1, such a blocking matrix can be obtained by the SVD method and the CCD method (Applebaum and Chapman, 1976; Buckley and Griffith, 1986; Goldstein and Reed, 1997; Jablon, 1986), which we will review briefly under the constraints specified in Section 2.2.1. We will find later in Chapter 3 that the blocking matrix could be constructed quite differently from the approach of these two methods with the aim of satisfying certain specific characteristics, which can be exploited for reducing the complexity of the subband adaptive GSC (Liu *et al.*, 2003a, 2004a) and the transform-domain GSC (An and Champagne, 1994; Chen and Fang, 1992; Goldstein *et al.*, 1992; Liu *et al.*, 2002a).

2.4.3.1 Singular Value Decomposition

The singular value decomposition theorem states that, given a matrix $\boldsymbol{A}$, there exist two unitary matrices $\boldsymbol{U}$ and $\boldsymbol{V}$, such that we have:

$$\boldsymbol{A} = \boldsymbol{U}\begin{bmatrix} \boldsymbol{\Sigma}_r & \boldsymbol{0} \\ \boldsymbol{0} & \boldsymbol{0} \end{bmatrix}\boldsymbol{V}^H \tag{2.87}$$

where $\boldsymbol{\Sigma}_r$ is an $r \times r$ diagonal matrix containing the ordered positive definite singular values of $\boldsymbol{A}$. The variable r is the rank of $\boldsymbol{A}$ and represents the number of linearly independent columns in this matrix.

Let us separate matrix $\boldsymbol{U}$ into two parts as follows:

$$\boldsymbol{U} = \left[\boldsymbol{U}_r \tilde{\boldsymbol{U}}_r\right] \tag{2.88}$$

where $\boldsymbol{U}_r$ holds the first r columns of the matrix $\boldsymbol{U}$, whereas $\tilde{\boldsymbol{U}}_r$ holds the remaining columns of $\boldsymbol{U}$; then it is easy to see that:

$$\tilde{\boldsymbol{U}}_r^H \boldsymbol{A} = \boldsymbol{0} \tag{2.89}$$

i.e. $\tilde{\boldsymbol{U}}_r$ forms a basis for the null space of $\boldsymbol{A}$. If we replace the matrix $\boldsymbol{A}$ by the constraint matrix $\boldsymbol{C}$ or $\tilde{\boldsymbol{C}}$ in the SVD decomposition, the resultant $\tilde{\boldsymbol{U}}_r$ will be our desired blocking matrix $\boldsymbol{B}$ or $\tilde{\boldsymbol{B}}$.

Note the SVD approach is not limited to the broadside constraints case and can be applied to any constraint matrix $\boldsymbol{C}$. Therefore it is a general approach for obtaining the blocking matrix.

2.4.3.2 Cascaded Columns of Differencing

In the CCD method, the blocking matrix is formed by S cascaded columns of differencing operations as shown in Figure 2.16 (Jablon, 1986).

In matrix form, the blocking matrix can be formulated as (Jablon, 1986):

$$\tilde{\boldsymbol{B}} = \boldsymbol{B}_M \cdot \boldsymbol{B}_{M-1} \cdots \boldsymbol{B}_{M-S+1} \tag{2.90}$$

where we have:

$$\boldsymbol{B}_i = \begin{bmatrix} 1 & -1 & & & \boldsymbol{0} \\ & \ddots & \ddots & & \\ & & \ddots & \ddots & \\ & & & \ddots & \ddots \\ \boldsymbol{0} & & & 1 & -1 \end{bmatrix}^T \in \mathbf{C}^{i \times i-1} \tag{2.91}$$

with $i = M, M-1, \ldots, M-S+1$. Clearly, if the signal of interest comes from the broadside, it will not be able to pass through such a blocking matrix. The zero response formed by this blocking matrix at the broadside will have a wider and wider lobe width with increasing S.

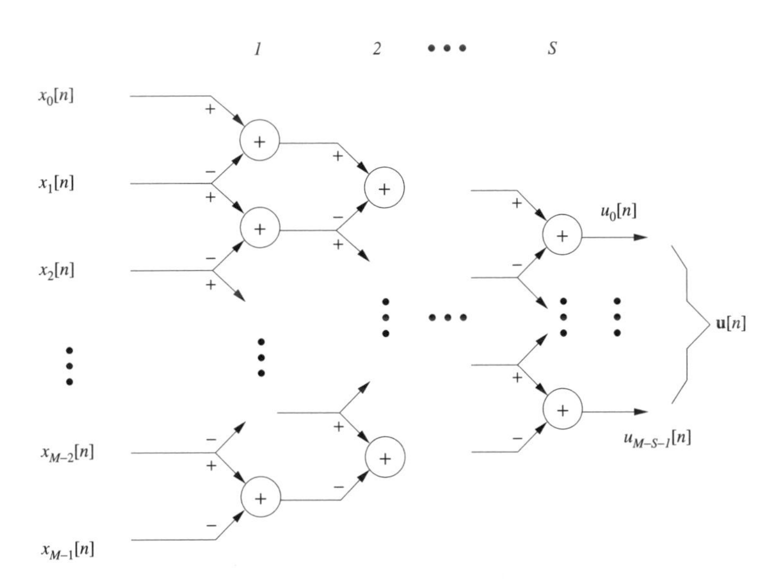

Figure 2.16 The blocking matrix obtained by S cascaded columns of differencing (Jablon, 1986).

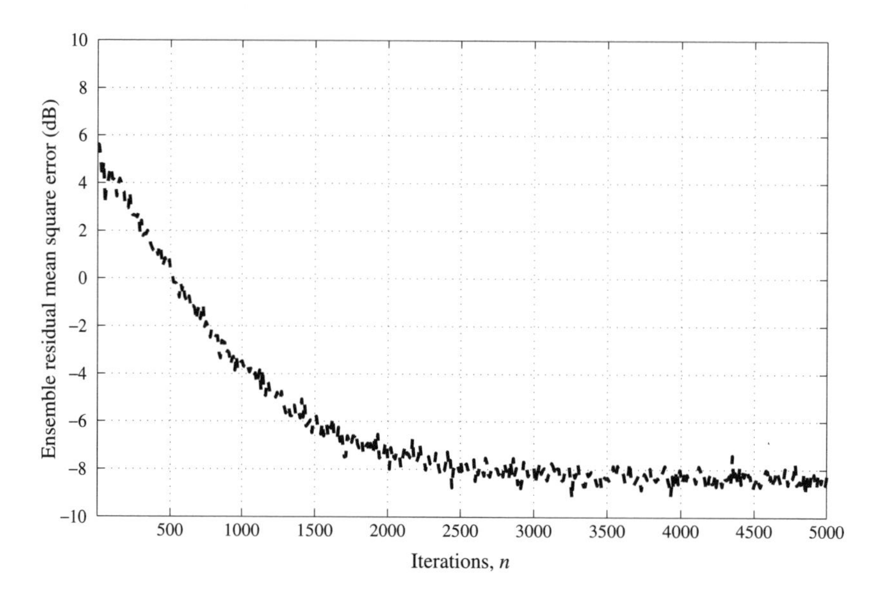

Figure 2.17 An averaged learning curve for the GSC

2.4.4 Simulations

In this part, we provide a simulation result based on the GSC structure. There are $M = 10$ sensors and the FIR filter length J for each sensor is set to $J = 48$. The aim is to receive a signal of interest from the broadside ($\theta = 0^\circ$) and adaptively suppress two wideband interfering signals with a normalized frequency bandwidth $\Omega \in [0.4\pi\ \pi]$ arriving from DOA angles $\theta = 30^\circ$ and -40°, respectively. The SIR for each interfering signal is -20 dB and the SNR is 20 dB. This has the same settings as the one in Section 2.2.4. Since the signal of interest is from the broadside, the constraint matrix is sparse and the GSC can be implemented by the simplified structure with TDLs. The blocking matrix is formulated by the CCD method with $S = 1$. For the adaptive part of the GSC, we use the normalized LMS algorithm with a step size of 0.1.

The ensemble mean square residual error $e[n]$ obtained by 500 independent runs is shown in Figure 2.17 and the resultant beam pattern for this scenario is shown in Figure 2.18 over the bandwidth of $\Omega \in [0.4\pi\ \pi]$, where the two nulls at $\theta = -30^\circ$ and 40° are clearly visible.

2.5 Other Minimum Variance Beamformers

In addition to the traditional LCMV beamformer and the GSC introduced in the previous sections, there are also some other minimum variance based beamformers and in this section we briefly review two of them. The first one is the soft constrained minimum variance (SCMV) beamformer (Ahmed and Evans, 1984; Er, 1993; Er and Cantoni, 1985, 1990; Kaneda and Ohga, 1986; Van Veen, 1991) and the second one is the correlation constrained minimum variance beamformer (CCMV) (Er, 1993; Er and Cantoni, 1990; Kikuma and Takao, 1989).

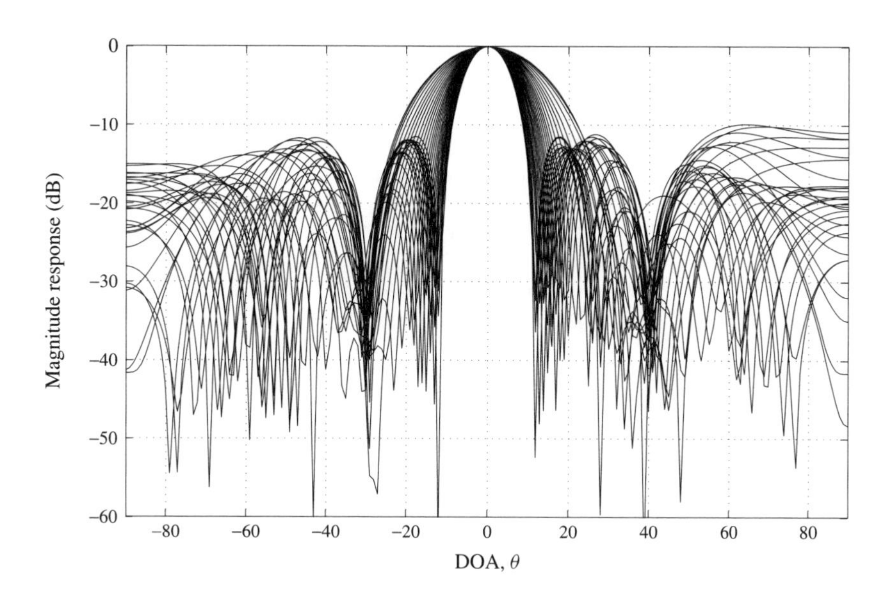

Figure 2.18 The resultant beam pattern for the GSC

2.5.1 *Soft Constrained Minimum Variance Beamformer*

In the classical LCMV beamformer, the response of the array to the desired signals and directions is specified exactly as given by the constraint equation and no deviation from the equation is permitted for the final solution. Although these kinds of 'hard' constraints guarantee a pre-determined response to the desired signals and directions in the resultant beamformer, it also reduces the degrees of freedom available in the weight vector for suppressing the interfering signals.

As a compromise, we can adopt a 'soft' constraint approach, i.e. allowing some error/tolerance in the desired response specified by the constraint equation, so that the beamformer can exploit this flexibility to suppress the interfering signals further, at the expense of a slightly distorted response to the desired signals and directions (Van Veen, 1991). The distortion can be calculated after obtaining the optimum weight vector and equalized at the beamformer output with an appropriately chosen temporal filter.

For a single point constraint, corresponding to the LCMV formulation in Equation (2.25), the SCMV beamforming problem can be expressed as (Ahmed and Evans, 1984):

$$\mathbf{w} = \arg\min_{\mathbf{w}} \mathbf{w}^H \boldsymbol{R}_{xx}\mathbf{w} \qquad \text{subject to} \qquad |\mathbf{d}^H(\theta_0, \omega_0)\mathbf{w} - G_0^*| \leq \varepsilon_0 \tag{2.92}$$

where ε_0 is the maximum tolerable error between the real response and the desired one for the point (θ_0, ω_0).

For a wide bandwidth of $\omega \in [\omega_l \ \omega_u]$, we have the following formulation:

$$\mathbf{w} = \arg\min_{\mathbf{w}} \mathbf{w}^H \boldsymbol{R}_{xx}\mathbf{w} \quad \text{subject to} \quad |\mathbf{d}^H(\theta_0, \omega)\mathbf{w} - G_\omega^*| \leq \varepsilon_\omega \quad \forall \omega \in [\omega_l \ \omega_u] \tag{2.93}$$

where G_ω is the desired response at the frequency ω and ε_ω the corresponding error parameter. For real signals, we need to consider the constraints on both the negative and positive frequencies.

An alternative implementation in the mean square error sense of the constraint in Equation (2.92) takes the following quadratic form:

$$|\mathbf{d}^H(\theta_0, \omega_0)\mathbf{w} - G_0^*|^2 \leq \varepsilon_0^2 \tag{2.94}$$

and for the wideband signal $\omega \in [\omega_l \ \omega_u]$, we have (Cox *et al.*, 1987; Er, 1993; Er and Cantoni, 1985; Er and Ng, 1990; Kaneda and Ohga, 1986; Van Veen, 1991):

$$\int_{\omega_l}^{\omega_u} |\mathbf{d}^H(\theta_0, \omega)\mathbf{w} - G_\omega^*|^2 \mathrm{d}\omega \leq \varepsilon \tag{2.95}$$

Suppose the weight vector $\mathbf{w}_q$ will implement the corresponding hard constraint exactly, namely:

$$\mathbf{d}^H(\theta_0, \omega)\mathbf{w}_q - G_\omega^* = 0 \qquad \text{for } \omega \in [\omega_l \ \omega_u] \tag{2.96}$$

which is equivalent to the constraint given in Equation (2.26) in theory. This is why we use the same symbol $\mathbf{w}_q$ as the one used for the quiescent vector in the GSC structure.

Then with $\mathbf{v} = \mathbf{w}_q - \mathbf{w}$, Equation (2.95) changes to:

$$\mathbf{v}^H \boldsymbol{A}\mathbf{v} \leq \varepsilon \tag{2.97}$$

with:

$$\boldsymbol{A} = \int_{\omega_l}^{\omega_u} \mathbf{d}(\theta_0, \omega)\mathbf{d}^H(\theta_0, \omega)\mathrm{d}\omega \tag{2.98}$$

Then the new SCMV problem formulation can be expressed as (Van Veen, 1991):

$$\mathbf{v} = \arg\min_{\mathbf{v}} (\mathbf{w}_q - \mathbf{v})^H \boldsymbol{R}_{xx}(\mathbf{w}_q - \mathbf{v}) \qquad \text{subject to} \qquad \mathbf{v}^H \boldsymbol{A}\mathbf{v} \leq \varepsilon \tag{2.99}$$

Its solution can be obtained by the method of Lagrange multipliers and is given by:

$$\mathbf{v}_{opt} = (\boldsymbol{R}_{xx} + \lambda\boldsymbol{A})^{-1}\boldsymbol{R}_{xx}\mathbf{w}_q \tag{2.100}$$

where λ is the Lagrange multiplier. By substituting $\mathbf{v}_{opt}$ into the constraint $\mathbf{v}^H \boldsymbol{A}\mathbf{v} \leq \varepsilon$, we have:

$$[(\boldsymbol{R}_{xx} + \lambda\boldsymbol{A})^{-1}\boldsymbol{R}_{xx}\mathbf{w}_q]^H \boldsymbol{A}(\boldsymbol{R}_{xx} + \lambda\boldsymbol{A})^{-1}\boldsymbol{R}_{xx}\mathbf{w}_q \leq \varepsilon \tag{2.101}$$

λ and ε are inversely related and we should choose an as large as possible value for λ (Van Veen, 1991). It is shown that the output SINR of the SCMV beamformer is a nondecreasing function of the permitted distortion parameter ε assuming that the direction of arrival of the signal of interest and its spectrum are known.

For details of various structures and adaptive algorithms to implement the SCMV beamformer, please refer to (Ahmed and Evans, 1984; Cox *et al.*, 1987; Er and Cantoni, 1985, 1990; Kaneda and Ohga, 1986; Van Veen, 1991).

2.5.2 Correlation Constrained Minimum Variance Beamformer

The correlation constrained minimum variance beamformer is another class of minimum variance beamformer, where the constraint is formed based on the correlation of the signal of interest between a reference point inside the beamformer and its final output (Er, 1993; Er and Cantoni, 1990; Kikuma and Takao, 1989).

Suppose the signal of interest comes from the direction θ_d and the corresponding received array signal vector is $\mathbf{x}_d(t)$. Then the component $y_d(t)$ for the signal of interest at the output is given by:

$$y_d(t) = \mathbf{w}^H \mathbf{x}_d(t) \tag{2.102}$$

If we take the centre of the array as the reference point and the received signal of interest for this reference point is $s_d(t)$, then the correlation constraint can be formulated as:

$$E\{s_d^*(t) y_d(t)\} = \mathbf{w}^H E\{s_d^*(t)\mathbf{x}_d(t)\} = \mathbf{w}^H \mathbf{r}_d = G_d \tag{2.103}$$

where $\mathbf{r}_d = E\{s_d^*(t)\mathbf{x}_d(t)\}$ is the correlation vector between $s_d(t)$ and $\mathbf{x}_d(t)$, and G_d is the desired correlation value.

Then the CCMV beamforming problem can be formulated as:

$$\mathbf{w} = \arg\min_{\mathbf{w}} \mathbf{w}^H \boldsymbol{R}_{xx} \mathbf{w} \qquad \text{subject to} \qquad \mathbf{r}_d^H \mathbf{w} = G_d^* \tag{2.104}$$

which is also a linearly constrained minimum variance problem and can be solved using the standard LCMV approach introduced before. However, a difference is, in this CCMV problem we need to find the correlation vector $\mathbf{r}_d$, which will require information of the spectrum of the desired signal, in addition to its bandwidth and DOA angle. Therefore, more information is needed to solve the CCMV problem than the traditional LCMV one.

Suppose the desired signal $s_d(t)$ with a unity power has a flat power spectral density of $2\pi/\Delta\omega$ over a limited bandwidth $\Delta\omega = \omega_u - \omega_l$ as shown in Figure 2.19. Then the correlation function $r_d(\tau)$ of the desired signal can be obtained by the inverse Fourier transform of the power spectral density as follows:

$$r_d(\tau) = \text{sinc}\left(\frac{\Delta\omega\tau}{2}\right) e^{j\frac{\omega_l+\omega_u}{2}\tau} \tag{2.105}$$

With $r_d(\tau)$, the correlation vector $\mathbf{r}_d$ in the constraint equation can be obtained by replacing the parameter τ by the corresponding delays. Note for the correlation function of a real-valued signal, we simply take the real part of $r_d(\tau)$.

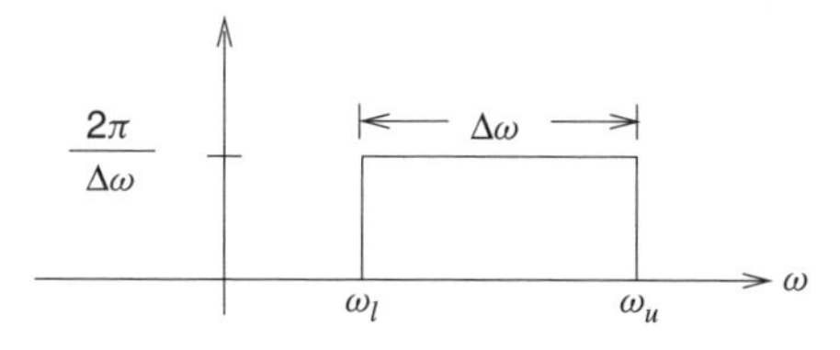

Figure 2.19 Power spectral density of the desired signal with a unit power

In practice, the power spectrum density of the signal of interest may not be known and in that case the beamformer will not achieve the desired response as specified by the constraint. One solution is to add the correlation matrix $\boldsymbol{R}_{x_d x_d} = E\{\mathbf{x}_d \mathbf{x}_d^H\}$ of the modelled desired signal to the original data correlation matrix $\boldsymbol{R}_{xx}$ so that a signal is artificially generated with the presumed power spectrum density in the direction of the signal of interest. The formulation for this modified problem is:

$$\mathbf{w} = \arg\min_{\mathbf{w}} \mathbf{w}^H(\boldsymbol{R}_{xx} + \alpha \boldsymbol{R}_{x_d x_d})\mathbf{w} \qquad \text{subject to} \qquad \mathbf{r}_d^H \mathbf{w} = G_d^* \tag{2.106}$$

where α is a scalar with a small positive value and needs to be changed according to the specific situation.

2.6 Robust Adaptive Beamforming

The minimum variance beamformers with the previously introduced constraints in the directions of interest can efficiently suppress sources of interference from other directions and achieve a satisfactory output signal-to-interference-plus-noise ratio (SINR) (Compton, 1988b; Mani and Base, 2008; Vook and Compton, 1992; Yu *et al.*, 2007). However, their performance is very sensitive to array calibration errors, especially the error in the DOA angle of the signal of interest. If the desired signal does not come exactly from the designed look direction of the array, it will be considered as an interfering signal by the array and the beamformer will tend to null out the desired signal at its output. In the GSC structure, due to this angle mismatch, we will see some leakage of the signal of interest into the blocking matrix output at the lower path.

Many methods have been proposed to improve the robustness of the beamformer to the mismatch error between the real DOA angle of the desired signal and the designed look direction of the array (Brandstein and Ward, 2001; Li and Stoica, 2005). For example, we can use a calibration signal to find the quiescent vector and the blocking matrix in a GSC tuned to the real DOA angle of the signal (Fudge and Linebarger, 1994) or we can employ some target tracking methods to estimate the true DOA angle or the signal subspace to reduce the mismatch error (Affes and Grenier, 1997; Affes *et al.*, 1996; Er, 1994). Similar to the calibration method, if we can have a period when only the signal of interest is present, we can use this set of data to tune the array to the right direction (Hoshuyama *et al.*, 1997, 1999a,b). Robustness of the wideband beamformer can also be achieved by applying a pre-processing focusing transform to the received signals (Wang and Kaveh, 1985; Zhang and Er, 1997). Most recently, there have been some approaches proposed based on convex optimization using the interior-point method (El-Keyi *et al.*, 2005; Rübsamen and Gershman, 2008; Slavakis and Yamada, 2007).

In the following we will focus on two classes of widely recognized robust beamforming approaches: one is to extend the constraints spatially to try to cover the possible signal directions; the other one is to restrain the norm of the resultant beamformer weights either explicitly or implicitly.

2.6.1 Spatially Extended Constraints

Due to the uncertainty of the DOA angle of the signal of interest, instead of imposing a single set of constraints in the presumed DOA angle θ_0, we can extend the constraints spatially and add more constraints to cover the angle range $\theta \in [\theta_l\ \theta_u]$, where the actual DOA angle of the signal falls.

To implement this idea, for the LCMV beamformer, we can sample the range $[\theta_l\ \theta_u]$ and for each angle point, a set of constraints can be formed using the eigenvector approach introduced in Section 2.3 (Buckley, 1987). The final constraint matrix can be formed by directly combining these constraints for different angles together or we can perform an SVD operation to this combined matrix and find an efficient low-rank approximation to it to reduce the number of constraints and increase the degrees of freedom of the beamformer weights for interference suppression.

For the SCMV beamformer, the soft constraint in Equation (2.95) will also include an integration over the angle range $\theta \in [\theta_l\ \theta_u]$, given by (Er and Cantoni, 1985):

$$\int_{\theta_l}^{\theta_u}\int_{\omega_l}^{\omega_u} |\mathbf{d}^H(\theta,\omega)\mathbf{w} - G_\omega^*|^2 \mathrm{d}\omega \mathrm{d}\theta \leq \varepsilon \tag{2.107}$$

For the CCMV beamformer, the signal model will be two dimensional and we can assume a flat power spectrum density over both the frequency range and the angle range when we calculate the correlation function $r_d(\tau)$ in Equation (2.105) (Kikuma and Takao, 1989).

Another approach is to impose derivative constraints on the beamformer coefficients by setting the derivatives of the squared response $|P(\theta,\omega)|^2$ in Equation (1.33) to the assumed direction to be zero so that a flat beam response can be achieved at this direction and any small deviation of the real DOA angle from the assumed direction will not lead to a severe attenuation to the desired signal at the beamformer output (Buckley and Griffiths, 1986; Er and Cantoni, 1983, 1986b; Er and Ng, 1990; Thng *et al.*, 1993, 1995; Zhang and Thng, 2002).

2.6.1.1 Derivative Constraints

To impose derivative constraints on the beamformer, the derivatives of the squared response $|P(\theta,\omega)|^2$ evaluated at the assumed direction θ_0 up to the order of n are set to be zero as follows:

$$\frac{\mathrm{d}^i(|P(\theta,\omega)|^2)}{\mathrm{d}\theta^i} = 0 \quad \text{for all} \quad \omega \in [\omega_{\min}\ \omega_{\max}] \tag{2.108}$$

with $i = 1, 2, \ldots, n$.

For the general case, the beamformer response will be a function of both the elevation angle and the azimuth angle and the derivative constraints in Equation (2.108) will become

$$\frac{\partial^{i_1+i_2}(|P(\theta,\phi,\omega)|^2)}{\partial\theta^{i_1}\partial\phi^{i_2}} = 0 \quad \text{for all} \quad \omega \in [\omega_{\min}\ \omega_{\max}] \tag{2.109}$$

where i_1 and i_2 are the corresponding partial derivative orders.

To illustrate the idea of derivative constraints, we consider an example based on a uniformly spaced linear array with:

$$|P(\theta,\omega)|^2 = \mathbf{w}^H\mathbf{d}(\theta,\omega)\mathbf{d}^H(\theta,\omega)\mathbf{w} \tag{2.110}$$

where:

$$\mathbf{d}(\theta,\omega) = [1 \ \dots \ \mathrm{e}^{-j(M-1)\mu\Omega\sin\theta} \ \mathrm{e}^{-j\Omega} \ \dots \ \mathrm{e}^{-j\Omega((M-1)\mu\sin\theta+1)} \\ \dots \ \mathrm{e}^{-j(J-1)\Omega} \ \dots \ \mathrm{e}^{-j\Omega((M-1)\mu\sin\theta+J-1)}]^T \tag{2.111}$$

as given in Equations (1.33) and (1.35) in Section 1.3.

For the first-order derivative constraint, we have:

$$\frac{\mathrm{d}(|P(\theta,\omega)|^2)}{\mathrm{d}\theta} = \frac{\mathrm{d}(\mathbf{w}^H\mathbf{d}(\theta,\omega))}{\mathrm{d}\theta}\mathbf{d}^H(\theta,\omega)\mathbf{w} + \mathbf{w}^H\mathbf{d}(\theta,\omega)\frac{\mathrm{d}(\mathbf{d}^H(\theta,\omega)\mathbf{w})}{\mathrm{d}\theta} \\ = 0 \tag{2.112}$$

A sufficient condition for Equation (2.112) is to set the derivative of both the real part and the imaginary part of $P(\theta,\omega) = \mathbf{w}^H\mathbf{d}(\theta,\omega)$ with respect to θ to be zero, namely:

$$\frac{\mathrm{d}(P(\theta,\omega))}{\mathrm{d}\theta} = \mathbf{w}^H\frac{\mathrm{d}(\mathbf{d}(\theta,\omega))}{\mathrm{d}\theta} = 0 \tag{2.113}$$

The element in the steering vector $\mathbf{d}(\theta,\omega)$ is in the form of $\mathrm{e}^{-j\Omega(m\mu\sin\theta+k)}$, which corresponds to the element $w_{m,k}$ in the weight vector $\mathbf{w}$, with $m = 0, 1, \dots, M-1$, and $k = 0, 1, \dots, J-1$. Then Equation (2.113) changes to:

$$(-j\mu\Omega\cos\theta)\sum_{m=0}^{M-1}\sum_{k=0}^{J-1} m\mathrm{e}^{-j\Omega(m\mu\sin\theta+k)} \times w_{m,k}^* = 0 \\ \Rightarrow \sum_{m=0}^{M-1}\sum_{k=0}^{J-1} m\mathrm{e}^{-j\Omega(m\mu\sin\theta+k)} \times w_{m,k}^* = 0 \tag{2.114}$$

For the broadside case ($\theta = \theta_0 = 0$), Equation (2.114) changes to:

$$\sum_{m=0}^{M-1}\sum_{k=0}^{J-1} m\mathrm{e}^{-jk\Omega} \times w_{m,k}^* = 0 \tag{2.115}$$

A sufficient condition (which is also independent of Ω) for this equation to hold is:

$$\sum_{m=0}^{M-1} m\mathrm{e}^{-jk\Omega} \times w_{m,k}^* = 0 \\ \Rightarrow \sum_{m=0}^{M-1} m \times w_{m,k}^* = 0 \tag{2.116}$$

for $k = 0, 1, \dots, J-1$.

For the second-order constraint, we have:

$$\sum_{m=0}^{M-1}\sum_{k=0}^{J-1} m^2\mathrm{e}^{-j\Omega(m\mu\sin\theta+k)} \times w_{m,k}^* = 0 \tag{2.117}$$

and similarly a sufficient condition is:

$$\sum_{m=0}^{M-1} m^2 w_{m,k}^* = 0 \tag{2.118}$$

for $k = 0, 1, \ldots, J-1$.

The two sufficient conditions are derived by implicitly taking the position of the first sensor ($m = 0$) as the zero-phase reference point of the signal. For the general case, we will have:

$$\sum_{m=0}^{M-1} (m - d_0)^i w_{m,k}^* = 0 \tag{2.119}$$

for the ith order derivative constraint, where d_0 is the zero-phase reference point. If we choose the middle point along the array line as the reference point and set the signal phase at this point to be zero, then we have $d_0 = (M-1)/2$.

In vector form, the nth order derivative constraint is:

$$\mathbf{c}_n^H \mathbf{w}_k = 0, \quad k = 0, 1, \ldots, J-1 \tag{2.120}$$

where $\mathbf{w}_k$ is defined in Equation (1.28) and $\mathbf{c}_n$ is given by:

$$\mathbf{c}_n = [c_n[0] \;\; c_n[1] \;\; \ldots \;\; c_n[M-1]]^T \in \mathbf{C}^{M\times 1} \tag{2.121}$$

with:

$$c_n[m] = (m - d_0)^n, \quad m = 0, 1, \ldots, M-1 \tag{2.122}$$

In this context, we can consider the constraint in Equation (2.28) as the zeroth-order derivative constraint and now we obtain the general formulation in the following when we impose derivative constraints of zeroth up to the nth orders on the beamformer:

$$\boldsymbol{C}^H \mathbf{w} = \mathbf{f} \tag{2.123}$$

where:

$$\boldsymbol{C} = \left[\hat{\boldsymbol{C}}_0 \ldots \hat{\boldsymbol{C}}_n\right] \in \mathbf{C}^{MJ\times nJ} \quad \text{with} \quad \hat{\boldsymbol{C}}_i = \begin{bmatrix} \mathbf{c}_i & & \mathbf{0} \\ & \ddots & \\ \mathbf{0} & & \mathbf{c}_i \end{bmatrix} \in \mathbf{C}^{MJ\times J} \tag{2.124}$$

and:

$$\mathbf{f} = [f[0] \;\; f[1] \;\cdots\; f[J-1] \;\; 0 \;\; 0 \;\cdots\; 0]^T \in \mathbf{C}^{nJ\times 1} \tag{2.125}$$

As an example, let us consider a uniformly spaced linear array with $M = 10$ sensors and a TDL length of $J = 96$. The signal of interest comes from the broadside. Figure 2.20 shows the response of the quiescent vector at the normalized frequency $\Omega = \pi$ based on the zeroth-order (solid line) and the third-order (dotted line) derivative constraints, respectively. As expected, a flatter response is achieved for the third-order case compared to the zeroth-order one.

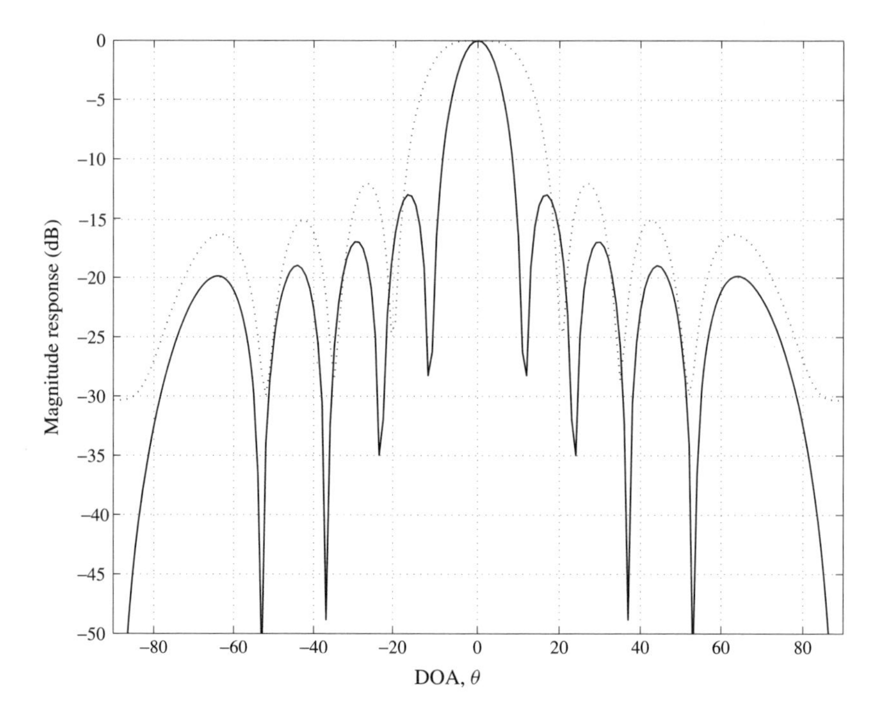

Figure 2.20 The beam response of the quiescent vector at the normalized frequency $\Omega = \pi$ for zero-order and third-order derivative constraints, where the solid line is the zeroth-order and the dotted line is the third-order

With this form of constraints, we can implement the LCMV beamformer in the simplified GSC structure with tapped delay-lines given in Section 2.4.2. Note the derivative constraints introduced here are based on the idea of sufficient conditions and their performance is dependent on the choice of the zero-phase reference point d_0 (Buckley and Griffiths, 1986). A set of constraints based on the necessary and sufficient conditions to meet Equation (2.109) was derived in Er and Ng (1990), which ensures a beam response independent of the choice of the phase reference point.

When implementing the derivative constraints using the GSC structure, the corresponding blocking matrix will have a flatter response at the assumed direction θ_0, so that the signal leakage to the lower path of a GSC can be reduced due to angle mismatch and significant cancellation of the signal of interest at the upper path can be avoided. Figure 2.21 shows an example of the beam response of a column of the blocking matrix corresponding to the example in Figure 2.20. We can see that a flatter zero response at the broadside direction has been achieved for the case with the third-order derivative constraints. Similar to this derivative constraint idea, we can deliberately design the blocking matrix with a relatively flat response at the direction θ_0 so that signal leakage to the lower path of the GSC can be minimized to reduce signal cancellation to the upper path (Claesson and Nordholm, 1992).

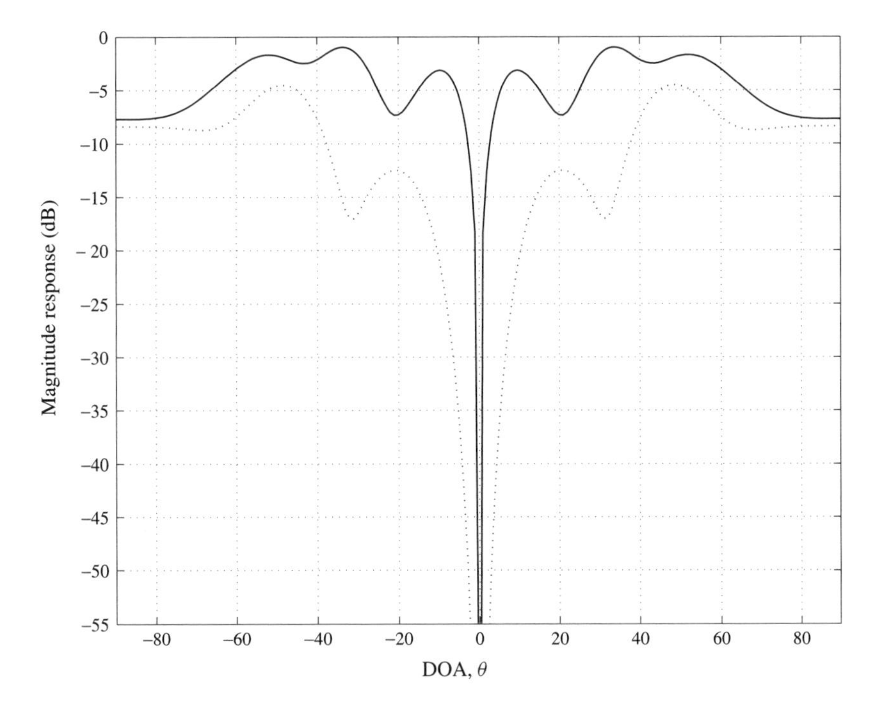

Figure 2.21 The beam response of a column of the blocking matrix corresponding to the example in Figure 2.20 at the normalized frequency $\Omega = \pi$ for the zeroth-order and third-order derivative constraints, where the solid line is the zeroth-order and the dotted line is the third-order

2.6.2 *Norm-Restrained Approaches*

For this class of robust adaptive beamformers, the norm of the beamformer coefficients is restrained either explicitly or implicitly.

To understand this idea, we first consider the derivative of the squared response $|P(\theta, \omega)|^2$ with respect to the DOA angle θ. From Equation (2.112), we can have:

$$\begin{aligned}\frac{\mathrm{d}(|P(\theta,\omega)|^2)}{\mathrm{d}\theta} &= \frac{\mathrm{d}(\mathbf{w}^H\mathbf{d}(\theta,\omega))}{\mathrm{d}\theta}\mathbf{d}^H(\theta,\omega)\mathbf{w} + \mathbf{w}^H\mathbf{d}(\theta,\omega)\frac{\mathrm{d}(\mathbf{d}^H(\theta,\omega)\mathbf{w})}{\mathrm{d}\theta}\\ &= \frac{\mathrm{d}(\mathbf{w}^H\mathbf{d}(\theta,\omega))}{\mathrm{d}\theta}\mathbf{d}^H(\theta,\omega)\mathbf{w} + \left(\frac{\mathrm{d}(\mathbf{w}^H\mathbf{d}(\theta,\omega))}{\mathrm{d}\theta}\mathbf{d}^H(\theta,\omega)\mathbf{w}\right)^H\\ &= 2\Re\left(\mathbf{w}^H\frac{\mathrm{d}(\mathbf{d}(\theta,\omega))}{\mathrm{d}\theta}\mathbf{d}^H(\theta,\omega)\mathbf{w}\right) \end{aligned} \tag{2.126}$$

where $\Re\{\}$ denotes the real part of a complex-valued variable.

Suppose the constraint matrix of the LCMV beamformer is designed to maintain a distortionless response at the look direction θ_0 and the real direction of the signal is $\theta_0 + \varepsilon$, where the value of ε is very small and represents the mismatch between the assumed DOA and the true DOA of the signal of interest. Since the signal of interest does not come from the assumed direction θ_0, it will be treated as interfering signals and suppressed at the beamformer output. As a result, the beam pattern of the resultant beamformer will have an almost zero response at the direction $\theta_0 + \varepsilon$. On the other hand, due to the imposed constraints, the beamformer has to maintain a distortionless response at θ_0, which leads to a very large value of the derivative $[\mathrm{d}(|P(\theta, \omega)|^2)]/\mathrm{d}\theta$ at $\theta = \theta_0$. Since both $\mathrm{d}[\mathbf{d}(\theta, \omega)]/\mathrm{d}\theta$ and $\mathbf{d}(\theta, \omega)$ are bounded in Equation (2.126), the only way to have a very large derivative value at θ_0 is to increase the norm of the weight vector significantly. Therefore, one way to avoid the suppression of the signal of interest in case of angle mismatch is to restrain the norm of the weight vector during the adaptation.

2.6.2.1 Diagonal Loading

For temporally and spatially white noise, its power at the output of the wideband beamformer is amplified by a factor of $\mathbf{w}^H\mathbf{w}$ and when we minimize the output power of an LCMV beamformer, this noise component also tends to be reduced and hence the norm of the weight vector. Therefore, if we deliberately add some white noise signal to the originally received array signals, the increase of the norm of the weight vector during adaptation will be limited by the minimization criterion of the LCMV beamformer and the complete cancellation of the signal of interest due to angle mismatch will then be avoided.

Mathematically, this is equivalent to modifying the received signal covariance matrix $\boldsymbol{R}_{xx}$ by adding a small value δ to its diagonal elements as follows (Cox *et al.*, 1987; Dogan and Mendel, 1994):

$$\hat{\boldsymbol{R}} = \boldsymbol{R}_{xx} + \delta \boldsymbol{I} \tag{2.127}$$

where δ is the perturbation factor and needs to be adjusted according to the original noise level and the angle mismatch error. This technique is called diagonal loading and more references with all kinds of extensions can be found in Li and Stoica (2005).

Now the LCMV beamforming problem changes to:

$$\mathbf{w} = \arg\min_{\mathbf{w}} \mathbf{w}^H(\boldsymbol{R}_{xx} + \delta\boldsymbol{I})\mathbf{w} \qquad \text{subject to} \qquad \boldsymbol{C}^H\mathbf{w} = \mathbf{f} \tag{2.128}$$

or:

$$\mathbf{w} = \arg\min_{\mathbf{w}} \mathbf{w}^H\boldsymbol{R}_{xx}\mathbf{w} + \delta\mathbf{w}^H\mathbf{w} \qquad \text{subject to} \qquad \boldsymbol{C}^H\mathbf{w} = \mathbf{f} \tag{2.129}$$

As given in Equation (2.36), the new optimum solution $\mathbf{w}_{opt}$ to this modified problem is:

$$\mathbf{w}_{opt} = (\boldsymbol{R}_{xx} + \delta\boldsymbol{I})^{-1}\boldsymbol{C}(\boldsymbol{C}^H(\boldsymbol{R}_{xx} + \delta\boldsymbol{I})^{-1}\boldsymbol{C})^{-1}\mathbf{f} \tag{2.130}$$

From Equation (2.129), we can see that the diagonal loading technique can be considered as adding a penalty term based on the norm of the weight vector $\mathbf{w}^H\mathbf{w}$ to the original cost function and therefore is a norm-restrained approach to robust adaptive beamforming.

2.6.2.2 Leaky LMS Adaptive Algorithm

The LCMV beamformer can be implemented by a GSC, where the weight vector $\mathbf{w}$ is decomposed into:

$$\begin{aligned}\mathbf{w} &= \mathbf{w}_q - \mathbf{v} \\ &= \mathbf{w}_q - \boldsymbol{B}\mathbf{w}_a\end{aligned} \tag{2.131}$$

as already shown in Figure 2.11.

Since both the quiescent vector $\mathbf{w}_q$ and the blocking matrix $\boldsymbol{B}$ are fixed, the diagonal loading approach can be implemented indirectly by restraining the norm of the unconstrained adaptive filter vector $\mathbf{w}_a$. The unconstrained optimisation problem in the GSC structure is:

$$\mathbf{w}_a = \arg\min_{\mathbf{w}_a} E\{|y[n]|^2\} \qquad \text{with} \qquad y[n] = d[n] - \hat{y}[n] \tag{2.132}$$

where:

$$\hat{y}[n] = \mathbf{w}_a^H \mathbf{u} \tag{2.133}$$

To restrain the norm of $\mathbf{w}_a$, we add a penalty term $\delta \mathbf{w}_a^H \mathbf{w}_a$ to the cost function and the new optimization problem is given by:

$$\mathbf{w}_a = \arg\min_{\mathbf{w}_a}(E\{|y[n]|^2\} + \delta \mathbf{w}_a^H \mathbf{w}_a) \tag{2.134}$$

Following the approaches in Section 2.1.1, we can obtain the optimum solution:

$$\mathbf{w}_{a,opt} = (\boldsymbol{R}_{uu} + \delta \boldsymbol{I})^{-1}\mathbf{p} \tag{2.135}$$

where:

$$\begin{aligned}\boldsymbol{R}_{uu} &= E\{\mathbf{u}[n]\mathbf{u}^H[n]\} \\ \mathbf{p} &= E\{\mathbf{u}[n]d[n]^*\}\end{aligned} \tag{2.136}$$

We can see this modified problem is equivalent to changing the covariance matrix $\boldsymbol{R}_{uu}$ of the tap input vector $\mathbf{u}[n]$ to $\hat{\boldsymbol{R}}_{uu} = \boldsymbol{R}_{uu} + \delta \boldsymbol{I}$.

Employing the standard stochastic gradient technique, we can derive the following update equation for $\mathbf{w}_a$:

$$\mathbf{w}[n+1] = (1 - \delta\mu)\mathbf{w}_a[n] + \mu y^*[n]\mathbf{u}[n] \tag{2.137}$$

This is called the leaky LMS adaptive algorithm (Haykin, 1996), and can be employed in the unconstrained adaptive part of the GSC for robust beamforming (Claesson and Nordholm, 1992; Hoshuyama *et al.*, 1997, 1999b).

2.6.2.3 Norm-Constrained Adaptive Algorithm

Another way to restrain the norm of the adaptive filter vector $\mathbf{w}_a$ in the GSC structure is to constrain its norm at each update as follows (Hoshuyama *et al.*, 1999a):

$$\begin{aligned}\mathbf{w}_a[n+1] &= \mathbf{w}_a[n] + \frac{\mu}{\mathbf{u}[n]^H\mathbf{u}[n]} y^*[n]\mathbf{u}[n] \\ \mathbf{w}_a[n+1] &= N_{w_a,\max}\frac{\mathbf{w}_a[n+1]}{N_{w_a}} \quad \text{if} \quad N_{w_a} > N_{w_a,\max}\end{aligned} \tag{2.138}$$

where N_{w_a} is the norm of the weight vector at $n+1$:

$$N_{w_a} = \sqrt{\mathbf{w}_a[n+1]^H \mathbf{w}_a[n+1]} \tag{2.139}$$

and $N_{w_a,\max}$ is the maximum norm value allowed for the weight vector $\mathbf{w}_a$ and should be determined according to the specific signal environment and the angle mismatch error considered.

This is a norm-constrained NLMS adaptive algorithm since compared to the NLMS adaptive algorithm introduced in Section 2.1.1, the only difference is the additional operation constraining the norm of the weight vector $\mathbf{w}_a$.

The two classes of robust adaptive beamforming approaches introduced in Sections 2.6.1 and 2.6.2 are focused on different parts of the beamformer and can be combined together to increase the robustness of the system. For example, in the GSC structure, the blocking matrix can be designed to have a relatively flat zero response at the assumed direction θ_0 and at the same time the leaky LMS adaptive algorithm can be adopted for updating $\mathbf{w}_a$ (Claesson and Nordholm, 1992).

2.7 Summary

In this chapter, we have studied a range of basic approaches to adaptive wideband beamforming.

Beamforming can be achieved by a standard adaptive filtering structure when a reference signal is available to the system, where the classic adaptive algorithms can be adopted, such as the LMS and RLS algorithms. When we know the DOA angle of the signal of interest, a linearly constrained minimum variance beamformer can be constructed and realized by either a constrained adaptive algorithm or an unconstrained one through an orthogonal decomposition of the constraint matrix. Such a decomposition leads to the well-known generalized sidelobe canceller structure, which can be further simplified to the GSC with tapped delay-lines when the desired signal comes from the broadside. Constraint design is a key issue in the implementation of an LCMV beamformer. An eigenvector based approach was reviewed in detail and its application to the wideband DOA estimation problem was introduced as an example.

In addition to the standard LCMV beamformer, two other minimum variance beamformers were studied, including the soft constrained beamformer and the correlation constrained beamformer. To improve the robustness of the beamformer in the presence of angle mismatch error for the signal of interest, at the end of this chapter we introduced several methods for robust adaptive beamforming, including spatially extended constraints, with the derivative constraint as an example, diagonal loading, a leaky LMS adaptive algorithm and a norm-constrained adaptive algorithm.

3

Subband Adaptive Beamforming

Given the wide bandwidth of the received array signal, various subband decomposition techniques can be employed in the beamforming process for an improved performance. In this chapter, we will introduce some subband structures for wideband adaptive beamforming. The two main advantages for subband adaptive beamforming are a reduced computational complexity due to a lower sampling rate at the decimated subbands and an increased convergence speed due to the prewhitening effect of the subband decomposition.

We will first give a brief overview of the fundamentals of filter banks and subband adaptive beamforming and then introduce several subband structures for wideband beamforming. Since DFT can be considered as a special class of filter banks, we will discuss the frequency-domain beamforming structure at the end of this chapter, with further extensions to the transform-domain based beamforming.

3.1 Fundamentals of Filter Banks

Subband techniques normally involve the use of two sets of filters. The first set is for subband decomposition so that the required processing, such as beamforming, can be performed at the resultant subbands; the second set is for fullband reconstruction, which combines the processed subband signals back into the original full band. Such a signal decomposition and reconstruction system is called a filter bank (Akansu and Haddad, 1992; Fliege, 1994; Strang and Nguyen, 1996; Vaidyanathan, 1993; Weiss and Stewart, 1998).

When the fullband signal is split into subbands, we can then sample it at a lower rate due to the reduced bandwidth. The resultant individual subbands may be treated separately during further processing such as audio coding (Crochiere, 1977, 1981; Crochiere *et al.*, 1976; Galand and Nussbaumer, 1984; Kim and Jones, 1991), image coding (Bamberger and Smith, 1992; Girod *et al.*, 1995; Woods, 1991; Woods and O'Neil, 1986) and adaptation (Gilloire, 1987; Gilloire and Vetterli, 1988, 1992; Harteneck *et al.*, 1999; Shynk, 1992; Weiss and Stewart, 1998). After processing, these subband signals can be reconstructed using a synthesis filter bank to obtain a fullband system output at the original sampling rate. Since typically different sampling rates are employed at different parts of the system, they are also referred to as multirate filter banks.

Figure 3.1 shows the general structure of a K-channel filter bank with a decimation factor of N, where the input signal $x[n]$ is decomposed into K subbands by an

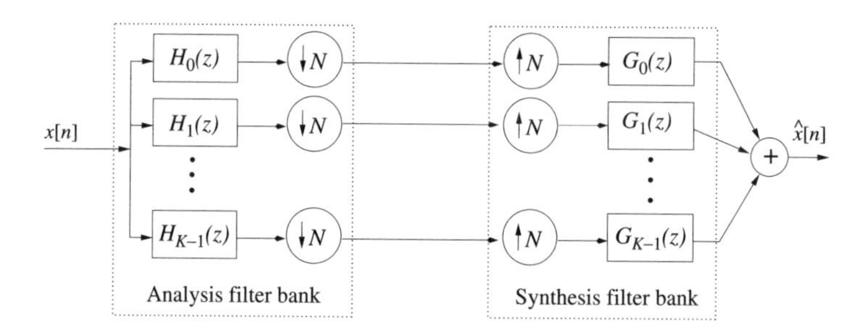

Figure 3.1 The general structure of a K-channel filter bank with a decimation factor N

analysis filter bank $H_0(z), H_1(z), \ldots, H_{K-1}(z)$ with each subband decimated or downsampled by a factor of $N \leq K$. After the required processing, which is not shown in Figure 3.1, the subband signals are then upsampled and recombined by a synthesis filter bank $G_0(z), G_1(z), \ldots, G_{K-1}(z)$ to yield the fullband output signal $\hat{x}[n]$. In general, we consider systems that are of perfect reconstruction (PR) (Mintzer, 1985; Smith and Barnwell III, 1986), where the output signal $\hat{x}[n]$ will be identical to the input $x[n]$ except for some delay.

In the following, we will first give an introduction to the basic operations in multirate filter banks and then analyse the PR conditions of filter banks. Finally, we will briefly discuss the design and implementation of a family of oversampled generalized DFTs (GDFTs) filter banks (Crochiere and Rabiner, 1983; Weiss and Stewart, 1998), which will be used in simulations for the introduced subband adaptive beamforming structures.

3.1.1 Basic Multirate Operations

3.1.1.1 Decimation and Interpolation

Decimation and interpolation are the two operations used to change the sampling rate in a discrete system (Akansu and Haddad, 1992; Vaidyanathan, 1993).

Decimation is the process of reducing the sampling rate of a discrete signal by an integer factor N, where the fullband signal $\tilde{x}[n]$ is first passed through an anti-aliasing filter $h[n]$, and then downsampled to a rate which is $1/N$ of the original one, as shown in Figure 3.2 (a).

A downsampler is also referred to as a subsampler and represented by a circle with $\downarrow N$ inside. This operation retains only every Nth sample of its input and then relabels the index axis. An example of a downsampling process by $N = 3$ is shown in Figure 3.2 (b).

In the time domain, the downsampling operation can be expressed as:

$$y[n] = x[Nn] \tag{3.1}$$

In the frequency domain, we have:

$$Y(e^{j\Omega}) = \frac{1}{N}\sum_{n=0}^{N-1} X(e^{j(\frac{\Omega-2n\pi}{N})}) \tag{3.2}$$

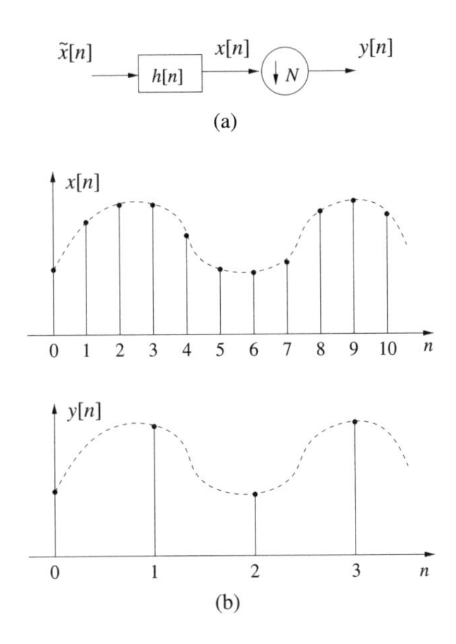

Figure 3.2 Decimation operation by a factor of N: (a) anti-aliasing filter and downsampler; (b) example for a downsampling operation by $N = 3$

where $X(e^{j\Omega})$ and $Y(e^{j\Omega})$ are the Fourier transforms of the input and output signals of the downsampler, respectively.

From the above equation we can see that the downsampling operation creates $N - 1$ aliased terms, $X(e^{j[(\Omega-2n\pi)/N]})$, $n = 1, 2, \ldots, N - 1$, at its output $Y(e^{j\Omega})$. If the spectra of these aliased terms overlap with each other, we will not be able to recover the original signal spectrum $X(e^{j\Omega})$ from $Y(e^{j\Omega})$, which will lead to loss of information, known as aliasing.

To avoid aliasing or spectrum overlap after downsampling, the bandwidth of the input signal has to be limited accordingly, which is the role of the anti-aliasing filter $h[n]$ in Figure 3.2 (a). As an example, for a low-pass input signal, its bandwidth should not exceed π/N.

Interpolation is the process of increasing the sampling rate of a discrete signal, and it is achieved by the combination of an upsampler and a follow-up filter $g[n]$ as shown in Figure 3.3 (a). An upsampler is also referred to as an expander, and the upsampling operation by an integer factor of N is achieved by inserting $N - 1$ zeros in between adjacent samples of the original discrete signal.

In the time domain, upsampling is represented by:

$$y[n] = \begin{cases} x\left[\frac{n}{N}\right] & : \quad n = 0, \pm N, \pm 2N, \cdots \\ 0 & : \quad \text{otherwise} \end{cases} \tag{3.3}$$

An example for an upsampling process by a factor of $N = 3$ is shown in Figure 3.3 (b).

In the frequency domain, upsampling is denoted as:

$$Y(e^{j\Omega}) = X(e^{j\Omega N}) \tag{3.4}$$

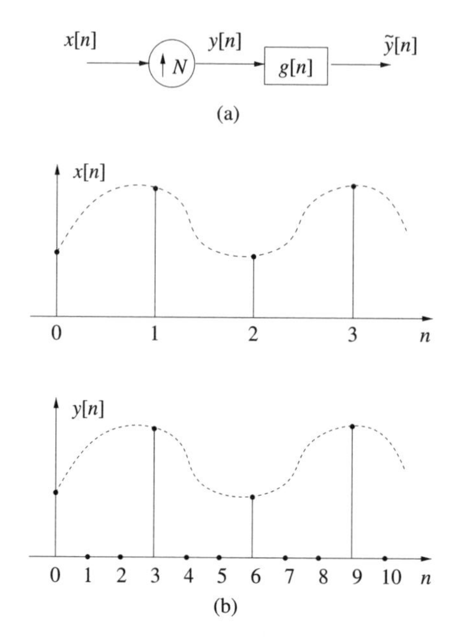

Figure 3.3 Interpolation operation by a factor of N: (a) upsampler and interpolation filter; (b) example for an upsampling operation by $N = 3$

which means that the output spectrum $Y(e^{j\Omega})$ is related to the input spectrum by a compression factor N. Then N copies of the original input signal spectrum $X(e^{j\Omega})$ will be packed into the range of $[-\pi \ \pi]$ of the output signal $Y(e^{j\Omega})$. The role of the filter $g[n]$ is to filter out the $N-1$ copies and leave only one copy at the final output signal $\tilde{y}[n]$. Normally the filter $g[n]$ is low-pass to leave the copy around the origin $\omega = 0$.

Finally, we give the time-domain representations for both the decimation and the interpolation operations, which include the filtering process in Figure 3.2 (a) and Figure 3.3 (a), respectively:

$$\text{Decimation:} \quad y[n] = \sum_k h[Nn-k]\tilde{x}[k]$$

$$\text{Interpolation:} \quad \tilde{y}[n] = \sum_k g[n-Nk]x[k] \tag{3.5}$$

3.1.1.2 Multirate Identities

There are several equivalent multirate building blocks, which are known as multirate identities (Akansu and Haddad, 1992; Vaidyanathan, 1993). They can be applied to simplify the derivation of the perfect reconstruction condition for a filter banks system.

Figure 3.4 shows two commonly used multirate identities (Akansu and Haddad, 1992; Vaidyanathan, 1993). In Figure 3.4 (a), an upsampler by a factor N followed by a filter $H(z^N)$ is equivalent to the structure with a filter $H(z)$ followed by the same upsampler; on the other hand, a filter $H(z^N)$ followed by a downsampler of a factor N has the same

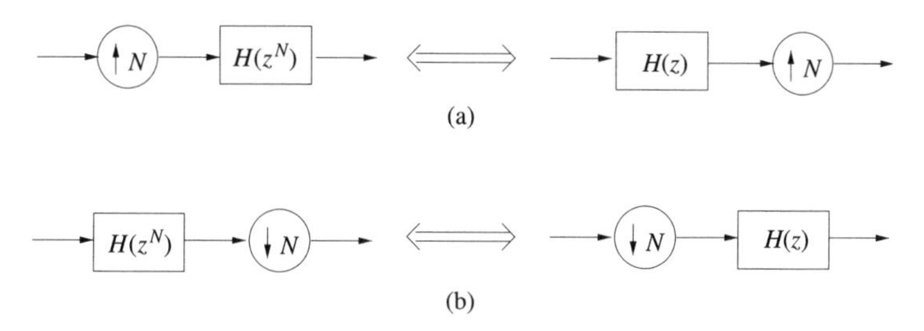

Figure 3.4 Equivalent structures for (a) upsampling and (b) downsampling, where $H(z)$ is the z-transform of a filter $h[n]$

function as the structure with the downsampler first, but followed by the filter $H(z)$, as shown in Figure 3.4 (b).

3.1.1.3 Polyphase Decomposition

Polyphase decomposition is an important tool in multirate signal processing and can simplify the analysis and implementation of a filter banks system significantly (Bellanger *et al.*, 1976; Vary, 1979). In order to derive the PR condition of filter banks in matrix form, we need to decompose the analysis and synthesis filters into their polyphase components. In the following, we introduce two types of polyphase decompositions.

For any discrete sequence $h[n]$, we can always define a series of subsequences $h_i[n]$ as follows:

$$h_i[n] = h[Nn + i], \quad i = 0, 1, \ldots, N-1 \tag{3.6}$$

The z-transform $H_i(z)$ of a subsequence $h_i[n]$ is given by:

$$H_i(z) = h[i] + h[i+N]z^{-1} + h[i+2N]z^{-2} + \cdots \tag{3.7}$$

Then the z-transform $H(z)$ of $h[n]$ can be expressed as:

$$H(z) = \sum_{i=0}^{N-1} z^{-i} H_i(z^N) \tag{3.8}$$

This is referred to as the Type-I polyphase decomposition of $H(z)$ and $H_i(z)$, $i = 0, 1, \ldots, N-1$, are the N Type-I polyphase components of $H(z)$.

A similar decomposition, referred to as the Type-II decomposition, is given by:

$$\tilde{h}_i[n] = h[Nn + N - 1 - i], \quad i = 0, 1, \cdots, N-1 \tag{3.9}$$

and:

$$H(z) = \sum_{i=0}^{N-1} z^{-(N-1-i)} \tilde{H}_i(z^N) \tag{3.10}$$

where $\tilde{H}_i(z)$ is the z-transform of $\tilde{h}_i[n]$.

The two types of polyphase components $H_i(z)$ and $\tilde{H}_i(z)$ are related by:

$$\tilde{H}_i(z) = H_{N-1-i}(z) \tag{3.11}$$

3.1.2 Perfect Reconstruction Condition for Filter Banks

We now study the PR condition of the K-channel filter banks system with a decimation factor N as shown in Figure 3.1. When $N = K$, the corresponding scheme is referred to as critically decimated filter banks or maximally decimated filter banks, whereas when $N < K$, it becomes oversampled filter banks. The PR condition discussed here is applicable to both cases.

In Figure 3.1, $H_k(z)$ and $G_k(z)$, $k = 0, 1, \ldots, K-1$, are the z-transforms of the analysis filters $h_k[n]$ and synthesis filters $g_k[n]$, respectively. We first place them into two vectors $\mathbf{h}(z)$ and $\mathbf{g}(z)$ as follows:

$$\mathbf{h}(z) = \begin{bmatrix} H_0(z) & H_1(z) & \cdots & H_{K-1}(z) \end{bmatrix}^T$$

$$\mathbf{g}(z) = \begin{bmatrix} G_0(z) & G_1(z) & \cdots & G_{K-1}(z) \end{bmatrix}^T$$

Then we decompose the analysis filters into their type-I polyphase components:

$$H_k(z) = \sum_{n=0}^{N-1} z^{-n} H_{k,n}(z^N) \tag{3.12}$$

and the synthesis filters into their type-II polyphase components:

$$G_k(z) = \sum_{n=0}^{N-1} z^{-(N-1-n)} G_{k,n}(z^N) \tag{3.13}$$

where $H_{k,n}(z)$ and $G_{k,n}(z)$ are the nth polyphase components of the kth analysis filter and synthesis filter, respectively.

Now the vectors $\mathbf{h}(z)$ and $\mathbf{g}(z)$ can be expressed in their polyphase form as:

$$\mathbf{h}(z) = \underbrace{\begin{bmatrix} H_{0,0}(z^N) & H_{0,1}(z^N) & \cdots & H_{0,N-1}(z^N) \\ H_{1,0}(z^N) & H_{1,1}(z^N) & \cdots & H_{1,N-1}(z^N) \\ \vdots & & \ddots & \vdots \\ H_{K-1,0}(z^N) & H_{K-1,1}(z^N) & \cdots & H_{K-1,N-1}(z^N) \end{bmatrix}}_{\boldsymbol{E}(z^N)} \cdot \underbrace{\begin{bmatrix} 1 \\ z^{-1} \\ \vdots \\ z^{-(N-1)} \end{bmatrix}}_{\mathbf{e}_N}$$

$$= \boldsymbol{E}(z^N)\mathbf{e}_N \tag{3.14}$$

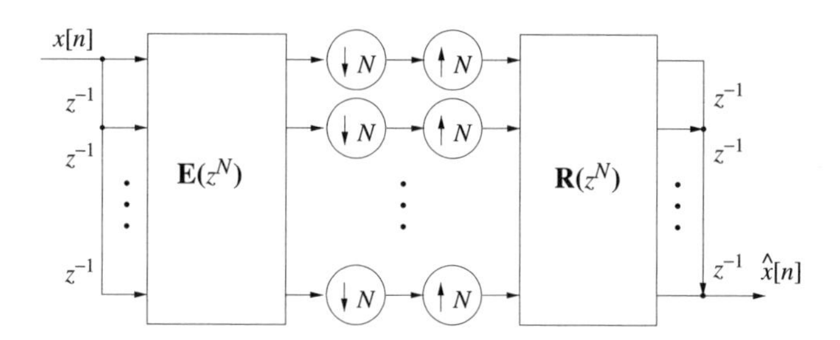

Figure 3.5 Polyphase representation of Figure 3.1

and:

$$\mathbf{g}^T(z) = \underbrace{\left[z^{-(N-1)} \; z^{-(N-2)} \; \cdots \; 1\right]}_{\tilde{\mathbf{e}}_N} \underbrace{\begin{bmatrix} G_{0,0}(z^N) & G_{1,0}(z^N) & \cdots & G_{K-1,0}(z^N) \\ G_{0,1}(z^N) & G_{1,1}(z^N) & \cdots & G_{K-1,1}(z^N) \\ \vdots & & \ddots & \vdots \\ G_{0,N-1}(z^N) & G_{1,N-1}(z^N) & \cdots & G_{K-1,N-1}(z^N) \end{bmatrix}}_{\boldsymbol{R}(z^N)}$$

$$= \tilde{\mathbf{e}}_N \boldsymbol{R}(z^N) \quad (3.15)$$

The matrix $\boldsymbol{E}(z)$ in Equation (3.14) is referred to as the polyphase analysis matrix, while $\boldsymbol{R}(z)$ in Equation (3.15) is called the polyphase synthesis matrix.

Using Equations (3.14) and (3.15), we can redraw Figure 3.1 in its polyphase representation, as shown in Figure 3.5. Using the multirate identities introduced before, we can shift the downsamplers to the left-hand side of the analysis polyphase matrix and replace z^N by z in the argument of $\boldsymbol{e}(z^N)$. Similarly, we can shift the upsamplers to the right-hand side of the synthesis polyphase matrix and obtain the structure shown in Figure 3.6, which can be further simplified to the form shown in Figure 3.7 with $\boldsymbol{P}(z) = \boldsymbol{R}(z)\boldsymbol{E}(z)$.

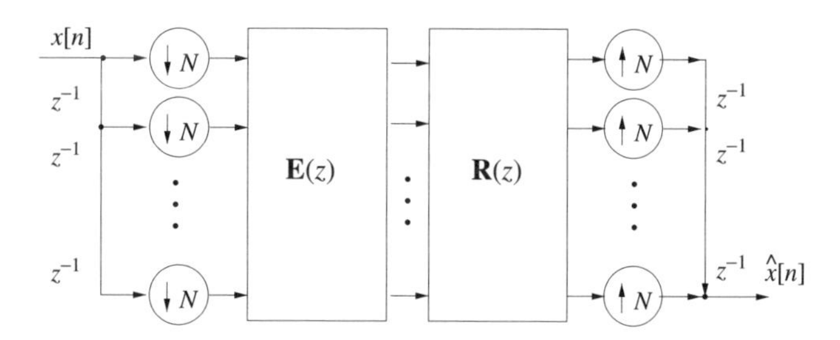

Figure 3.6 Polyphase representation using multirate identities

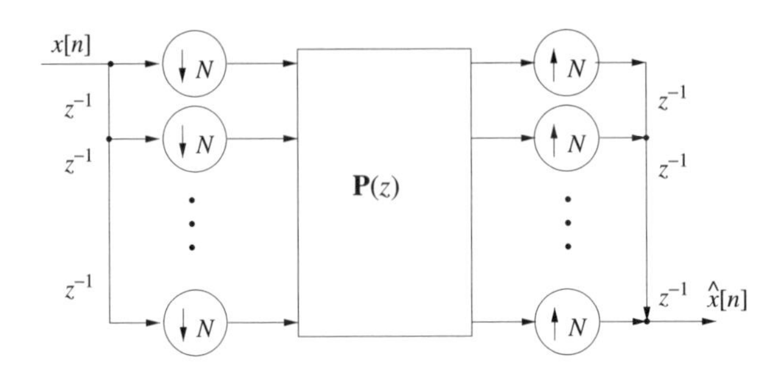

Figure 3.7 Equivalent structure of Figure 3.6 with $\boldsymbol{P}(z) = \boldsymbol{R}(z)\boldsymbol{E}(z)$

A general PR condition is given by (Vaidyanathan and Nguyen, 1987):

$$\boldsymbol{P}(z) = cz^{-\alpha}\begin{bmatrix} \mathbf{0} & \boldsymbol{I}_{N-r} \\ z^{-1}\boldsymbol{I}_r & \mathbf{0} \end{bmatrix}_{N\times N} \tag{3.16}$$

where c is a nonzero constant, and α and r are integers with $0 \leq r \leq (N-1)$. Under this condition the input–output relationship becomes:

$$\hat{x}[n] = cx[n - n_0] \tag{3.17}$$

where $n_0 = N\alpha + r + N - 1$ is the overall delay of the system.

A useful sufficient condition is:

$$\boldsymbol{P}(z) = z^{-\alpha}\boldsymbol{I}_N \tag{3.18}$$

which corresponds to the case where $r = 0$. When $\alpha = 0$, the equivalent form in the time domain for Equation (3.18) is given by (Steffen *et al.*, 1993):

$$\sum_n h_i[n]g_j[-lN - n] = \delta[l]\delta[i-j], \quad 0 \leq i, j \leq K-1 \tag{3.19}$$

where $\delta[l]$ is the Kronecker delta function. With this condition, the system in Figure 3.6 is reduced to a multiplexer and a demultiplexer in the analysis and synthesis banks, respectively, and we have:

$$\hat{x}[n] = x[n - n_0] \text{ with } n_0 = N\alpha + N - 1 \tag{3.20}$$

3.1.3 Oversampled Modulated Filter Banks

In the K-channel filter banks, the analysis filters $h_k[n]$ and the synthesis filters $g_k[n]$, $k = 0, 1, \ldots, K-1$, have different characteristics and need to be designed carefully to satisfy the PR condition. In order to reduce the design and implementation complexity, many modulated filter banks have been proposed, where both the analysis and the synthesis filters are derived from a prototype lowpass filter by some suitable modulation.

Popular modulation schemes include the cosine modulation (Koilpillai and Vaidyanathan, 1992), the extended lapped transforms (Malvar, 1992) and the discrete Fourier transform (DFT) (Bölcskei *et al.*, 1995; Crochiere and Rabiner, 1983). By this approach, the design of PR filter banks is reduced to that of a prototype filter and the system can be realized by the coefficients of the prototype filter and the modulation block.

Here we focus on a class of oversampled modulated filter banks with $N < K$, which are referred to as generalized DFT (GDFT) filter banks (Crochiere and Rabiner, 1983). In such filter banks, the analysis filters are derived from a real-valued lowpass prototype FIR filter $p[n]$ by a GDFT according to:

$$h_k[n] = \mathrm{e}^{j\frac{2\pi}{K}(k+k_0)(n+n_0)} p[n], \quad \text{with } k = 0, 1, \ldots, K-1, \quad n = 0, \ldots, l_p - 1 \qquad (3.21)$$

where l_p is the length of the prototype filter: k_0 and n_0 are two offsets introduced into the frequency and time indices of the DFT, leading to a generalized version of the DFT, hence the name GDFT.

According to Equation (3.21), the spectrum of the resultant analysis filters $h_k[n]$ are shifted versions of the prototype filter $p[n]$ along the frequency axis by $2\pi(k + k_0)/K$. Specifically, with $k_0 = 1/2$, for a real-valued input signal $x[n]$ it is sufficient to only process the first $K/2$ subbands covering the frequency interval $[0\ \pi]$ as the remaining subbands are the complex conjugate versions of these subbands. For the full effect of the frequency offset k_0, please refer to Crochiere and Rabiner (1983). In some applications such as image coding (Bamberger and Smith, 1992; Girod *et al.*, 1995; Woods, 1991; Woods and O'Neil, 1986), maintaining a linear phase is required, which can be fulfilled by choosing:

$$n_0 = -\frac{l_p - 1}{2} \qquad (3.22)$$

along with a real-valued linear phase prototype filter $p[n]$ of even length.

If the prototype filter is designed such that the polyphase analysis matrix $\boldsymbol{E}(z)$ is paraunitary, namely:

$$\boldsymbol{E}^H(z^{-1})\boldsymbol{E}(z) = c\boldsymbol{I} \qquad (3.23)$$

then we can simply choose the polyphase synthesis matrix as $\boldsymbol{R}(z) = z^{-\alpha}\boldsymbol{E}^H(z^{-1})$ in order to comply with the PR requirement of Equation (3.18). In this case, the impulse responses of the analysis and synthesis filters are time-reversed, complex conjugate versions of each other:

$$g_k[n] = h_k^*[l_p - 1 - n] \qquad (3.24)$$

If the analysis filters have a linear phase, then we further have:

$$g_k[n] = h_k[n], \quad n = 0, 1, \ldots, l_p - 1 \qquad (3.25)$$

With the GDFT modulation approach, the memory required for storing the filter coefficients can be reduced significantly and its implementation simplified by efficient polyphase decomposition and FFT (Cvetković and Vetterli, 1998; Weiss and Stewart, 1998). According to Weiss and Stewart (2000), the number of real multiplications required to implement the GDFT filter banks is:

$$C_{GDFT}^{real} = \frac{1}{N}\left(l_p + 4K\log_2 K + 4K\right) \tag{3.26}$$

for a real-valued input signal and:

$$C_{GDFT}^{complex} = \frac{1}{N}\left(2l_p + 4K\log_2 K + 8K\right) \tag{3.27}$$

for a complex-valued input signal. Note the term $4K\log_2 K$ in the above two equations is derived from the computational complexity when we implement the oversampled GDFT filter banks using FFTs based on K being a power of two. For other values of K, a similar result can be obtained and the above two equations are for guidance only.

Furthermore, the filter bank design problem is reduced to that of the prototype filter, which has to satisfy two requirements. Firstly, its stopband attenuation for $\Omega \in [\pi/N\ \pi]$ has to be large enough to avoid the distortion of the subband signal since any signal component of the input signal $x[n]$ with a frequency $\Omega \in [\pi/N\ \pi]$ will be aliased into the baseband $[-\pi/N\ \pi/N]$ after filtering and downsampling. Secondly, the choice of the prototype filter should ensure that the whole system is of perfect reconstruction. If the stopband attenuation of the prototype filter is sufficiently large so that the aliasing problem can be ignored, then the PR condition is reduced to the following power complementarity requirement (Vaidyanathan, 1993):

$$\sum_{k=0}^{K-1} |H_k(\mathrm{e}^{j\Omega})|^2 = 1, \forall\, \Omega \tag{3.28}$$

Many methods have been proposed for the design of such prototype filters, such as the iterative least-squares methods (Harteneck *et al.*, 1999; Weiss *et al.*, 1998c) and the methods based on dyadically iterated halfband filters (Fliege, 1993; Neuvo *et al.*, 1984; Weiss and Stewart, 1998).

Figure 3.8 shows a prototype filter with length of $l_p = 448$ designed by an iterative least-squares method according to Harteneck *et al.* (1999) for a $K = 16$ channel GDFT filter bank with a decimation factor $N = 14$.

3.2 Subband Adaptive Filtering

Subband adaptive filtering (SAF) techniques (Gilloire and Vetterli, 1988; Shynk, 1992; Weiss and Stewart, 1998; Weiss *et al.*, 2001; Yamada *et al.*, 1994; Yang *et al.*, 1995) have been widely applied to problems such as acoustic echo cancellation (Gilloire, 1987; Gilloire and Vetterli, 1992; Kellermann, 1988), identification of room acoustics (Schönle *et al.*, 1993), or equalization (Weiss *et al.*, 1998a,b, 1999a), where a large number of adaptive parameters have to be adjusted and as a result, the computational complexity can be very high and the convergence rate of the adaptive filter can be slow using standard techniques.

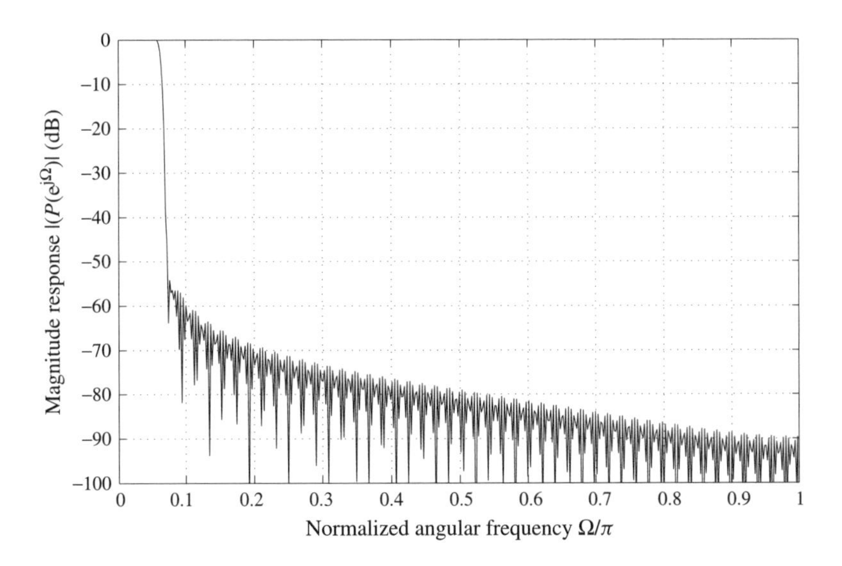

Figure 3.8 Frequency response $P(e^{j\Omega})$ of a prototype filter with length $l_p = 448$ designed by the iterative least-squares method in Harteneck *et al.* (1999) ($K = 16$, $N = 14$)

Similar problems arise in wideband beamforming where arrays with a large number of sensors and filter coefficients have to be employed to perform beamforming with high interference rejection and resolution. Therefore subband adaptive filtering techniques can also be applied to the wideband beamforming area to reduce the computational complexity and increase the convergence speed of the system.

A general SAF system is shown in Figure 3.9, where both the input signal and the desired signal are split into decimated subbands by analysis filter banks and then the subband adaptive filters, which run at a much lower rate compared to the original fullband system, are employed to estimate the subband desired signals using the subband input signals. The resultant subband error signals are then reconstructed into a fullband error signal by a synthesis filter bank.

Depending on the filter banks employed, the subband adaptive filters can have different structures. In the case of critically decimated filter banks ($N = K$), some cross-terms at least between adjacent subbands have to be employed to compensate for the information loss in the spectrum overlap region (Gilloire and Vetterli, 1992), or some gap filter banks can be employed to remove the spectrum overlap region to avoid the aliasing problem (Tanrikulu *et al.*, 1997; Yamada *et al.*, 1994). All these measures have their own drawbacks. For example, the introduced cross-terms will reduce the convergence speed and increase the computational cost, while the gap filter banks may lead to unacceptable signal distortion for some applications.

Another choice of the filter banks in the SAF system is to employ the oversampled filter banks. Since the aliasing level in subbands after decimation has been kept sufficiently low in the design of the oversampled filter banks, the cross-terms in the critically decimated case can be avoided and only one independent subband adaptive filter can be operated in each of the corresponding subbands as shown in Figure 3.10. The coefficients of the oversampled filter banks can be either real-valued or complex-valued. For real-valued

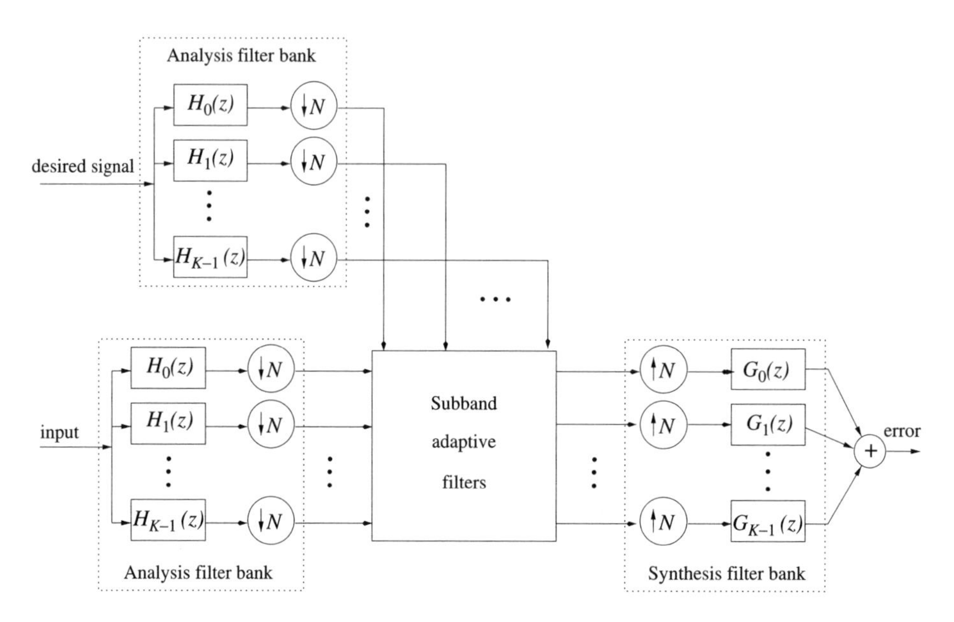

Figure 3.9 A general SAF structure, where subband decomposition and fullband error reconstruction is performed by filter banks

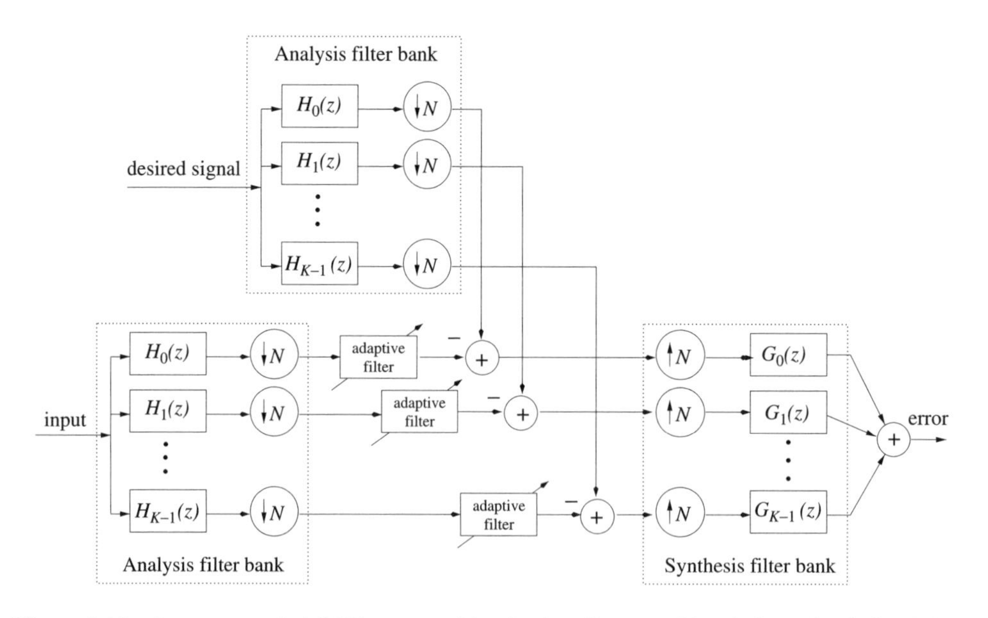

Figure 3.10 An oversampled SAF system with adaptive filters working independently in K decimated subbands

systems, to meet the bandpass sampling theorem (Vaughan *et al.*, 1991), the bandwidth of the analysis filters and the corresponding decimation ratio has to be chosen very carefully, which normally leads to a non-uniform filter banks system (Harteneck and Stewart, 1997; Harteneck *et al.*, 1998; Somayazulu *et al.*, 1989). Since it is more complicated to design and implement non-uniform filter banks than the modulated uniform filter banks system, in the following work employing SAF techniques, we only focus on those based on the previously introduced oversampled GDFT filter banks (Weiss and Stewart, 1998).

In SAF systems, the adaptive filter length at each subband can be much shorter compared to a fullband adaptive filter in accordance with the sampling rate reduction by a factor of $N \leq K$. However, the relationship between a fullband adaptive filter of length l_{full} and the corresponding adaptive filter length l_{sub} at each subband with similar modelling capabilities is complicated and it is determined not only by the decimation factor N, but also an additional offset term due to the transients caused by the filter banks (Gilloire and Vetterli, 1992; Weiss and Stewart, 1998, 1999). The following is an approximation (Weiss and Stewart, 1998):

$$l_{sub} = \frac{l_{full} + l_p}{N} \tag{3.29}$$

where l_p is the length of the prototype filter used in the oversampled GDFT filter banks. If we consider a large adaptive system, in general $l_p \ll l_{full}$ and hence the approximation $l_{sub} = l_{full}/N$ arises. Since Equation (3.29) is a heuristic result and also for the sake of simplicity, we will assume $l_{sub} = l_{full}/N$ unless otherwise specified.

There are three major advantages by performing adaptive filtering in decimated subbands and we list them briefly in the following:

1. **Reduction in Computational Complexity**

   Firstly, as mentioned, the adaptive filter length required in each subband can be shortened approximately by a factor of $N < K$ compared to a fullband adaptive filter. Secondly, updating of the adaptive filters is carried out at a much lower rate. Therefore, if we ignore the extra computational overhead involved in the subband decomposition part, significant reduction in computational complexity can be achieved by subband implementation.

   However, considering subband decomposition and adaptation as a whole, the SAF method does not necessarily save computations. If the adaptive filter has a relatively short length, the extra computational load of the filter banks is likely to exceed any reduction of computational complexity achieved in the subband adaptive part. With an increasing adaptive filter length, reduction of the computational complexity achieved by the subband adaptive part will become more substantial. Since the complexity of the filter banks is fixed, the overall SAF system will have a lower computational complexity than the fullband one. When the length of the adaptive filter is sufficiently high, the computations involved in the filter banks part can be ignored and the reduction ratio for the whole subband adaptive system will be approximately $\mathcal{O}(K/N^2)$ for LMS-type algorithms and $\mathcal{O}(K/N^3)$ for RLS algorithms.

   An illustrative example for the computational complexity of both the subband and fullband methods is provided in Figure 3.11.

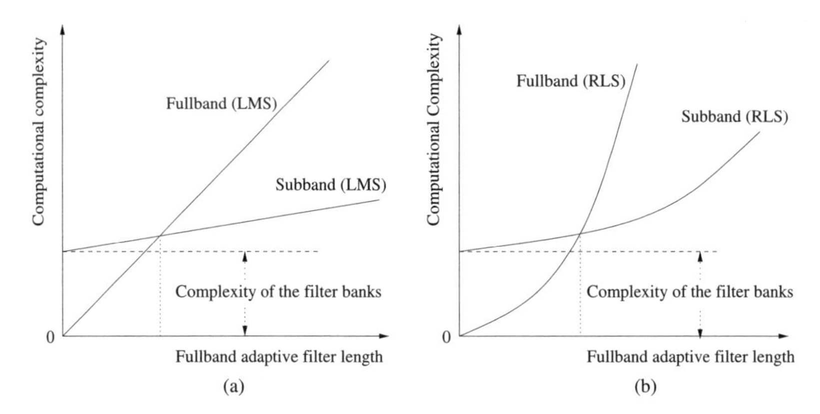

Figure 3.11 A qualitative description of the computational complexities of fullband and subband implementations for (a) LMS and (b) RLS adaptations

2. **Spectral Prewhitening**

   For coloured input signals, such as speech, after subband decomposition and decimation, the spectral dynamics of the resultant subband signals will be reduced and a flatter overall spectrum can be achieved for each subband. In this sense, the filter banks system has carried out a spectral prewhitening operation, although the spectrum is not totally whitened. Since the convergence speed of the LMS-type adaptive algorithms is controlled by the eigenvalue spread or condition number of the input covariance matrix (Haykin, 1996), a reduced spectral dynamic range, hence a reduced eigenvalue spread, will lead to a much increased convergence speed.

3. **Parallel Processing**

   The filter banks system splits the original fullband adaptive filter into K parallel subband filters and each subband adaptive filtering task can be performed on different processors if required (Gilloire and Vetterli, 1992), especially when the reduction of the computational complexity achieved by the subband implementation alone is not sufficient for enabling a real-time processing on a single processor.

In the next section, we will apply the SAF techniques to wideband beamforming and introduce a general subband adaptive beamforming structure.

3.3 General Subband Adaptive Beamforming

In the past, the application of subband methods to adaptive beamforming such as microphone or antenna arrays has been studied by a number of researchers (Chau *et al.*, 2002; De Haan *et al.*, 2002; Khalab and Ibrahim, 1994; Khalab and Woolfson, 1994; Liu and Langley, 2007; Liu *et al.*, 2004a; Lorenzelli *et al.*, 1996; Neo and Farhang-Boroujeny, 2002; Weiss and Proudler, 2002; Weiss *et al.*, 1999b,c; Zhang *et al.*, 2001, 2005). The basic idea of subband adaptive beamforming (SAB) is to first split the received sensor signals into different subbands and then operate an independent beamformer in each of them, with the subband beamformer selected according to the specific applications.

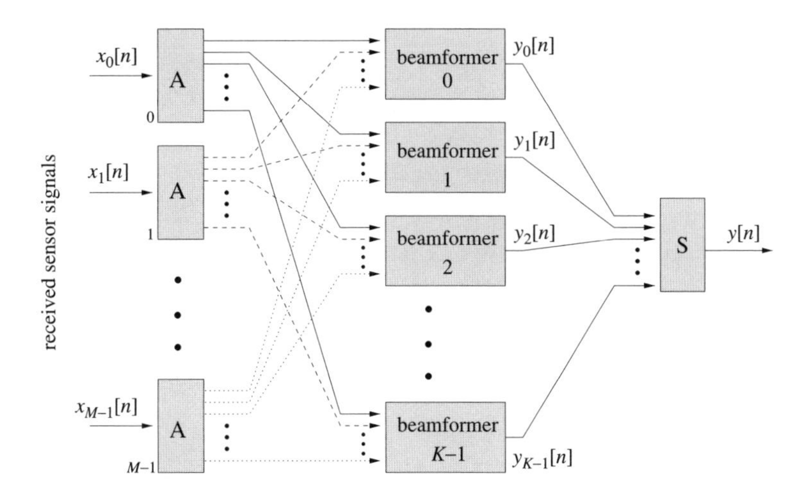

Figure 3.12 A general subband adaptive beamforming structure

Figure 3.12 shows a general SAB structure, where each of the M received array signals $x_m[n]$, $m = 0, 1, \ldots, M - 1$ is decomposed into K subbands by a K-channel analysis filter bank and a beamformer is then set up at each set of the M corresponding decimated subband signals. The output $y_k[n]$, $k = 0, 1, \ldots, K - 1$, of these K subband beamformers are then combined by a synthesis filter bank to form the fullband output $y[n]$. In Figure 3.12, the blocks labelled 'A' are the analysis filter banks (including the downsampling operators) and the block labelled 'S' is the synthesis filter bank (including the upsampling operators). In total there are M analysis filter banks and one synthesis filter bank.

Depending on the specific applications, we can choose a reference signal based beamformer, or a generalized sidelobe canceller, or any other appropriate ones. In the following, we will consider two examples.

3.3.1 Reference Signal Based Beamformer

The fullband reference signal based beamformer has a multichannel structure, as shown in Figure 2.1. To implement this structure in subbands, we need to split both the reference signal $r[n]$ and the sensor signals $x_m[n]$, $m = 0, 1, \ldots, M - 1$ into subbands and the resultant subband adaptive reference signal based beamformer is shown in Figure 3.13 (Liu and Langley, 2009), where the block labeled 'MCAF' is the multichannel adaptive filtering part with M subband channels.

We now give a detailed analysis of the computational complexity of the subband structure. There are a total of $M + 1$ analysis filter banks and one synthesis filter bank. Assume that we employ the oversampled GDFT filter banks with a prototype filter of length l_p. Then for each filter bank $1/N\left(l_p + 4K\log_2 K + 4K\right)$ real multiplications are required for a real-valued input and $1/N\left(2l_p + 4K\log_2 K + 8K\right)$ real multiplications are required for a complex-valued input, as given in Equations (3.26) and (3.27). Thus, the total number of real multiplications for the filter banks part is $(M+2)/N\left(l_p + 4K\log_2 K + 4K\right)$ or $(M+2)/N\left(2l_p + 4K\log_2 K + 8K\right)$ for real- and complex-valued inputs, respectively.

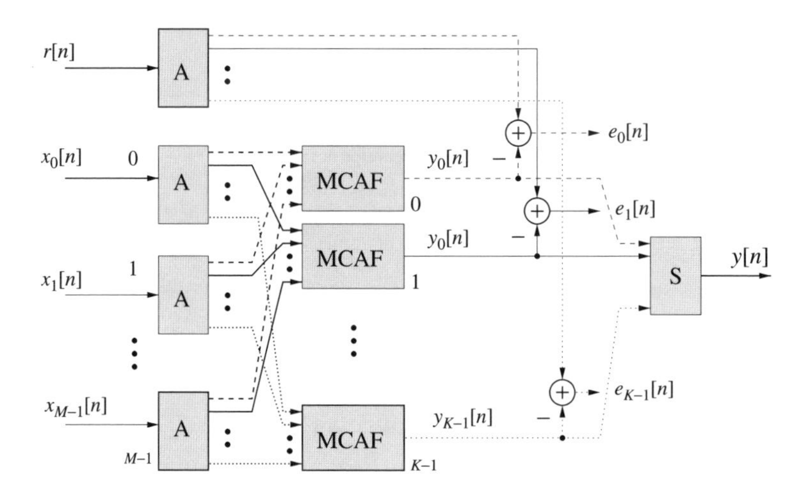

Figure 3.13 A general subband adaptive beamforming structure with a reference signal

For the adaptive part, in general, it yields a reduction in computational complexity by a factor of $\mathcal{O}(K/N^2)$ for LMS-type algorithms and $\mathcal{O}(K/N^3)$ for RLS algorithms, as mentioned in Section 3.2. More specifically, suppose that the subband adaptive filter length for each channel in each of the K MCAF blocks is J_{sub}, which is given by:

$$J_{sub} = \frac{J + l_p}{N} \tag{3.30}$$

according to Equation (3.29), where J is the TDL length of the fullband beamformer for each sensor. Then the total number of adaptive filter coefficients for each MCAF block is MJ_{sub}.

Since the subband signals in the oversampled GDFT filter banks are complex-valued, according to Table 2.2, in the real-valued input sensor signal case, it requires $K/(2N)(8MJ_{sub} + 2)$ (LMS), $K/(2N)(8MJ_{sub} + 2M + 3)$ (NLMS) or $K/(2N)(12M^2J_{sub}^2 + 12J_{sub} + 4)$ (RLS) real multiplications, because the number of adaptive filter coefficients required for each of the subband MCAF is MJ_{sub} and only $K/2$ number of complex subbands decimated by $N < K$ have to be processed. If the array input signals are complex-valued, then the computational complexity will be $K/N(8MJ_{sub} + 2)$ (LMS), $K/N(8MJ_{sub} + 2M + 3)$ (NLMS) or $K/N(12M^2J_{sub}^2 + 12J_{sub} + 4)$ (RLS) real multiplications, because we have to process all of the K subband MCAFs.

The above results are summarised in Table 3.1. As a comparison, it also includes the computational complexities of the corresponding fullband beamformer.

3.3.2 Generalized Sidelobe Canceller

When we know the DOA angle of the signal of interest, we can employ a GSC at each set of subbands for beamforming and the new structure is shown in Figure 3.14.

In this implementation, the fullband linear constraints of the beamformer and in particular the response vector **f**, have to be decomposed into subband-based constraints in

Table 3.1 Computational complexities for the fullband and subband adaptive beamformers with a reference signal

Fullband	Real multiplications (real-valued)	Real multiplications (complex-valued)
LMS	$2MJ+1$	$8MJ+2$
NLMS	$2MJ+M+2$	$8MJ+2M+3$
RLS	$3M^2J^2+3MJ$	$12M^2J^2+12MJ+4$
Subband	**Real multiplications (real-valued)**	**Real multiplications (complex-valued)**
LMS	$(M+2)/N(l_p + 4K\log_2 K+4K)+$ $K/(2N)(8MJ_{sub}+2)$	$(M+2)/N(2l_p + 4K\log_2 K+8K)+$ $K/N(8MJ_{sub}+2)$
NLMS	$(M+2)/N(l_p + 4K\log_2 K+4K)+$ $K/(2N)(8MJ_{sub}+2M+3)$	$(M+2)/N(2l_p + 4K\log_2 K+8K)+$ $K/N(8MJ_{sub}+2M+3)$
RLS	$(M+2)/N(l_p + 4K\log_2 K+4K)+$ $K/(2N)(12M^2J_{sub}^2+12J_{sub}+4)$	$(M+2)/N(2l_p + 4K\log_2 K+8K)+$ $K/N(12M^2J_{sub}^2+12J_{sub}+4)$

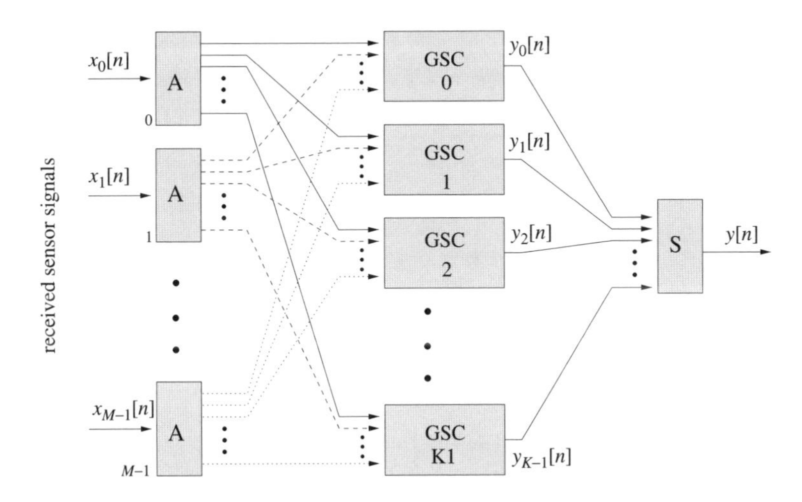

Figure 3.14 A general SAB structure with a GSC at each set of the subbands

order to construct the GSC for each subband (Weiss *et al.*, 1999c). If the reconstruction error of the filter banks is too large, this projection can incur inaccuracies which cannot be neglected. However, for the broadside constraints discussed in Section 2.4.2, if the response vector is a simple delay, as shown in Figure 2.15, and the delay is a multiple of the decimation ratio N of the filter banks, then the constraint at each subband GSC will have a similar form as the fullband one. More specifically, if the fullband constraint is given as shown at the beginning of Section 2.4.2, with:

$$\mathbf{f}_0 = [0\ \ldots\ 1\ \ldots\ 0]^T \in \mathbf{C}^{J\times 1} \tag{3.31}$$

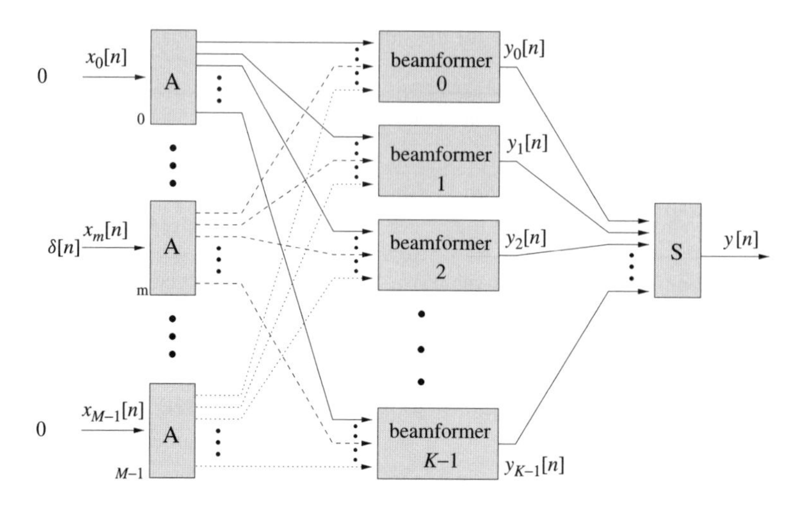

Figure 3.15 Reconstruction of the mth branch filter (mth TDL) of the equivalent fullband beamformer from the SAB structure

then at each subband, the constraint becomes:

$$\boldsymbol{C}_{sub} = \left[\hat{\boldsymbol{C}}_{sub,0} \ldots \hat{\boldsymbol{C}}_{sub,S-1}\right] \in \mathbf{C}^{MJ_{sub} \times SJ_{sub}} \tag{3.32}$$

with:

$$\hat{\boldsymbol{C}}_{sub,i} = \begin{bmatrix} \mathbf{c}_i & & \mathbf{0} \\ & \ddots & \\ \mathbf{0} & & \mathbf{c}_i \end{bmatrix} \in \mathbf{C}^{MJ_{sub} \times J_{sub}} \tag{3.33}$$

and:

$$\mathbf{f}_{sub} = [a_0\mathbf{f}_{sub,0}^T \; a_1\mathbf{f}_{sub,0}^T \; \ldots \; a_{S-1}\mathbf{f}_{sub,0}^T]^T \in \mathbf{C}^{SJ_{sub} \times 1} \tag{3.34}$$

with:

$$\mathbf{f}_{sub,0} = [0 \; \ldots \; 1 \; \ldots \; 0]^T \in \mathbf{C}^{J_{sub} \times 1} \tag{3.35}$$

The difference between the fullband response vector $\mathbf{f}_0$ and the subband one $\mathbf{f}_{sub,0}$ is the position of the element 1. If the element 1 is located at the mNth position of $\mathbf{f}_0$, then it will be at the mth position of $\mathbf{f}_{sub,0}$, where m is an integer, N is the decimation ratio of the filter banks and $mN \leq J$.

Now the constraint equation for each subband can be formulated as:

$$\boldsymbol{C}_{sub}^H \mathbf{w}_{sub} = \mathbf{f}_{sub} \tag{3.36}$$

where the weight vector $\mathbf{w}_{sub}$ is formed in the same way as in the fullband $\mathbf{w}$ with a reduced dimension $MJ_{sub} \times 1$.

In this case, the subband GSC can also be implemented in its simplified form as described in Figure 2.15 and the analysis of the computational complexity of the subband and fullband GSC is based on this simplified implementation.

For the computational complexity of this subband beamformer, it can also be split into two parts: the filter banks part and the subband GSC. There are M analysis filter banks and one synthesis filter bank. Then the total number of real multiplications for the filter banks is $(M+1)/N\left(l_p+4K\log_2 K+4K\right)$ or $(M+1)/N\left(2l_p+4K\log_2 K+8K\right)$ for real- and complex-valued inputs, respectively.

For the subband GSCs, since the subband signals in the oversampled GDFT filter banks are complex-valued, the number of real multiplications required for the real-valued quiescent vector $\tilde{\mathbf{w}}_q$ and the real-valued blocking matrix $\tilde{\boldsymbol{B}}$ is $2M+2(M-S)M=2M(M-S+1)$ for each subband GSC. For the adaptive part of the GSC, it is an $M-S$ channel adaptive filter with $(M-S)J_{sub}$ coefficients, which requires $(8(M-S)J_{sub}+2)$ (LMS), $(8(M-S)J_{sub}+2(M-S)+3)$ (NLMS) or $(12(M-S)^2J_{sub}^2+12J_{sub}+4)$ (RLS) real multiplications. Note for real-valued sensor signals, we only process the first $K/2$ subband GSCs and the update rate is $1/N$. Then for real-valued sensor signals, the total number of real multiplications for the subband GSCs is $K/(2N)(2M(M-S+1)+(8(M-S)J_{sub}+2))$ (LMS), $K/(2N)(2M(M-S+1)+(8(M-S)J_{sub}+2(M-S)+3))$ (NLMS) or $K/(2N)(2M(M-S+1)+(12(M-S)^2J_{sub}^2+12J_{sub}+4))$ (RLS), respectively; for complex-valued sensor signals, it is $K/N(2M(M-S+1)+(8(M-S)J_{sub}+2))$ (LMS), $K/N(2M(M-S+1)+(8(M-S)J_{sub}+2(M-S)+3))$ (NLMS) or $K/N(2M(M-S+1)+(12(M-S)^2J_{sub}^2+12(M-S)J_{sub}+4))$ (RLS).

The above results are summarized in Table 3.2. As a comparison, it also includes the computational complexities of the corresponding fullband GSC.

3.3.3 Reconstruction of the Fullband Beamformer

When the subband adaptive system reaches its steady state, we may find it useful to calculate the fullband beamformer equivalent to the subband-based beamformer in order to operate the beamformer in the fullband or to calculate its beam pattern. To reconstruct the m–th branch filter of the equivalent fullband beamformer shown in Figure 1.5, the m–th sensor signal $x_m[n]$ is excited by the Kronecker delta function $\delta[n]$, while all the other $M-1$ sensors receive a zero input (Weiss *et al.*, 1999c), as shown in Figure 3.15. Assuming perfect reconstruction of the filter banks, the measured impulse response at the output of the subband beamformer $y[n]$ will be the equivalent fullband coefficients $w_{m,0}$, $w_{m,1}, \ldots, W_{m,J-1}$.

3.3.4 Simulations

In this part, we will run some simulations to show the increased convergence rate of the subband adaptive beamformer based on a GSC structure.

We use a beamformer with $M=10$ sensors and $J=128$ coefficients for each attached filter. The system is constrained to receive a signal of interest from the broadside and the received array signals are corrupted by additive white Gaussian noise at an SNR of 20 dB. The beamformer aims to suppress an interfering signal coming from a DOA of $\theta=20^\circ$, which is composed of six sinusoidal signals with normalized frequencies of 0.1π, 0.3π 0.5π, 0.7π and 0.9π, respectively. The SIR is -40 dB and a simple broadside constraint is imposed ($S=1$).

Table 3.2 Computational complexities for the fullband and subband adaptive beamformers based on the simplified GSC structure in Figure 2.15

Fullband	Real multiplications (real-valued)
LMS	$M(M-S+1)+(2(M-S)J+2)$
NLMS	$M(M-S+1)+(2(M-S)J+M-S+2)$
RLS	$M(M-S+1)+(3(M-S)^2J^2+3(M-S)J)$
Subband	**Real multiplications (real-valued)**
LMS	$(M+1)/N(l_p+4K\log_2 K+4K)+$ $K/(2N)(2M(M-S+1)+(8(M-S)J_{sub}+2))$
NLMS	$(M+1)/N(l_p+4K\log_2 K+4K)+$ $K/(2N)(2M(M-S+1)+(8(M-S)J_{sub}+2(M-S)+3))$
RLS	$(M+1)/N(l_p+4K\log_2 K+4K)+$ $K/(2N)(2M(M-S+1)+(12(M-S)^2J_{sub}^2+12J_{sub}+4))$
Fullband	**Real multiplications (complex-valued)**
LMS	$2M(M-S+1)+(8(M-S)J+2)$
NLMS	$2M(M-S+1)+(8(M-S)J+2(M-S)+3)$
RLS	$2M(M-S+1)+(12(M-S)^2J^2+12J+4)$
Subband	**Real multiplications (complex-valued)**
LMS	$(M+1)/N(2l_p+4K\log_2 K+8K)+$ $K/N(2M(M-S+1)+(8(M-S)J_{sub}+2))$
NLMS	$(M+1)/N(2l_p+4K\log_2 K+8K)+$ $K/N(2M(M-S+1)+(8(M-S)J_{sub}+2(M-S)+3))$
RLS	$(M+1)/N(2l_p+4K\log_2 K+8K)+$ $K/N(2M(M-S+1)+(12(M-S)^2J_{sub}^2+12(M-S)J_{sub}+4))$

The subband structure is based on the 10-channel ($K=10$) oversampled GDFT filter banks with a decimation ratio $N=8$. The prototype filter length of the filter banks is $l_p=110$. $J_{sub}=(J+l_p)/N=29$ coefficients are used for each subband MCAF channel. The MCAFs for both the fullband and subband cases are updated by a normalized LMS algorithm.

Because the interference is the sum of a finite number of narrowband signals, in principle, we do not really need a wideband beamformer with TDLs to suppress it. The aim of this set of simulations is mainly to show the significant improvement in convergence speed by the subband method when the input signal is highly coloured.

The learning curves of the fullband and the subband adaptive beamformers based on the GSC with its blocking matrix formed by the CCD method are shown in Figure 3.16. They are obtained by averaging the results of 200 independent runs. The stepsize for the fullband GSC is 0.46 and 0.1 for the subband GSC, which is chosen empirically for both systems to achieve roughly the same steady-state value.

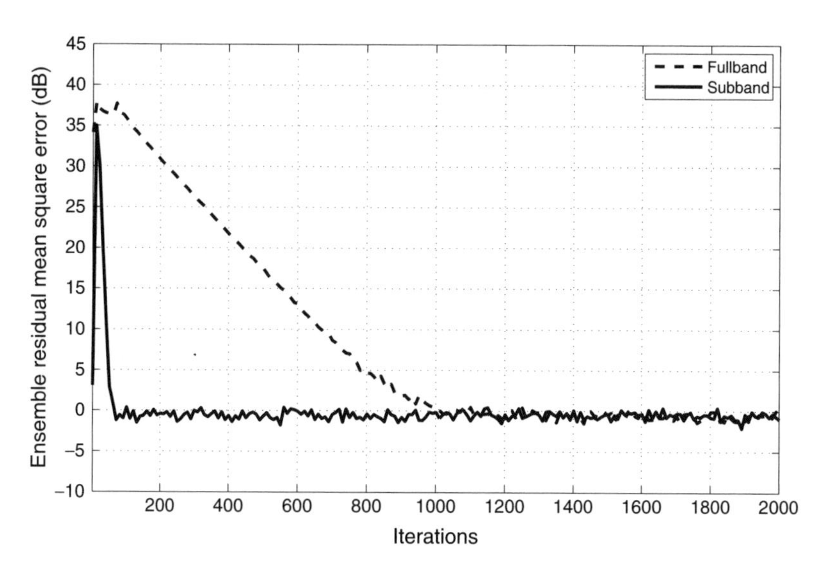

Figure 3.16 Fullband and subband learning curves for the GSC using an NLMS adaptive algorithm

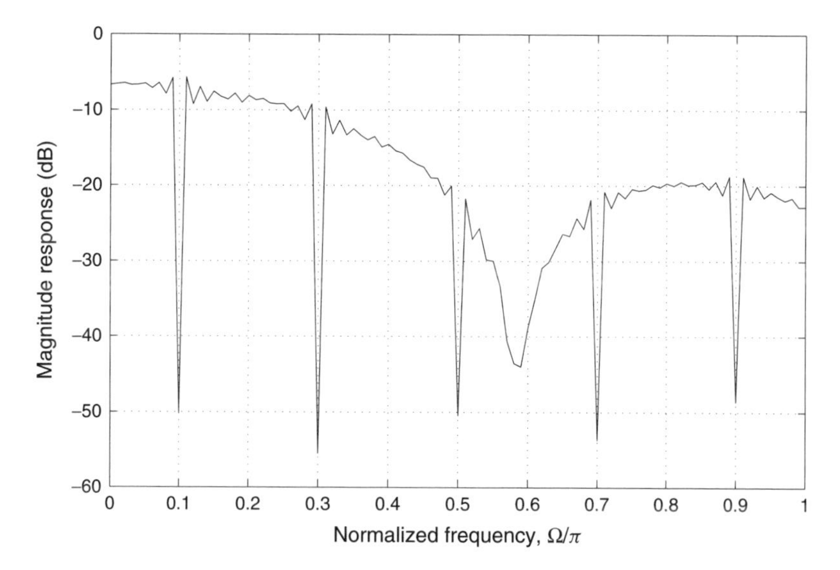

Figure 3.17 A slice of the beam pattern of the subband beamformer at the interference direction $\theta = 20^{\circ}$

It can be seen that the subband adaptive structure converges much faster than the fullband one because of its pre-whitening effect. A slice of the beam pattern of the subband beamformer after convergence at the direction $\theta = 20^{\circ}$ is shown in Figure 3.17, where the nulls at the five frequency components of the interfering signal can be seen clearly.

3.4 Subband Adaptive GSC

In the subband adaptive beamforming structure introduced in Section 3.3 based on a GSC, we are restricted to use the same number of analysis filter banks as the sensor number M, and also the same number of GSCs as the subband number K, because we have to split each of the sensor signals into subbands and apply a GSC to each of the corresponding subbands. When the number of sensors and subbands is high, these operations will impose a high computational load on the system. Moreover, in this method, the fullband constraints of the beamformer have to be decomposed into subband-based constraints in order to construct a GSC for each of the subbands. This projection can incur inaccuracies because of the non-PR property of the filter banks and the limited number of weights to represent the constraints in each subband.

To overcome these problems, we can replace the MCAF block of the fullband GSC in Figure 2.14 by the subband MCAF structure directly, leading to a subband adaptive GSC (Liu *et al.*, 2001b). In contrast to Figure 3.14, the sensor signals are directly fed into the GSC, and the outputs of its blocking matrix are split into subbands, in which independent adaptation is performed.

By this approach, we use a single GSC and apply it to the fullband sensor signals directly, thus bypassing any constraint decomposition operations. In addition, the number of analysis filter banks used is the same as the number of blocking matrix outputs, which can be considerably lower than the sensor number, especially for partially adaptive GSCs (Chapman, 1976; Van Veen and Roberts, 1987; Wang and Fang, 2000; Wang *et al.*, 1999; Yang and Ingram, 1997). Therefore, the computational complexity of this subband adaptive GSC is reduced further than the previous method.

3.4.1 Structure

Considering the multichannel characteristics of the GSC adaptation, when applying SAF techniques to the MCAF in the GSC structure in Figure 2.14, the subband setup as shown in Figure 3.18 arises, where the block with $\hat{\mathbf{f}}$ is the filtering operation defined by the simplified response vector $\hat{\mathbf{f}}$ in Section 2.4.2. Both the blocking matrix output signals $u_m[n]$, $m = 0, 1, \ldots, M - S - 1$ and the upper path output $d[n]$ are split into subbands and in each set of corresponding subbands, an MCAF system is set up to perform an unconstrained optimization based on the subband error signal $y_k[n]$, $k = 0, 1, \ldots, K - 1$. These subband error signals are then reconstructed into a fullband beamformer output $y[n]$ by a synthesis filter bank.

Comparing Figure 3.18 with Figure 3.14, we can see that as long as the constraints are decomposed into each subband without any error, especially when the response vector $\mathbf{f}$ takes the form described in Section 3.3.2, these two structures are equivalent and there should not be much performance difference between them except for the reduced computational complexity.

3.4.2 Analysis of the Computational Complexity

For the computational complexity of this subband adaptive GSC, assuming that the response vector $\hat{\mathbf{f}}$ is a simple delay, the number of real multiplications required for

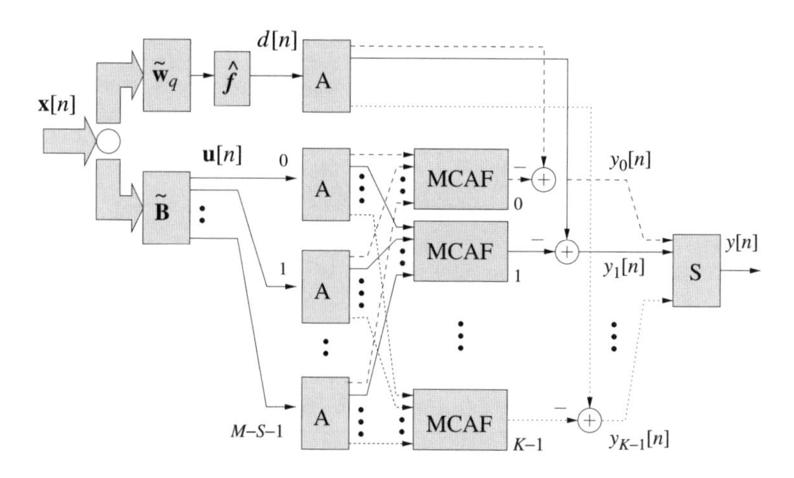

Figure 3.18 A general subband adaptive generalized sidelobe canceller

the real-valued quiescent vector $\tilde{\mathbf{w}}_q$ and the real-valued blocking matrix $\tilde{\boldsymbol{B}}$ is $M + (M - S)M = M(M - S + 1)$ for real-valued sensor signals and $2M(M - S + 1)$ for complex-valued sensor signals. For the adaptive part of the GSC, it is an $(M - S)$-channel subband adaptive filter and there are $M - S + 1$ analysis filter banks and one synthesis filter bank. Then the total number of real multiplications for the filter banks is $(M - S + 2)/N\left(l_p + 4K\log_2 K + 4K\right)$ or $(M - S + 2)/N\left(2l_p + 4K\log_2 K + 8K\right)$ for real- and complex-valued inputs, respectively.

For the following adaptive part after the analysis filter banks, we can use the result in Section 3.3.1 to calculate the number of real multiplications. That is, in the real-valued sensor signals case, this part requires $K/(2N)(8(M - S)J_{sub} + 2)$ (LMS), $K/(2N)(8(M - S)J_{sub} + 2(M - S) + 3)$ (NLMS) or $K/(2N)(12(M - S)^2 J_{sub}^2 + 12J_{sub} + 4)$ (RLS) real multiplications; for complex-valued sensor signals, it is $K/N(8(M - S)J_{sub} + 2)$ (LMS), $K/N(8(M - S)J_{sub} + 2(M - S) + 3)$ (NLMS) or $K/N(12(M - S)^2 J_{sub}^2 + 12J_{sub} + 4)$ (RLS) real multiplications.

The above results are summarized in Table 3.3.

3.4.3 Reconstruction of the Fullband Beamformer

The reconstruction of the fullband beamformer can be performed in the same way as in Section 3.3.3, i.e. in order to obtain the coefficients of the mth branch TDL, the mth sensor signal is excited by $\delta[n]$, while leaving all the other $M - 1$ sensors with a zero input, as shown in Figure 3.19.

3.4.4 Simulations

In this part, we give a simple comparison between the performance of the subband adaptive GSC and the fullband one. The setup is the same as in Section 3.3.4 except that now there is an additional interfering signal with a bandwidth $\Omega \in [0.3\pi\ 0.9\pi]$, an SIR of -20 dB

Table 3.3 Computational complexities for the subband adaptive GSC

Subband	Real multiplications (real-valued)	Real multiplications (complex-valued)
LMS	$M(M-S+1)+$ $(M-S+2)/N(l_p + 4K\log_2 K + 4K)+$ $K/(2N)(8(M-S)J_{sub}+2)$	$2M(M-S+1)+$ $(M-S+2)/N(2l_p + 4K\log_2 K + 8K)+$ $K/N(8(M-S)J_{sub}+2)$
NLMS	$M(M-S+1)+$ $(M-S+2)/N(l_p + 4K\log_2 K + 4K)+$ $K/(2N)(8(M-S)J_{sub}+2(M-S)+3)$	$2M(M-S+1)+$ $(M-S+2)/N(2l_p + 4K\log_2 K + 8K)+$ $K/N(8(M-S)J_{sub}+2(M-S)+3)$
RLS	$M(M-S+1)+$ $(M-S+2)/N(l_p + 4K\log_2 K + 4K)+$ $K/(2N)(12(M-S)^2 J_{sub}^2 + 12J_{sub}+4)$	$2M(M-S+1)+$ $(M-S+2)/N(2l_p + 4K\log_2 K + 8K)+$ $K/N(12(M-S)^2 J_{sub}^2 + 12J_{sub}+4)$

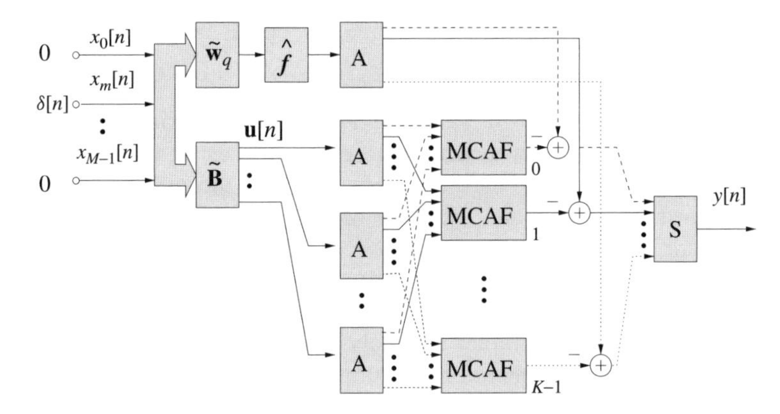

Figure 3.19 Reconstruction of the mth TDL coefficients of the equivalent fullband beamformer from the subband adaptive GSC

and from the direction $\theta = -30°$. The resultant learning curves are given in Figure 3.20, which shows a significantly increased convergence speed.

3.5 Temporally/Spatially Subband-Selective Beamforming

In this section, we introduce a class of GSCs employing a spatially/temporally subband-selective blocking matrix for partially adaptive beamforming (Liu *et al.*, 2001a, 2002b,c, 2003a, 2004a), where the impulse responses hosted by the columns of the blocking matrix constitute a series of bandpass filters. These filters select signals with specific DOAs and frequencies and result in bandlimited spectra of the blocking matrix outputs. When we apply such a blocking matrix to the subband adaptive GSC described in Section 3.4, the system complexity may be reduced. Moreover, due to its combined decorrelation in both the spatial and temporal domains, a faster convergence rate is also achieved.

In Section 3.5.1, we will give a brief introduction to partially adaptive beamforming based on the GSC structure. Thereafter we focus on the role of the blocking matrix and

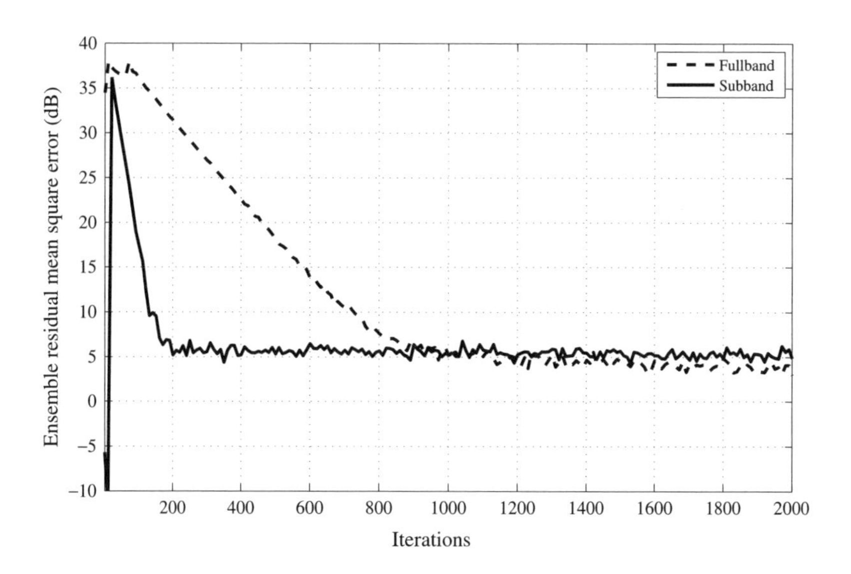

Figure 3.20 Learning curves of the fullband and the subbband adaptive GSCs

its new construction exploiting spatial/temporal filtering properties in Section 3.5.2. In Section 3.5.3, an alternative formation of this blocking matrix by a subband-selective transformation matrix is provided. The application of this blocking matrix to the subband adaptive GSC is given in Section 3.5.4 with extensions to the general beamforming structure described in Section 3.5.5. Some simple simulation results are provided in Section 3.5.6.

3.5.1 Partially Adaptive GSC

As mentioned before, to perform beamforming with high interference rejection and resolution, arrays with a large number of sensors and filter coefficients have to be employed and the computational complexity of a fully adaptive beamformer thus becomes considerable. One way to reduce this complexity is through partially adaptive beamforming techniques, which employ only a subset of available degrees of freedom (DOFs) in the weight update process at the expense of a somewhat reduced performance (Chapman, 1976).

Partial adaptation techniques have been studied widely in the past, especially in the narrowband beamforming area. Chapman (1976) performed one of the earliest works, which reduces the number of adaptive channels by means of a transformation. In Van Veen and Roberts (1987), a fixed transformation was applied to achieve a weight reduction by minimizing the output power over a set of interference scenarios. A 'power-space method', which uses singular value decomposition to obtain a reduced-rank transformation was developed in Yang and Ingram (1997).

In wideband beamforming, a broadband focusing adaptive beamformer was proposed in Simanapalli and Kaveh (1994), where a pre-processing focusing operation is applied to the received array signals, followed by a narrowband beamforming structure with a single set of adaptive coefficients for the focused frequency; the wavelet-based beamformer, introduced in Wang and Fang (2000) and Wang *et al.* (1999), reduces the dimension of the blocking matrix by utilizing a set of wavelet filters, which are considered as a series of spatial filters. Thereafter, a dynamic selection of the blocking matrix outputs is performed by a prescribed statistical hypothesis test. The main problem with this approach is that the design of wavelet filters with a good band-selection property is not easy when their lengths are relatively short, and such filters would sacrifice too many DOFs of the beamformer.

For a brief description of the partially adaptive beamformer based on the GSC structure, we consider the simplified system shown in Figures 2.14 and 2.15. There the only constraint imposed on the blocking matrix $\tilde{\boldsymbol{B}}$ is Equation (2.75), expressed as:

$$\tilde{\boldsymbol{C}}^H \tilde{\boldsymbol{B}} = \mathbf{0} \qquad \text{where} \quad \tilde{\boldsymbol{C}} = \begin{bmatrix} \mathbf{c}_0 & \dots & \mathbf{c}_{S-1} \end{bmatrix} \tag{3.37}$$

with $\tilde{\boldsymbol{C}} \in \mathbf{C}^{M \times S}$ and $\tilde{\boldsymbol{B}} \in \mathbf{C}^{M \times M-S}$. However, the column dimension of $\tilde{\boldsymbol{B}}$ is not restricted to be $L = M - S$, which is only the maximum possible value. Assume $\tilde{\boldsymbol{B}} \in \mathbf{C}^{M \times L}$ with:

$$\tilde{\boldsymbol{B}} = \begin{bmatrix} \mathbf{b}_0 & \mathbf{b}_1 & \dots & \mathbf{b}_{L-1} \end{bmatrix} \qquad \text{and} \tag{3.38}$$

$$\mathbf{b}_l = \begin{bmatrix} b_{l,0} & b_{l,1} & \dots & b_{l,M-1} \end{bmatrix}^H \tag{3.39}$$

where $l = 0, 1, \dots, L-1$. With this new definition, we redraw the simplified GSC for MVDR beamforming in Figure 3.21, where the output signal vector $\mathbf{u}[n]$ of the blocking matrix is obtained by $\mathbf{u}[n] = \tilde{\boldsymbol{B}}^H \mathbf{x}[n]$ with:

$$\begin{aligned} \mathbf{u}[n] &= \begin{bmatrix} u_0[n] & u_1[n] & \cdots & u_{L-1}[n] \end{bmatrix}^T \\ \mathbf{x}[n] &= \begin{bmatrix} x_0[n] & x_1[n] & \cdots & x_{M-1}[n] \end{bmatrix}^T \end{aligned} \tag{3.40}$$

Since $\tilde{\boldsymbol{C}}$ is an $M \times S$ matrix, the dimension L can be selected arbitrarily with $L \leq M - S$. The maximum value $M - S$ corresponds to the fully adaptive GSC. When a large number of sensors are employed, we can choose a smaller value for L, i.e. $L < M - S$, resulting in a partially adaptive GSC (Van Veen and Roberts, 1987; Wang and Fang, 2000). By partial adaptation, the number of DOFs, i.e. the number of adaptive weights, is reduced, which offers reduced complexity traded off against a potentially somewhat inferior performance. In the next section, we will trade the loss of DOFs against a specific design of the blocking matrix.

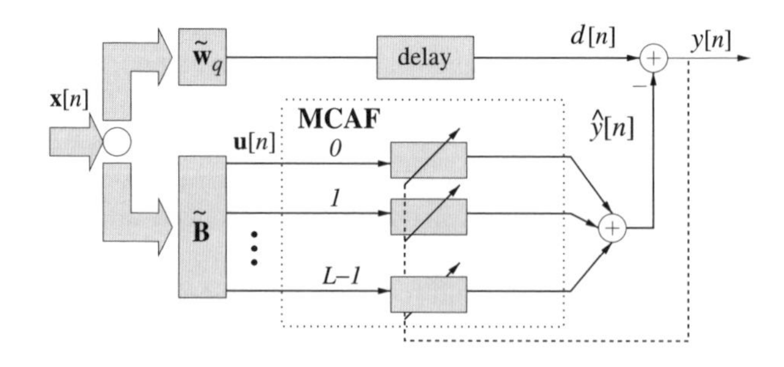

Figure 3.21 A general structure for the partially adaptive GSC

3.5.2 *Temporally/Spatially Subband-Selective Blocking Matrix*

3.5.2.1 Blocking Matrix with Temporal/Spatial Subband Selectivity

In Figure 3.21, the lth output $u_l[n]$ of the blocking matrix is given by:

$$u_l[n] = \sum_{m=0}^{M-1} b_{l,m} x_m[n] \tag{3.41}$$

Assume that the system is a uniformly spaced linear array. In order to attain an interpretation of the spatially and temporally subband selective filters constituting the blocking matrix, consider a signal with an angular frequency ω and a DOA angle θ. Then the response of the column vector $\mathbf{b}_l$ to this impinging signal can be expressed as:

$$R_l(\Omega, \theta) = \sum_{m=0}^{M-1} b_{l,m} \mathrm{e}^{-jm\mu\Omega \sin\theta} \text{ with } \mu = \frac{d}{cT_s} \tag{3.42}$$

where $\Omega = \omega T_s$ and T_s is the sampling period, as also given in Equation (1.36) with $J = 1$. As discussed before, when the sampling frequency is twice the highest frequency component of the signal and the array spacing is half the wavelength of the highest frequency component, we have $\mu = 1$. Without loss of generality, we will always assume $\mu = 1$ in the following.

With $B_l(\mathrm{e}^{j\Psi}) = \sum_{m=0}^{M-1} b_{l,m} \mathrm{e}^{-jm\Psi}$, i.e. $B_l(\mathrm{e}^{j\Psi})$ is the frequency response of the filter with impulse responses $b_{l,0}, b_{l,1}, \ldots, b_{l,M-1}$, we have:

$$R_l(\Omega, \theta) = B_l(\mathrm{e}^{j\Psi}) \tag{3.43}$$

with $\Psi = \Omega \sin\theta$. For a fixed $\Omega \in [-\pi\ \pi]$, when θ changes from $-90°$ to $90°$, the value of Ψ changes from $-\Omega$ to Ω.

Now the columns of the blocking matrix $\tilde{\boldsymbol{B}}$ can be regarded as a set of spatial filters. If the beamformer is constrained to receive signals from the broadside, then the blocking matrix has to suppress any component impinging from $\theta = 0$. Therefore, at $\Psi = 0$ the response of any column vector $\mathbf{b}_l$ has to be zero. Now we arrange the design of $\mathbf{b}_l$ to yield spatial bandpass filters on the interval $\Psi \in [0\ \pi]$ as shown in Figure 3.22, whereby ideally:

$$|B_l(\mathrm{e}^{j\Psi})| = \begin{cases} 1 \text{ for } & \Psi \in [\Psi_{l,\text{lower}}\ \Psi_{l,\text{upper}}] \\ 0 \text{ otherwise.} & \end{cases} \tag{3.44}$$

Here we assume that $\mathbf{b}_l$ is real-valued and has a symmetric response $|B_l(\mathrm{e}^{j\Psi})|$ for the part $\Psi \in [-\pi\ 0]$.

In the arrangement of Figure 3.22, all values of $\Psi \in (0\ \pi]$ except for $\Psi = 0$ have to be covered by the filters to ensure that the lower path of the GSC contains all possible interfering signals. If the interfering signals impinge only from a certain set of angles with a certain bandwidth, then the appeal is that only some outputs of the blocking matrix and therefore some branches of the subsequent multichannel adaptive filter will contain significant contributions. Then, a design of the blocking matrix columns according to Figure 3.22 will lead to a spatial decomposition or decorrelation of the array data.

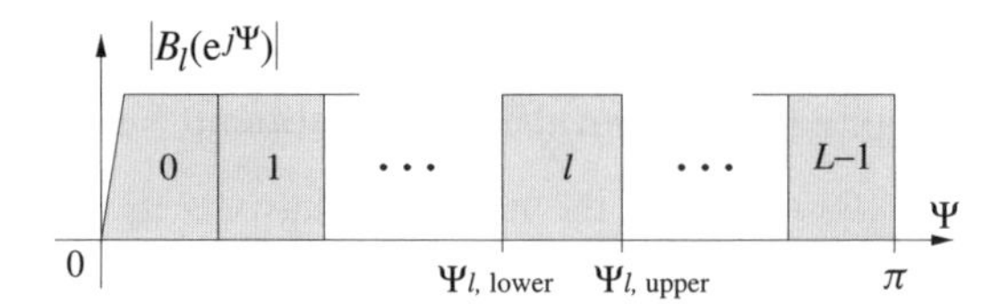

Figure 3.22 Characteristics of the L column vectors in $\tilde{\boldsymbol{B}}$

To avoid redundancy in the blocking matrix outputs, we would like the column vectors of $\tilde{\boldsymbol{B}}$ to be linearly independent, i.e. none of them can be expressed as a linear combination of the others. It can be proved that these column vectors $\mathbf{b}_l$ are orthogonal under the ideal arrangement of Figure 3.22. If an overlap occurs between the neighbouring bands in Figure 3.22, these column vectors $\mathbf{b}_l$ can also be proved to be at least linearly independent. To show this, we consider the linear combination of all these vectors in the following form:

$$\mathbf{0} = \alpha_0\mathbf{b}_0 + \alpha_1\mathbf{b}_1 + \cdots + \alpha_{L-1}\mathbf{b}_{L-1} \tag{3.45}$$

where $\alpha_0, \ldots, \alpha_{L-1}$ are scalars to be found for this equation to hold. Taking the Hermitian transpose of both sides and then multiplying the equation with the vector $[1\ e^{-j\Psi}\ \cdots\ e^{-j(M-1)\Psi}]^T$, we arrive at:

$$0 = \alpha_0 B_0(e^{j\Psi}) + \alpha_1 B_1(e^{j\Psi}) + \cdots + \alpha_{L-1} B_{L-1}(e^{j\Psi}) \tag{3.46}$$

As it should hold for all values of Ψ, we choose the middle point of each of the passbands of the column vectors. Since at these points only one of the L Fourier transforms $B_l(e^{j\Psi})$, $l = 0, 1, \ldots, L-1$, has a value of unity and all the others are zeros, we find that all the scalars $\alpha_0, \ldots, \alpha_{L-1}$ must be zero for Equation (3.45) to hold, i.e. the column vectors are linearly independent and the proof is complete.

From the above discussion we can also see that as long as every vector has at least one point, where its Fourier transform has a nonzero value and coincides with zeros of all the other vectors' Fourier transforms, these vectors will be linearly independent.

We now focus on the temporal filtering effect of the arrangement in Equation (3.44). Consider the range $\Omega \in [0\ \pi]$. As $|\sin\theta| \in [0\ 1]$ for $\theta \in [-\pi/2\ \pi/2]$, the possible maximum frequency component of the lth output $u_l[n]$ is $\Omega = \pi$, which corresponds to $|\sin\theta| = (\Psi_{l,\text{upper}})/\pi$, while the possible minimum frequency component is $\Omega = \Psi_{l,\text{lower}}$, which corresponds to $|\sin\theta| = 1$. Then we have the response of $B_l(e^{j\Psi})$ as a function of $\Omega = \Psi/\sin\theta$:

$$|B_l(e^{j\Omega\sin\theta})| = \begin{cases} 1 & \text{for } \Omega \in [\Psi_{l,\text{lower}}\ \pi] \\ 0 & \text{otherwise} \end{cases} \quad \forall\,\theta \tag{3.47}$$

This temporal filtering characteristic of the lth column vector of $\tilde{\boldsymbol{B}}$ is shown in Figure 3.23. If the interfering signals have components with $\Omega > \pi$, then as long as they appear in the directions of $|\sin\theta| \in [\Psi_{l,\text{lower}}/\Omega\ \Psi_{l,\text{upper}}/\Omega]$, they will be received by the lth column vector. However, with the assumption that the sampling rate is twice the maximum signal frequency, there will be no signal existing with $\Omega > \pi$.

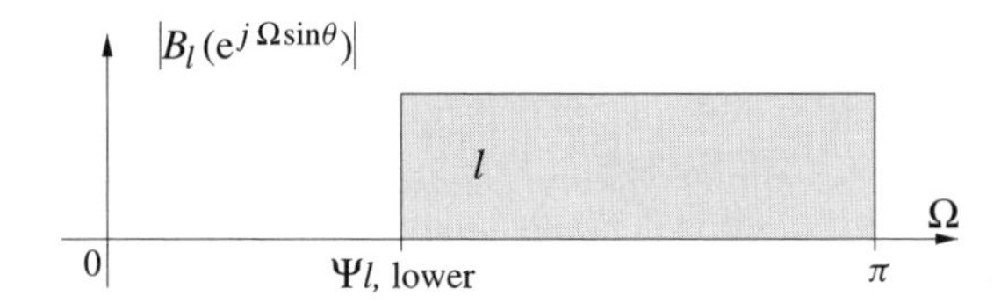

Figure 3.23 Temporal filtering effect of the lth spatial filter in $\tilde{\boldsymbol{B}}$

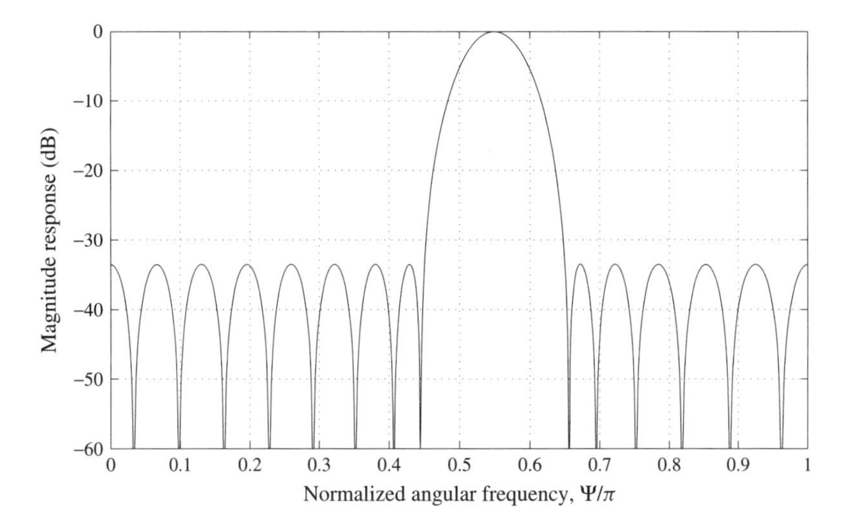

Figure 3.24 Frequency response of a bandpass filter as an example for the column vectors of blocking matrix $\tilde{\boldsymbol{B}}$

Example: To show this temporal filtering effect, we here give a simple example. Figure 3.24 shows the magnitude response of a 30-tap bandpass filter designed by the MATLAB© function *remez* (Mat, 2001). If the filter coefficients are employed as a column vector in $\tilde{\boldsymbol{B}}$, then a gain response or beam pattern to signals with different frequencies and DOAs can be calculated according to Equation (3.43) and the result is shown in Figure 3.25. To see its highpass filtering effect more clearly, the figure can be rotated and inspected in term of its frequency dependence only. The resultant 2-D response with respect to different frequencies is shown in Figure 3.26, and clearly exhibits the noted highpass characteristic.

Thus, the blocking matrix is capable of decomposing the received signals and interferences not only in the spatial domain, but also in the temporal domain, i.e. the column vectors simultaneously perform a spatial selection and a temporal highpass filtering operation. With increasing l, these filters are associated with tighter and tighter highpass spectra until the last output $u_{L-1}[n]$ only contains the ultimate highpass component.

In reality, $B_l(e^{j\Psi})$ will not have an ideal bandpass response and transition bands have to be allowed. However, a better design quality can be attained by reducing the number of columns, L, below the limit of $M - S$, thus yielding a partially adaptive beamformer by sacrificing some DOFs. As the blocking matrix covers all possible interfering signals, it

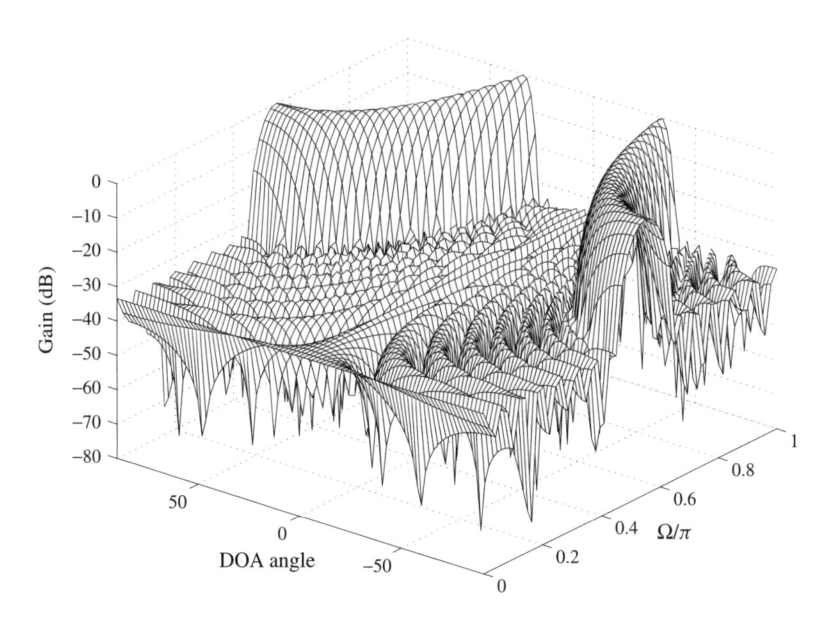

Figure 3.25 Three-dimensional response of a bandpass filter to signals with different frequencies and DOAs

can still suppress any incoming interferences, but the achievable maximum SINR would be potentially lower than a fully adaptive beamformer. Due to the loss of ideal responses as specified in Figure 3.22, the linear independence of the column vectors of $\tilde{\boldsymbol{B}}$ has to be inspected after design.

As the blocking matrix plays a central role in our following applications, a column vector design with a good band-selective property is of great importance. We will deal with this issue in Sections 3.5.2.2 and 3.5.2.3, where a full design and a cosine-modulated design of the blocking matrix will be described, respectively.

3.5.2.2 Full Design of the Blocking Matrix

From Equation (2.75), when considering the $S-1$ order derivative constraints with $d_0 = 0$, as defined in Equation (2.124), we can express the constraints to be satisfied by the lth column vector of the blocking matrix as:

$$\sum_{m=0}^{M-1} m^i b_{l,m} = 0\,, \quad \text{for} \quad i = 0, 1, \ldots, S-1, \;\; l = 0, 1, \ldots, L-1 \tag{3.48}$$

Subject to the above constraints, the objective function Φ_l to be minimized for the lth column vector is:

$$\Phi_l = \int_0^{\Psi_{l,\text{lower}}} \| B_l(\mathrm{e}^{j\Psi})\|^2 \mathrm{d}\Psi \;+\; \int_{\Psi_{l,\text{upper}}}^{\pi} \| B_l(\mathrm{e}^{j\Psi})\|^2 \mathrm{d}\Psi \tag{3.49}$$

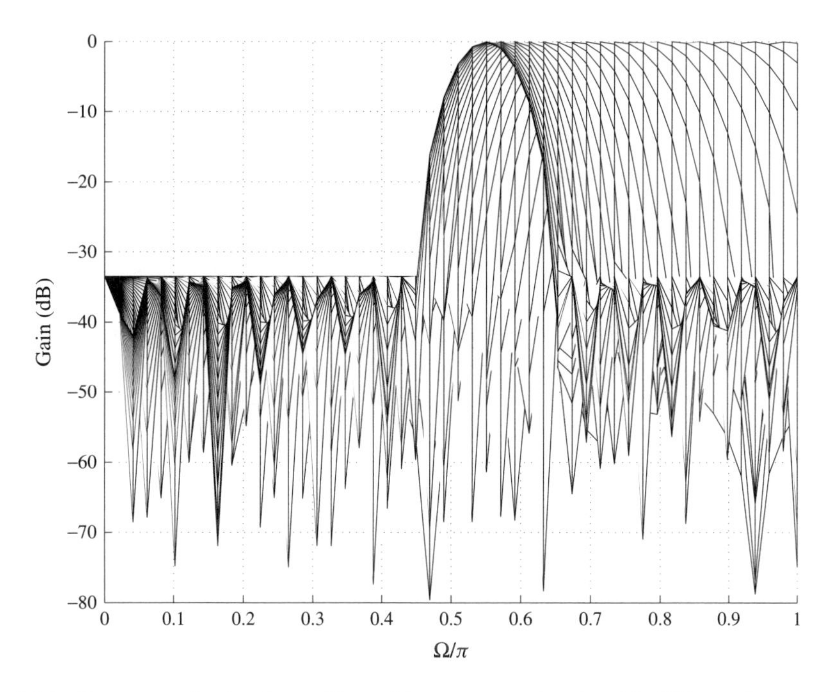

Figure 3.26 Response of a bandpass filter with respect to signals with different frequencies as a column vector of the blocking matrix $\tilde{\boldsymbol{B}}$

The design of the blocking matrix $\tilde{\boldsymbol{B}}$ can then be formulated as the following linearly constrained minimization problem:

$$\mathbf{b}_l = \arg\min_{\mathbf{b}_l} \Phi_l \quad \text{subject to Equation (3.48)}, \quad l = 0, 1, \ldots, L-1 \tag{3.50}$$

Various methods exist to solve such a linearly constrained optimization problem. One example is the Lagrange multipliers method. Here the minimization of Φ_l is accomplished by using the subroutines LCONF/DLCONF in the IMSL library (Vis, 2002). A design example for the blocking matrix with $M = 16$ sensors, first-order derivative constraints ($S = 2$) and $L = 8$ column filters is given in Figure 3.27. Displayed are the frequency responses $B_l(\mathrm{e}^{j\Psi})$, $l = 0, 1, \ldots, 7$, which exhibit a reasonably good bandpass characteristic.

If Equation (3.48) is used to express the first S parameters in each $\mathbf{b}_l$ by the remaining $M - S$ vector elements, an unconstrained optimization can be performed over those remaining parameters, for example by means of a genetic algorithm (GA) (Man *et al.*, 1997; Tang *et al.*, 1996). Using a GA, we can obtain a result with all elements in $\tilde{\boldsymbol{B}}$ in the form of sums of power of two (SOPOTs) (Chan *et al.*, 2001; Liu, 2000; Liu *et al.*, 2000a,b). By SOPOT representation, the arithmetic for $\tilde{\boldsymbol{B}}$ can be implemented by simple shifts and additions to further reduce its computational complexity (Lim and Parker, 1983; Lim *et al.*, 1983).

For a brief introduction to GAs, please refer to Appendix C.

Here, as an example, let us consider the case of $S = 2$, whereby the first two elements of $\mathbf{b}_l$ are fixed in dependency on the remaining coefficients:

$$b_{l,0} = -\sum_{m=1}^{M-1} b_{l,m}$$
$$b_{l,1} = -\sum_{m=2}^{M-1} m b_{l,m} \qquad (3.51)$$

Note that the constrained optimization problem has now been transformed into an unconstrained one over the remaining coefficients $b_{l,2}, \ldots, b_{l,M-1}$, which is straightforward to solve by means of a GA. In the optimization process of a GA yielding SOPOT parameters, each of the coefficients $b_{l,m}$, $m = 2, 3, \ldots, M-1$ is represented as:

$$b_{l,m} = \sum_{i=0}^{P[l,m]-1} a_i[l,m] \times 2^{L_i[l,m]} \qquad \text{with :}$$
$$a_i[l,m] \in \{-1, 1\}, \quad L_i[l,m] \in \{Q_1, Q_1+1, \ldots, Q_2-1, Q_2\} \qquad (3.52)$$

where $P[l,m]$ is the limit for the number of SOPOT terms for $b_{l,m}$, and Q_1 and Q_2 are integers determined by the range of the corresponding variable. Normally, $P[l,m]$ is limited to a small number. With the same parameters as those of the example shown in Figure 3.27, a GA design result in SOPOT notation with $P[l,m] = 3$, $Q_1 = -9$ and $Q_2 = 0$ can be found in Table C.1 in Appendix C. Its frequency response is displayed in Figure 3.28.

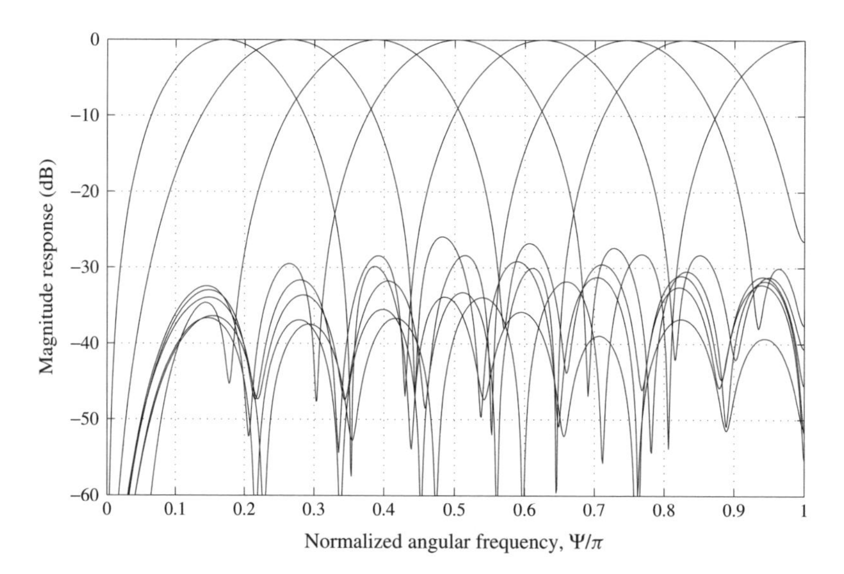

Figure 3.27 A design example for a 16 × 8 blocking matrix with first-order derivative constraints based on the constrained optimization

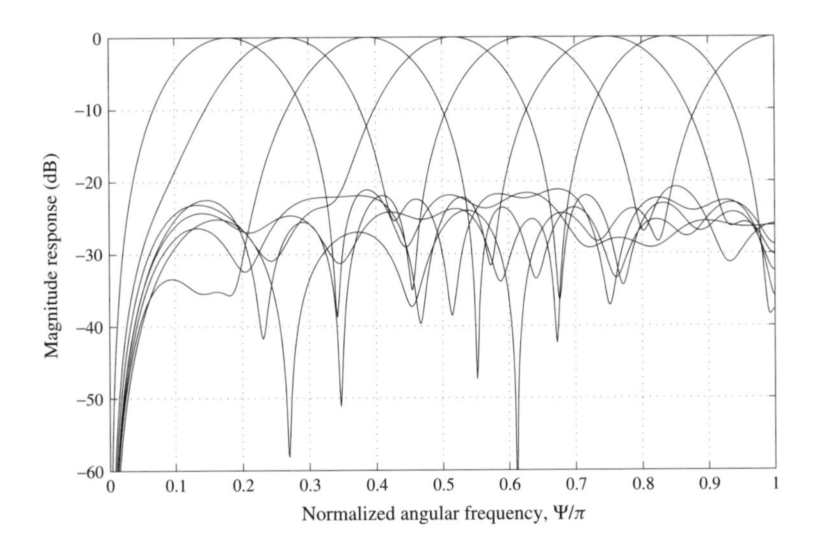

Figure 3.28 Characteristics of the columns vectors of the 16×8 blocking matrix using a GA design with SOPOT representation

3.5.2.3 Design Based on Prototype Modulation

In order to reduce the design and implementation complexity of the blocking matrix, different from the approach discussed in Section 3.5.2, the column vectors of $\tilde{\boldsymbol{B}}$ can also be derived from a prototype vector through various modulation schemes, e.g. a cosine modulation, whereby the broadside constraint is guaranteed by imposing spectral zeros appropriately on the prototype vector.

Assume the prototype vector is $h[m]$, $m = 0, 1, \ldots, M-1$. Based on $h[m]$, we employ the DCT-IV (discrete cosine transform type IV) modulation (Vaidyanathan, 1993) to obtain the lth column vector $\mathbf{b}_l$, $l = 0, 1, \ldots, L-1$, with its element $b_{l,m}$ given by:

$$b_{l,m} = h[m] \cos\left[\frac{\pi}{2L+2}(2l+3)\left(m - \frac{M-1}{2}\right) - (-1)^l \frac{\pi}{4}\right] \tag{3.53}$$

In the frequency domain, this modulation creates two copies of the prototype vector's frequency response shifted along the frequency axis by $(2l+3)\pi/(2L+2)$ and $-(2l+3)\pi/(2L+2)$, respectively and then adds them together. To comply with the zero-order broadside constraint $B_l(\mathrm{e}^{j\Psi})|_{\Psi=0} = 0$, the frequency response $H(z)$ of $h[m]$ should have one spectral zero at each frequency point $\Omega_l = \pm(2l+3)\pi/(2L+2)$, $l = 0, 1, \ldots, L-1$. If we factorize $H(z)$ into two parts

$$\begin{aligned} H(z) &= P(z)Q(z) \qquad \text{with :} \\ Q(z) &= \prod_{l=0}^{L-1}\left(1 - \mathrm{e}^{j\frac{2l+3}{2L+2}\pi} z^{-1}\right)\left(1 - \mathrm{e}^{-j\frac{2l+3}{2L+2}\pi} z^{-1}\right) \end{aligned} \tag{3.54}$$

then the broadside constraint will be automatically satisfied by $Q(z)$ for all the column vectors and the free parameters contained in $P(z)$ can be used to optimize its

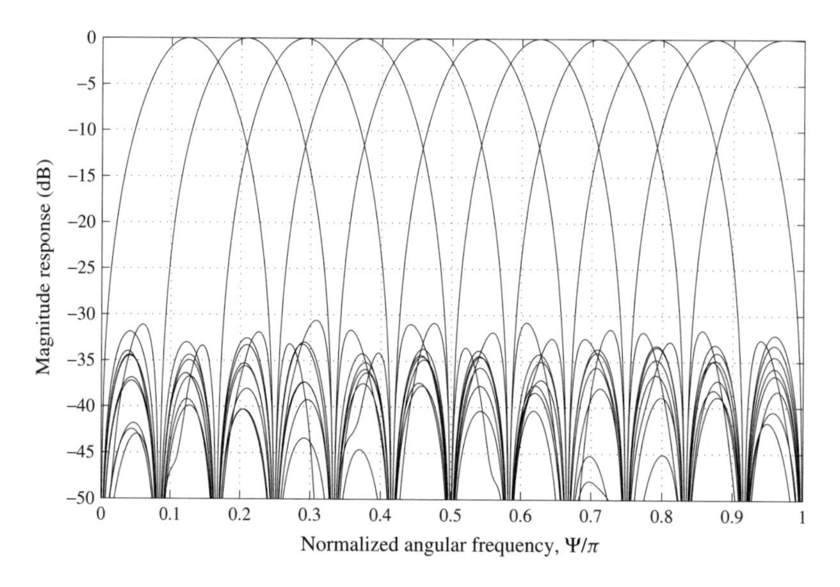

Figure 3.29 A design example for a 28 × 11 blocking matrix

frequency response. By this factorization, the design of the blocking matrix is reduced to an unconstrained optimization problem for the prototype vector. The objective function we minimize is:

$$\Phi = \int_{\Omega_s}^{\pi} \|H(e^{j\Psi})\|^2 d\Psi \tag{3.55}$$

where Ω_s is the stopband cutoff frequency. To solve this unconstrained optimization problem, we here employ the subroutines BCONF/DBCONF in the IMSL library used earlier (Vis, 2002). A design example for the blocking matrix with $M = 28$ sensors, and $L = 11$ column vectors is given in Figure 3.29.

Note that the length of the filter $Q(z)$ is $2L + 1$ and the minimum length of $P(z)$ is 1, thus we have:

$$L \leq \frac{M-1}{2} \tag{3.56}$$

i.e. the maximum value of the output dimension L achieved by this prototype modulation method is $(M-1)/2$, instead of the theoretical value $M - S$ as indicated in Equation (2.75).

For $S - 1$ order derivative constraints, we can replace $Q(z)$ in Equation (3.54) by $Q(z)^S$ and the maximum value of L achieved will be $(M-1)/2S$, which will sacrifice a considerable number of DOFs for $S > 1$ and thus a satisfying performance may not be achieved for small-size arrays. This reduction in DOFs presents a limitation of this design method.

3.5.3 *Temporally/Spatially Subband-Selective Transformation Matrix*

In the partially adaptive GSC introduced in Section 3.5.1, the blocking matrix plays two different roles: the first is to block the signal of interest from the broadside, while the second is to reduce the dimension of its output from $M - S$ to $L < M - S$ for partial adaptation. A blocking matrix designed according to these specifications can be decomposed into the product of two matrices, an $M \times (M - S)$-dimensional blocking matrix $\tilde{\boldsymbol{B}}$ for a fully adaptive GSC, which is set to block the broadside signal of interest, followed by an $(M - S) \times L$-dimensional transformation matrix $\boldsymbol{T}$, which can reduce the output dimension according to some specific requirements.

A partially adaptive GSC following this approach has been introduced in Van Veen and Roberts (1987) and Yang and Ingram (1997) and is shown in Figure 3.30. In this section, we will propose a temporally/spatially subband-selective transformation matrix. When this matrix is cascaded with a standard blocking matrix, they exhibit characteristics similar to the subband-selective blocking matrix proposed in Section 3.5.2.

3.5.3.1 Transformation Matrix with Temporal/Spatial Subband-Selectivity

As shown in Figure 3.30, the dimension of the blocking matrix output vector $\tilde{\mathbf{u}}[n] = \tilde{\boldsymbol{B}}^H \mathbf{x}[n]$ is $(M - S) \times 1$. After passing through the transformation matrix $\boldsymbol{T} \in \mathbf{C}^{L\times(M-S)}$, the data vector's dimension is reduced to $L \times 1$ by $\mathbf{u}[n] = \boldsymbol{T}\tilde{\mathbf{u}}[n]$, where $\mathbf{u}[n]$ is the final input to the following multi-channel adaptive filtering process and:

$$\begin{aligned} \boldsymbol{T} &= \begin{bmatrix} \mathbf{t}_0 & \mathbf{t}_1 & \cdots & \mathbf{t}_{L-1} \end{bmatrix}^T \\ \mathbf{t}_l &= \begin{bmatrix} t_{l,0} & t_{l,1} & \ldots & t_{l,M-S-1} \end{bmatrix}^T \end{aligned} \tag{3.57}$$

with $l = 0, 1, \ldots, L - 1$.

Consider the same signal environment and parameters as those in Section 3.5.2. Using the substitution $\Psi = \Omega \sin\theta$, the steering vector of the sensor array is given by:

$$\mathbf{d}(\Psi) = \begin{bmatrix} 1 & \mathrm{e}^{-j\Psi} & \cdots & \mathrm{e}^{-j(M-1)\Psi} \end{bmatrix}^T \tag{3.58}$$

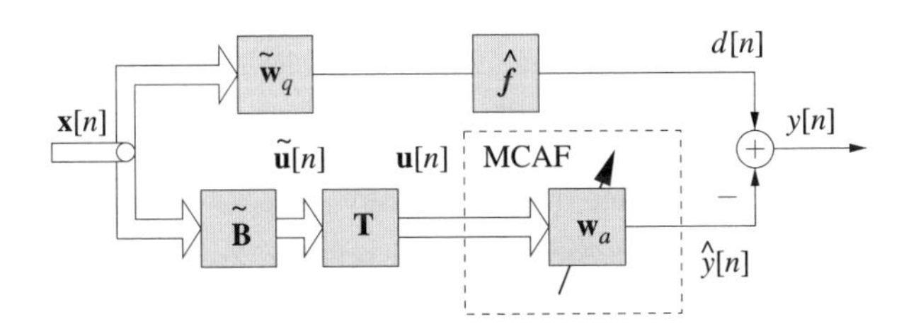

Figure 3.30 A partially adaptive GSC with a transformation matrix

Assume that the blocking matrix $\tilde{\boldsymbol{B}}$ is constructed by the CCD method as described in Section 2.4.3, namely:

$$\tilde{\boldsymbol{B}} = \boldsymbol{B}_M \cdot \boldsymbol{B}_{M-1} \cdots \boldsymbol{B}_{M-S+1} \tag{3.59}$$

where:

$$\boldsymbol{B}_i = \begin{bmatrix} 1 & -1 & & & \mathbf{0} \\ & \ddots & \ddots & & \\ & & \ddots & \ddots & \\ & & & \ddots & \ddots \\ \mathbf{0} & & & 1 & -1 \end{bmatrix}^T \in \mathbf{C}^{i\times(i-1)} \tag{3.60}$$

with $i = M, M-1, \ldots, M-S+1$. For an impinging signal $\mathrm{e}^{jn\Omega}$ from the direction θ, the output of the blocking matrix can be expressed as:

$$\tilde{\mathbf{u}}[n] = (1-\mathrm{e}^{-j\Psi})^S \times \begin{bmatrix}1 & \mathrm{e}^{-j\Psi} & \cdots & \mathrm{e}^{-j(M-S-1)\Psi}\end{bmatrix}^T \times \mathrm{e}^{jn\Omega} \tag{3.61}$$

We see that the blocking matrix has a zero response to the signal from the broadside as required. For a general blocking matrix $\tilde{\boldsymbol{B}}$, its output vector follows as:

$$\tilde{\mathbf{u}}[n] = [B_0(\mathrm{e}^{j\Psi}) \;\; B_1(\mathrm{e}^{j\Psi}) \;\; \cdots \;\; B_{M-S-1}(\mathrm{e}^{j\Psi})]^T \times \mathrm{e}^{jn\Omega} \tag{3.62}$$

Since a blocking matrix must have the desired zeros for the broadside signal of interest, the above vector response with elements $B_l(\mathrm{e}^{j\Psi})$, $l = 0, 1, \ldots, M-S-1$, can be decomposed into the product of $(1-\mathrm{e}^{-j\Psi})^S$ and a vector with polynomials in $\mathrm{e}^{j\Psi}$. Such a vector with polynomials in $\mathrm{e}^{j\Psi}$ can be further decomposed into the product of a matrix and the vector $[1 \;\; \mathrm{e}^{-j\Psi} \;\; \cdots \;\; \mathrm{e}^{-j(M-S-1)\Psi}]^T$. Therefore we can say that any blocking matrix for broadside constraints can be regarded as a product of the blocking matrix obtained by the CCD method and some other matrix, i.e. the CCD method forms the basis for any other blocking matrices with a broadside constraint.

The lth output of the transformation matrix, $u_l[n]$, $l = 0, 1, \ldots, L-1$, can be denoted as:

$$\begin{aligned} u_l[n] &= \mathbf{t}_l^T \cdot \tilde{\mathbf{u}}[n] \\ &= (1-\mathrm{e}^{-j\Psi})^S \sum_{m=0}^{M-S-1} t_{l,m}\mathrm{e}^{-jm\Psi}\mathrm{e}^{jn\Omega} \\ &= (1-\mathrm{e}^{-j\Psi})^S T_l(\mathrm{e}^{j\Psi})\mathrm{e}^{jn\Omega} \end{aligned} \tag{3.63}$$

with $T_l(\mathrm{e}^{j\Psi}) \leftrightarrow t_{l,m}$ being a Fourier transform pair.

Similar to Section 3.5.2, we arrange $T_l(\mathrm{e}^{j\Psi})$, $l = 0, \ldots, L-1$, on the interval $\Psi \in [0\ \pi]$ as shown in Figure 3.31, such that:

$$|T_l(\mathrm{e}^{j\Psi})| = \begin{cases} 1 & \text{for } \Psi \in [\Psi_{l,\text{lower}} \ \Psi_{l,\text{upper}}] \\ 0 & \text{otherwise} \end{cases} \tag{3.64}$$

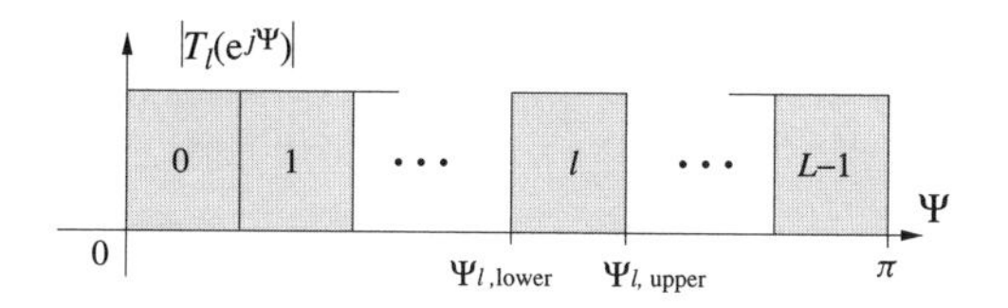

Figure 3.31 Arrangement of the L row vectors in $\boldsymbol{T}$

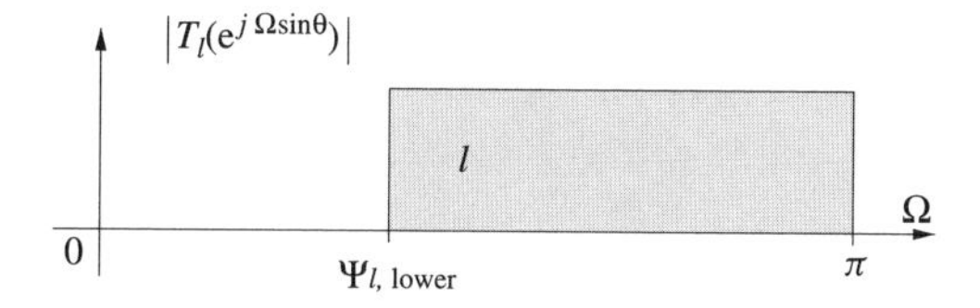

Figure 3.32 Temporal filtering effect of the lth spatial filter in $\boldsymbol{T}$

As $|\sin\theta| \in [0\ 1]$ when $\theta \in [-\pi/2\ \pi/2]$, the possible maximum frequency component of the lth output $u_l[n]$ is $\Omega = \pi$, which corresponds to $|\sin\theta| = (\Psi_{l,\text{upper}})/\pi$, while the possible minimum frequency component is $\Omega = \Psi_{l,\text{lower}}$, which corresponds to $|\sin\theta| = 1$. Therefore we have the same result as in Section 3.5.2, namely:

$$|T_l(e^{j\Omega\sin\theta})| = \begin{cases} 1 & \text{for } \Omega \in [\Psi_{l,\text{lower}}\ \pi] \\ 0 & \text{otherwise} \end{cases} \quad \forall\,\theta \tag{3.65}$$

as shown in Figure 3.32.

Obviously, if we consider the blocking matrix $\tilde{\boldsymbol{B}}$ jointly with the transformation matrix $\boldsymbol{T}$, a characteristic similar to that of the blocking matrix proposed in Section 3.5.2 is obtained. Alternatively, the resultant matrix could simply be regarded as a new realization of the method presented in the previous section. In the following applications, we will consider the concatenation of the blocking matrix and the subband-selective transformation matrix as one realization of the general subband-selective blocking matrix proposed in Section 3.5.2.

3.5.3.2 Design of the Transformation Matrix

The subband-selective transformation matrix introduced in the previous section does not have to fulfil any constraints other than the band-selectivity requirement. The design problem of the transformation matrix is therefore that of a series of general filter designs having cutoff frequencies specified by Equation (3.64). However, cosine modulation can also be employed to reduce the design and implementation complexity of the transformation matrix.

Assume that the prototype vector contains the elements $h[m]$, $m = 0, 1, \ldots, M - S - 1$. Then the elements $t_{l,m}$ of the row vectors $\mathbf{t}_l$, $l = 0, 1, \ldots, L - 1$, of the

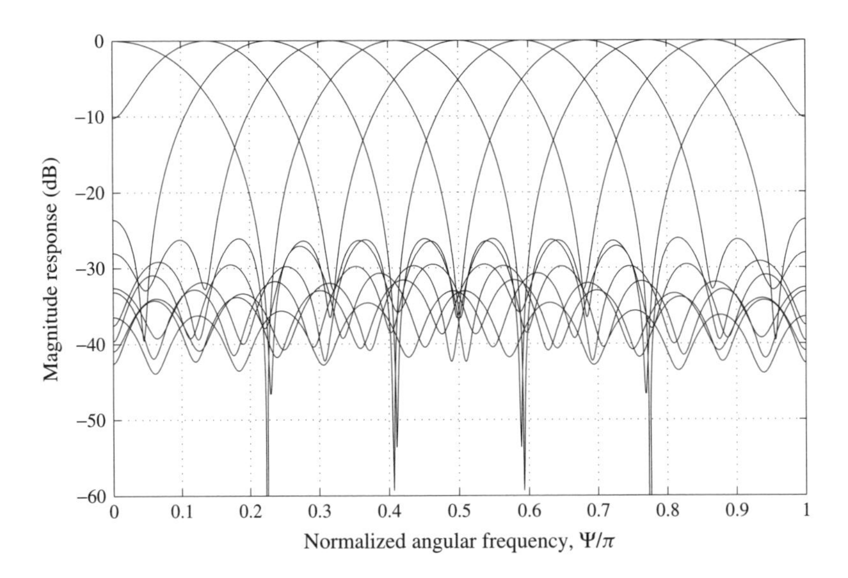

Figure 3.33 A design example for a 11 × 16 transformation matrix

transformation matrix can be obtained by (Koilpillai and Vaidyanathan, 1992):

$$t_{l,m} = h[m]\cos\left[\frac{\pi}{2L}(2l+1)\left(m - \frac{M-S-1}{2}\right) + (-1)^l\frac{\pi}{4}\right] \tag{3.66}$$

Thus, the design problem of the transformation matrix is simplified to the unconstrained design of a lowpass prototype filter $h[m]$, which can be readily solved by standard filter design algorithms, such as the *remez* function in MATLAB© (Mat, 2001). Many of these standard filter design routines are for obtaining linear phase filters. This however is not a requirement for the transformation matrix and ties down available DOFs. Here we opt for a less restrictive method, based again on the IMSL library (Vis, 2002). A result obtained by the subroutine BCONF of the IMSL library for a 11 × 16 transformation matrix is shown in Figure 3.33.

3.5.4 Application to Subband Adaptive GSC

In this part, we apply the subband-selective blocking matrix including the transformation method of Section 3.5.3, to the subband adaptive GSC introduced in Section 3.4 in order to reduce the computational complexity of the system (Liu *et al.*, 2001a, 2002b,c, 2003a, 2004a).

3.5.4.1 Structure

Sections 3.5.2 and 3.5.3 have introduced two schemes for the blocking matrix which permit a joint spatial/temporal filtering of the received array signals. With the bandlimited spectra of the blocking matrix outputs $u_l[n]$, $l = 0, 1, \ldots, L-1$, a further spectral

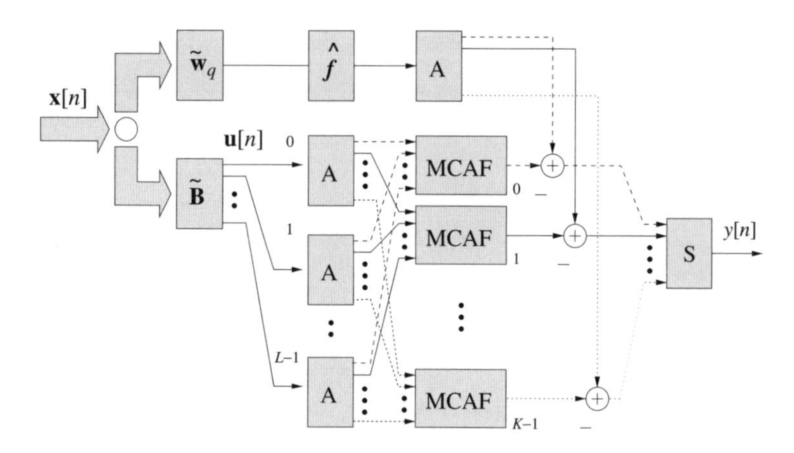

Figure 3.34 Subband decomposition applied to the output of the subband-selective blocking matrix

decomposition is applied to these L outputs to remove the sampling redundancy due to their reduced bandwidths and adaptive processing is then performed in subbands, as shown in Figure 3.34.

This subband setup is actually the same as the one in Figure 3.18 in Section 3.4 and it can be regarded as an application of the subband-selective blocking matrix to the subband adaptive GSC. Because $u_l[n]$, $l = 0, 1, \ldots, L-1$, is a highpass signal, the subband signals in the corresponding lowpass subbands at each MCAF will be zero and can be omitted from the subband adaptive processing. Therefore both the filter length and the number of channels are reduced, which together with the decreased update rates and the lower output dimension of $\tilde{\boldsymbol{B}}$ results in a substantial reduction of the beamformer's computational complexity.

Another advantage of this combination is that the subbands discarded in the adaptation can be determined a priori and independent of the array signals because both the blocking matrix and the filter banks are selected without knowledge of the operating environment. Additionally a further reduction of computational complexity can be achieved by monitoring the remaining subbands and dynamically discarding the processing in those subbands where the signal power falls below a given threshold.

3.5.4.2 Computational Complexity

Now we quantify the achievable computational complexity reduction introduced by the proposed temporally/spatially subband-selective blocking matrix.

Assume that we employ the oversampled GDFT filter banks in the subband decomposition and the response vector $\hat{\mathbf{f}}$ represents a simple delay. Compared with the subband adaptive GSC structure in Figure 3.18, the reduction of the computational complexity is achieved in two ways: firstly, the dimension of the blocking matrix output is reduced from $M - S$ to L by partial rather than full adaptation and secondly by discarding the corresponding lowpass subbands, which contain negligible signal power in each of the MCAFs, we achieve a further complexity reduction.

Table 3.4 Computational complexity for the subband adaptive GSC employing the subband-selective blocking matrix

Subband	Real multiplications (real-valued)	Real multiplications (complex-valued)
LMS	$M(L+1)+$ $(L+2)/N(l_p + 4K\log_2 K + 4K)+$ $K/(4N)(8(L)J_{sub}+2)$	$2M(L+1)+$ $(L+2)/N(2l_p + 4K\log_2 K + 8K)+$ $K/(2N)(8(L)J_{sub}+2)$
NLMS	$M(L+1)+$ $(L+2)/N(l_p + 4K\log_2 K + 4K)+$ $K/(4N)(8(L)J_{sub}+2(L)+3)$	$2M(L+1)+$ $(L+2)/N(2l_p + 4K\log_2 K + 8K)+$ $K/(2N)(8(L)J_{sub}+2(L)+3)$
RLS	$M(L+1)+$ $(L+2)/N(l_p + 4K\log_2 K + 4K)+$ $K/(4N)(12(L)^2J_{sub}^2+12J_{sub}+4)$	$2M(L+1)+$ $(L+2)/N(2l_p + 4K\log_2 K + 8K)+$ $K/(2N)(12(L)^2J_{sub}^2+12J_{sub}+4)$

Assuming that a sufficiently selective column vector $\mathbf{b}_l$ can be designed, the first MCAF indexed as $k = 0$ would be a single-channel adaptive filter, drawing its low frequency input solely from the first branch of $\tilde{\boldsymbol{B}}$. The second ($k = 1$) MCAF block in Figure 3.34 will only cover some of the lower outputs of $\tilde{\boldsymbol{B}}$, while finally only the last MCAF block ($K - 1$) consists of L non-sparse channels. Thus, a channel reduction in the MCAFs is achieved, yielding a considerably reduced complexity. This characteristic underlines the advantage of a combined spatial/temporal subband selection by subband processing in both the spatial and temporal domains. Under ideal conditions, the dimensions of the MCAFs can almost be halved, with roughly the same decrease in computational complexity.

Referring to Table 3.3, the detailed computational complexity figures derived under the ideal assumption are shown in Table 3.4 for the subband adaptive GSC employing the subband-selective blocking matrix.

3.5.5 Extension to the General Subband Adaptive Beamforming Structure

The idea of the temporally/spatially subband-selective blocking/transformation matrix can be generalized and applied to the general subband adaptive beamforming structure shown in Figure 3.12 (Liu and Langley, 2009).

In this new structure, as shown in Figure 3.35, the received M array signals $x_m[n]$, $m = 0, \ldots, M-1$, are transformed into a new set of signals $z_l[n]$, $l = 0, \ldots, L-1$ by a full-rank $L \times M$ transformation matrix $\boldsymbol{T}$, with $L \leq M$, as shown below:

$$\mathbf{z}[n] = \boldsymbol{T}\mathbf{x}[n] \quad \text{with :} \tag{3.67}$$

$$\mathbf{x}[n] = \left[x_0[n]\ x_1[n]\ \cdots\ x_{M-1}[n]\right]^{\mathrm{T}}$$

$$\mathbf{z}[n] = \left[z_0[n]\ z_1[n]\ \cdots\ z_{L-1}[n]\right]^{\mathrm{T}}$$

$$[\boldsymbol{T}]_{l,m} = t_{l,m} \tag{3.68}$$

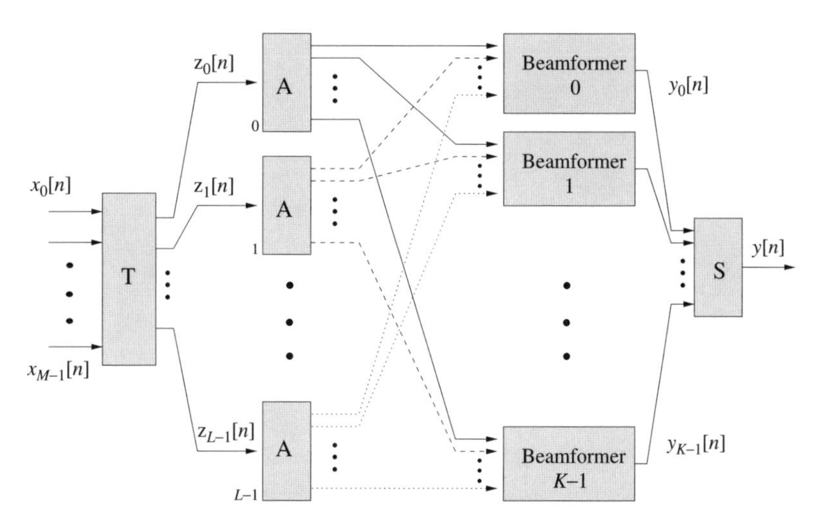

Figure 3.35 Extension of the subband-selective blocking/transformation matrix to the general subband adaptive beamforming structure

We can consider each of the row vectors of the transformation matrix as a simple beamformer with its output given by:

$$z_l[n] = \sum_{m=0}^{M-1} t_{l,m} x_m[n] \tag{3.69}$$

The response of this beamformer with coefficients $t_{l,m}$, $m = 0, 1, \ldots, M-1$ can be expressed as:

$$R_l(\Omega, \theta) = T_l(\mathrm{e}^{j\Psi}) \tag{3.70}$$

with $\Psi = \Omega \sin\theta$, as defined in Equation (3.63).

Now we can arrange the frequency responses $T_l(\mathrm{e}^{j\Psi})$ in the same way as shown in Figure 3.31. The arrangement there is based on the assumption that the elements of the matrix $\boldsymbol{T}$ are real-valued, so that $T_l(\mathrm{e}^{j\Psi})$ has a symmetric response with respect to the point $\Psi = 0$ over the full range $\Psi \in [-\pi\ \pi]$. However, the choice of $\boldsymbol{T}$ is not limited to the real-valued case and its elements can in general be complex-valued. Here we discuss an arrangement for this case with each $T_l(\mathrm{e}^{j\Psi})$ having a bandwidth of $2\pi/M$ and together they cover the whole signal band $[-\pi\ \pi]$, as shown in Figure 3.36 for an odd number M.

The bandpass characteristics of the row vectors will have a similar highpass filtering effect on the received array signals as discussed before. As an example, let us consider the lth row vector with a frequency response given in Figure 3.36, namely:

$$|T_l(\mathrm{e}^{j\Psi})| = \begin{cases} 1 & \text{for } \Psi \in [\Psi_{l,lower}\ \Psi_{l,upper}] > 0 \\ 0 & \text{otherwise} \end{cases} \tag{3.71}$$

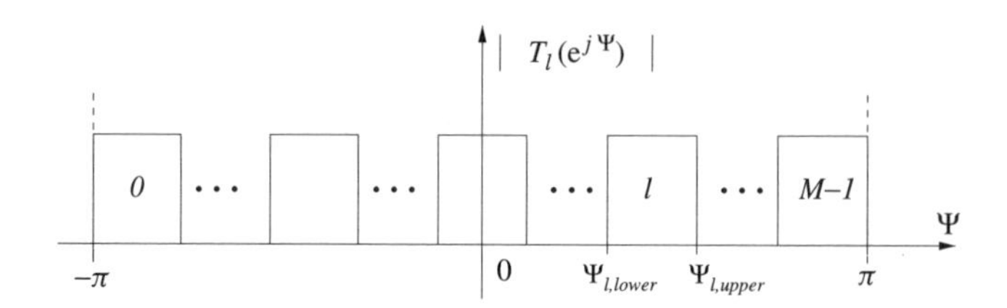

Figure 3.36 Characteristics of the M row vectors contained in $\boldsymbol{T}$

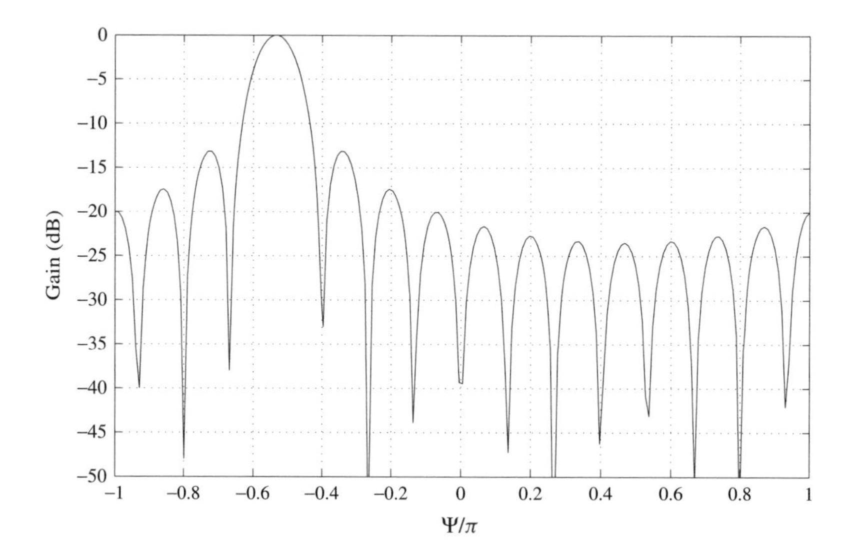

Figure 3.37 The frequency response of a bandpass filter with 15 coefficients

With such a response, all of the received array signal components with a frequency $\Omega \in (-\Psi_{l,lower}\ \Psi_{l,lower})$ will not be able to pass through this filter, since, no matter which direction it comes from, $\Psi = \Omega \sin\theta$ will never fall into the passband of $[\Psi_{l,lower}\ \Psi_{l,upper}]$. As a result, the frequency range of its output will be $|\Omega| \geq \Psi_{l,lower}$ and its lower bound is defined by the lower cutoff frequency $\Psi_{l,lower}$ when $\Psi_{l,lower} > 0$. On the contrary, when $\Psi_{l,lower} < \Psi_{l,upper} < 0$, the lower band will be decided by $|\Psi_{l,upper}|$.

Now consider the example in Figure 3.37, which shows the frequency response of a bandpass filter with 15 coefficients. If we replace one of the row vectors of the transformation matrix $\boldsymbol{T}$ by the coefficients of this bandpass filter, then its response $R_m(\Omega, \theta)$ with respect to signal frequency Ω and arrival angle θ will be highpass, which can be seen clearly by examining the response shown in Figure 3.38, where each curve represents a frequency response for a fixed angle θ.

To remove the redundancy in the transformed array signals $\mathbf{z}[n]$, each of the signals $z_m[n]$ is split into K subbands by a K-channel analysis filter bank and an independent adaptive beamformer can then be set up at each set of corresponding subband array signals. The outputs $y_k[n]$, $k = 0, 1, \ldots, K-1$, of these K subband beamformers are then combined by a synthesis filter bank to form the fullband beamforming output $y[n]$, as

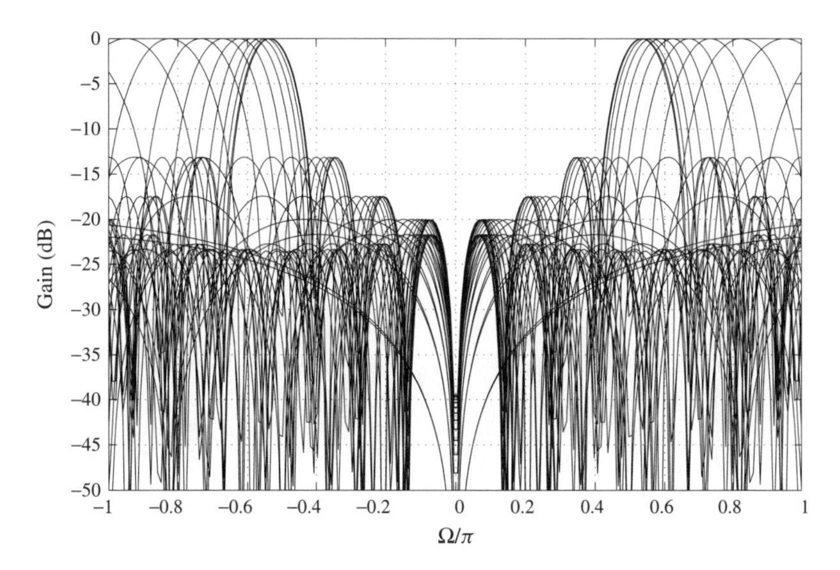

Figure 3.38 The response of the bandpass filter shown in Figure 3.37 with respect to signal frequency and arrival angle when applied to replace one of the row vectors of the transformation matrix $\boldsymbol{T}$

shown in Figure 3.35. Due to the highpass filtering effect, some of the lower subbands will not carry any signal, and therefore can be discarded during the beamforming process. For the example discussed before, any subbands with a frequency range between $-\Psi_{l,lower}$ and $\Psi_{l,lower} > 0$ for the lth row vector output can be discarded without affecting the beamformer's performance. Under ideal conditions, on average the total number of input channels for the K subband MCAFs can be reduced by half.

The row vectors of $\boldsymbol{T}$ can be designed one by one or by prototype modulation, as discussed in Section 3.5.3. For the complex-valued case, as an example, we can use the DFT matrix for $L = M$, as given by:

$$\boldsymbol{T} = \begin{bmatrix} w^{0\cdot 0} & w^{0\cdot 1} & \dots & w^{0\cdot(M-1)} \\ w^{1\cdot 0} & w^{1\cdot 1} & \dots & w^{1\cdot(M-1)} \\ \vdots & \vdots & \ddots & \ddots \\ w^{(M-1)\cdot 0} & w^{(M-1)\cdot 1} & \dots & w^{(M-1)\cdot(M-1)} \end{bmatrix} \tag{3.72}$$

where $w = \mathrm{e}^{-j\frac{2\pi}{M}}$.

However, the sidelobe attenuation of the DFT matrix is only about 15 dB. For a higher attenuation, we can try various window functions as the prototype filter, such as the hamming window (Oppenheim and Schafer, 1975).

3.5.6 Simulations

Extensive simulation results highlighting the benefits of the temporally/spatially subband-selective blocking/transformation matrix can be found in Liu (2003), Liu and Langley

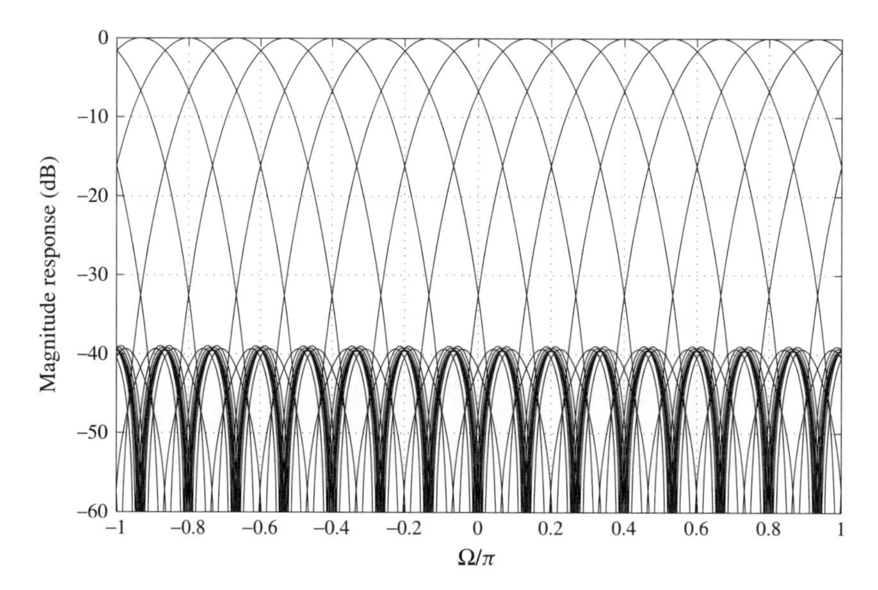

Figure 3.39 Magnitude responses of the 15 × 15 transformation matrix obtained by modulating a hamming window

(2007) and Liu *et al.* (2001a, 2002b,c, 2003a, 2004a) and here we only provide a simple example based on the structure in Figure 3.35 (Liu and Langley, 2009), where the reference signal based beamformer is used for subband beamforming. The sensor number is $M = 15$ and the fullband adaptive filter length J is 140. A 15×15 transformation matrix $\boldsymbol{T}$ is employed, which is obtained by modulating a hamming window with a length 15 and its frequency responses are shown in Figure 3.39. Each of its outputs is divided into $K = 16$ subband channels by oversampled GDFT filter banks with a decimation ratio $N = 14$. The prototype lowpass filter employed for the GDFT filter banks has a length of 448 with a bandwidth of $\pi/8$ and a stopband cutoff frequency $\Omega = \pi/14$. The length J_{sub} of the subband adaptive channel for each MCAF is 10. The signal of interest illuminates the array from $\theta = 40^\circ$ and four interfering signals impinge from $\theta = -80^\circ$, -60°, -20°, 80°, respectively, each of them with an SIR of -30 dB. All the signals have a bandwidth of $\Omega \in [0.30\pi \; 0.95\pi]$. Additionally, all sensors receive temporally and spatially uncorrelated noise at a 20 dB SNR.

For the subband-selective method, in each MCAF block, we have discarded channels with very low signal power according to the responses of the filter banks and the transformation matrix. In this case, there are in total $K = 16$ subband beamformers and 15 channels for each subband beamformer. The number of channels discarded was 66 so that only $15 \times 16 - 66 = 174$ channels were processed for each update.

Figure 3.40 shows the learning curves based on a normalized LMS algorithm with a step size of 0.06, averaged over 150 independent simulations. The convergence rate of the proposed method is much higher than the adaptive fullband beamformer due to the combined decorrelation effect of both the transformation matrix and the filter banks. In term

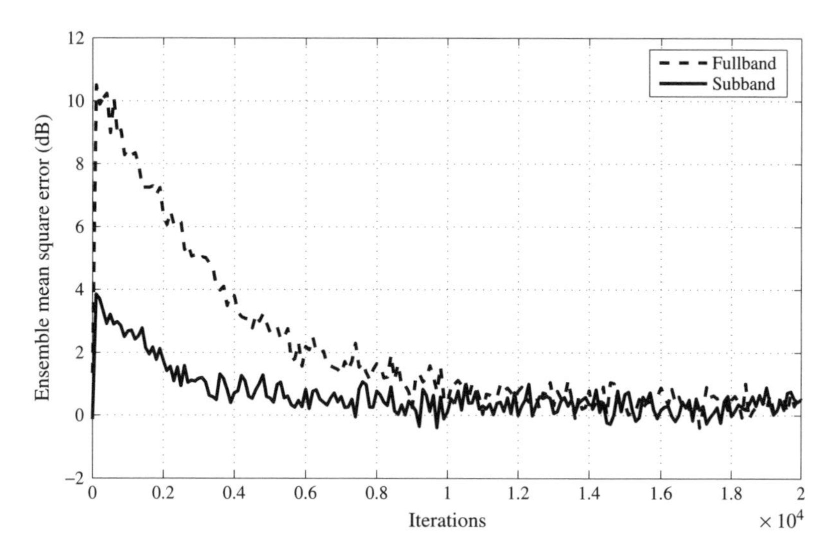

Figure 3.40 Learning curves of the fullband adaptive beamformer and the proposed subband-selective beamformer

of computational complexity, the proposed method only needs 21% real multiplications of its fullband counterpart.

3.6 Frequency-Domain Adaptive Beamforming

To reduce the computational complexity of a wideband adaptive beamformer, another choice is to employ frequency-domain adaptive algorithms (Benesty and Morgan, 2000; Kellermann and Buchner, 2003; Shynk, 1992). Various frequency-domain methods have been proposed and the well-known one is to transform the signals into the frequency domain by a DFT matrix as given in Equation (3.72) and after beamforming the results are then transformed back to the time domain by an inverse DFT (IDFT) matrix. If we consider each column vector of the DFT matrix as an analysis filter and that of the IDFT matrix as an synthesis filter, the whole DFT and IDFT processing can be considered as a maximally decimated filter banks system and such a frequency-domain adaptation structure can be considered as a special case of the previously introduced subband implementation.

In this section, we introduce a special class of frequency-domain adaptive algorithms applied to the LCMV beamformer (Liu *et al.*, 2004b), which can solve the LCMV problem with significantly reduced computational complexity, although with a slower convergence speed compared to the time-domain approach. Firstly, the frequency-domain formulation of the LCMV problem is given in Section 3.6.1: based on the similarity of the formulations in both the frequency domain and the time domain, we derive the constrained frequency-domain adaptive algorithm in Section 3.6.2, and its alternative implementation with the GSC structure in Section 3.6.3; finally, simulation results are presented in Section 3.6.4.

3.6.1 Frequency-Domain Formulation

For convenience, we rearrange the elements of both the weight vector $\mathbf{w}$ and the signal vector $\mathbf{x}[n]$ as follows:

$$\begin{aligned}
\mathbf{w} &= \left[\mathbf{w}_0^{\mathrm{T}}\ \mathbf{w}_1^{\mathrm{T}}\ \ldots\ \mathbf{w}_{M-1}^{\mathrm{T}}\right]^{\mathrm{T}} \\
\mathbf{w}_m &= \left[w_{m,0}\ w_{m,1}\ \cdots\ w_{m,J-1}\right]^{\mathrm{T}} \\
\mathbf{x}[n] &= \left[\mathbf{x}_0^{\mathrm{T}}[n]\ \mathbf{x}_1^{\mathrm{T}}[n]\ \ldots\ \mathbf{x}_{M-1}^{\mathrm{T}}[n]\right]^{\mathrm{T}} \\
\mathbf{x}_m[n] &= [x_m[n]\ x_m[n-1]\ \ldots\ x_m[n-J+1]]^{\mathrm{T}}
\end{aligned} \tag{3.73}$$

The fullband beamformer output $y[n]$ is still given by:

$$y[n] = \mathbf{w}^{\mathrm{H}}\mathbf{x}[n] \tag{3.74}$$

and the time-domain LCMV problem is formulated as:

$$\min_{\mathbf{w}} \mathbf{w}^{\mathrm{H}}\boldsymbol{R}_{xx}\mathbf{w} \qquad \text{subject to} \qquad \boldsymbol{C}^{\mathrm{H}}\mathbf{w} = \mathbf{f} \tag{3.75}$$

where $\boldsymbol{R}_{xx} = E\{\mathbf{x}\mathbf{x}^{\mathrm{H}}\}$ is the covariance matrix, $\boldsymbol{C} \in \mathbf{C}^{MJ\times r}$ the re-arranged constraint matrix and $\mathbf{f} \in \mathbf{C}^{r}$ the corresponding response vector.

Stacking the outputs $y^*[nJ]$, $y^*[nJ+1], \ldots, y^*[nJ+J-1]$ into a vector $\mathbf{y}[n]$, we have:

$$\mathbf{y}[n] = \sum_{m=0}^{M-1} \boldsymbol{X}_m^{\mathrm{H}}[n]\mathbf{w}_m \tag{3.76}$$

where:

$$\boldsymbol{X}_m[n] = [\mathbf{x}_m[nJ]\ \mathbf{x}_m[nJ+1]\ \cdots\ \mathbf{x}_m[nJ+J-1]] \tag{3.77}$$

We can expand the convolutional matrices $\boldsymbol{X}_m[n]$ to a circular form (Golub and Van Loan, 1996) and write the output of the beamformer as:

$$\begin{bmatrix} \mathbf{v} \\ \mathbf{y}[n] \end{bmatrix} = \sum_{m=0}^{M-1} \begin{bmatrix} \tilde{\boldsymbol{X}}_m^{\mathrm{H}}[n] & \boldsymbol{X}_m^{\mathrm{H}}[n] \\ \boldsymbol{X}_m^{\mathrm{H}}[n] & \tilde{\boldsymbol{X}}_m^{\mathrm{H}}[n] \end{bmatrix} \cdot \begin{bmatrix} \mathbf{w}_m \\ \underline{\mathbf{0}} \end{bmatrix} \tag{3.78}$$

where $\mathbf{v}$ is an arbitrary J element vector and $\tilde{\boldsymbol{X}}_m$ is a Toplitz matrix using the data samples of $\boldsymbol{X}_m$ except for an arbitrary element along the main diagonal. Defining a DFT domain vector $\underline{\mathbf{y}}[n] \in \mathbf{C}^{2J}$ with the help of a $2J$–point DFT matrix $\boldsymbol{T}$:

$$\begin{aligned}
\underline{\mathbf{y}}[n] &= \boldsymbol{T}\begin{bmatrix} \underline{\mathbf{0}} \\ \mathbf{y}[n] \end{bmatrix} \\
&= \underbrace{\boldsymbol{T}\begin{bmatrix} \mathbf{0}_{J\times J} & \mathbf{0}_{J\times J} \\ \mathbf{0}_{J\times J} & \boldsymbol{I}_{J\times J} \end{bmatrix}\boldsymbol{T}^{\mathrm{H}}}_{\boldsymbol{G}} \cdot \boldsymbol{T}\begin{bmatrix} \mathbf{v} \\ \mathbf{y}[n] \end{bmatrix}
\end{aligned}$$

$$= \boldsymbol{G} \sum_{m=0}^{M-1} \boldsymbol{T} \begin{bmatrix} \tilde{\boldsymbol{X}}_m^{\mathrm{H}}[n] & \boldsymbol{X}_m^{\mathrm{H}}[n] \\ \boldsymbol{X}_m^{\mathrm{H}}[n] & \tilde{\boldsymbol{X}}_m^{\mathrm{H}}[n] \end{bmatrix} \boldsymbol{T}^{\mathrm{H}} \boldsymbol{T} \cdot \begin{bmatrix} \mathbf{w}_m \\ \underline{\mathbf{0}} \end{bmatrix}$$

$$= \boldsymbol{G} \sum_{m=0}^{M-1} \underline{\underline{\boldsymbol{\Gamma}}}_m[n] \, \underline{\mathbf{w}}_m = \boldsymbol{G} \, \underline{\underline{\boldsymbol{\Gamma}}}[n] \, \underline{\mathbf{w}} \tag{3.79}$$

we obtain the diagonal matrices $\underline{\underline{\boldsymbol{\Gamma}}}_m[n]$, where their main elements are the Discrete Fourier transform of the first column of the corresponding circular matrices (Benesty and Morgan, 2000).

Now we consider the constraint equation. In the time domain, the constraints are expressed as:

$$\boldsymbol{C}^{\mathrm{H}} \mathbf{w} = \sum_{m=0}^{M-1} \boldsymbol{C}_m^{\mathrm{H}} \mathbf{w}_m = \mathbf{f} \tag{3.80}$$

whereby the original matrix equation can be separated into M additive components. Note that $\boldsymbol{C}_m \in \mathbf{C}^{J \times r}$ has an arbitrary form (in particular not Toeplitz), where r is the number of linearly independent constraints. In term of the DFT domain coefficients, we have:

$$\sum_{m=0}^{M-1} [\, \boldsymbol{C}_m^{\mathrm{H}} \;\; \mathbf{0}_{r \times J} \,] \begin{bmatrix} \mathbf{w}_m \\ \underline{\mathbf{0}} \end{bmatrix} = \mathbf{f} \tag{3.81}$$

or:

$$\sum_{m=0}^{M-1} [\, \boldsymbol{C}_m^{\mathrm{H}} \;\; \mathbf{0}_{r \times J} \,] \boldsymbol{T}^{\mathrm{H}} \underline{\mathbf{w}}_m = \underline{\underline{\boldsymbol{C}}}^{\mathrm{H}} \, \underline{\mathbf{w}} = \mathbf{f} \tag{3.82}$$

where $\underline{\underline{\boldsymbol{C}}} \in \mathbf{C}^{2MJ \times r}$ is the new constraint matrix:

$$\underline{\underline{\boldsymbol{C}}} = \left[\begin{bmatrix} \boldsymbol{C}_0 \\ \mathbf{0}_{J \times r} \end{bmatrix}^{\mathrm{H}} \boldsymbol{T}^{\mathrm{H}} \; \cdots \; \begin{bmatrix} \boldsymbol{C}_{M-1} \\ \mathbf{0}_{J \times r} \end{bmatrix}^{\mathrm{H}} \boldsymbol{T}^{\mathrm{H}} \right]^{\mathrm{H}} \tag{3.83}$$

applicable to the DFT domain coefficient vector $\underline{\mathbf{w}}$.

Now the mean squared frequency-domain output of the beamformer is given by:

$$E\{\underline{\mathbf{y}}^{\mathrm{H}}[n] \underline{\mathbf{y}}[n]\} = \underline{\mathbf{w}}^{\mathrm{H}} \boldsymbol{R} \underline{\mathbf{w}} \tag{3.84}$$

where $\boldsymbol{R} = E\{\underline{\underline{\boldsymbol{\Gamma}}}[n]^{\mathrm{H}} \boldsymbol{G}^{H} \boldsymbol{G} \underline{\underline{\boldsymbol{\Gamma}}}[n]\} = E\{\underline{\underline{\boldsymbol{\Gamma}}}[n]^{\mathrm{H}} \boldsymbol{G} \underline{\underline{\boldsymbol{\Gamma}}}[n]\}$.

Thus, the LCMV formulation in the frequency domain is given by:

$$\min_{\underline{\mathbf{w}}} \underline{\mathbf{w}}^{\mathrm{H}} \boldsymbol{R} \underline{\mathbf{w}} \qquad \text{subject to} \qquad \underline{\underline{\boldsymbol{C}}}^{\mathrm{H}} \underline{\mathbf{w}} = \mathbf{f} \tag{3.85}$$

3.6.2 Constrained Frequency-Domain Adaptive Algorithm

Since Equation (3.85) has a form similar to Equation (3.75), we can follow the derivation of the Frost algorithm to obtain an iterative update equation.

First, using the method of Lagrange multipliers, we form the new objective function $\Phi(\underline{\mathbf{w}})$:

$$\Phi(\underline{\mathbf{w}}) = \underline{\mathbf{w}}^H \boldsymbol{R}\underline{\mathbf{w}} + \boldsymbol{\lambda}^H(\underline{\underline{\boldsymbol{C}}}^H\underline{\mathbf{w}} - \mathbf{f}) + \boldsymbol{\lambda}^T(\underline{\underline{\boldsymbol{C}}}^T\underline{\mathbf{w}}^* - \mathbf{f}^*) \tag{3.86}$$

where $\boldsymbol{\lambda} \in \mathbf{C}^r$ is a vector with undetermined Lagrange multipliers.

Differentiating the function $\Phi(\underline{\mathbf{w}})$ with respect to $\underline{\mathbf{w}}^*$, we have:

$$\frac{\partial \Phi(\underline{\mathbf{w}})}{\partial \underline{\mathbf{w}}^*} = \boldsymbol{R}\underline{\mathbf{w}} + \underline{\underline{\boldsymbol{C}}}\boldsymbol{\lambda} \tag{3.87}$$

Then, for our constrained adaptive algorithm, we set the weight vector to $\underline{\mathbf{w}}[0] = \underline{\mathbf{w}}_q$ for initialization, which satisfies the constraint in Equation (3.82), with $\underline{\mathbf{w}}_q = \underline{\underline{\boldsymbol{C}}}(\underline{\underline{\boldsymbol{C}}}^{\mathrm{H}}\underline{\underline{\boldsymbol{C}}})^{-1}\mathbf{f}$. At each iteration, the vector $\underline{\mathbf{w}}$ is updated in the direction of the negative gradient given in Equation (3.87) by a step proportional to a scaling factor μ according to:

$$\underline{\mathbf{w}}[n+1] = \underline{\mathbf{w}}[n] - \mu\big(\boldsymbol{R}\underline{\mathbf{w}}[n] + \underline{\underline{\boldsymbol{C}}}\boldsymbol{\lambda}[n]\big) \tag{3.88}$$

Since $\underline{\mathbf{w}}[n+1]$ must satisfy the constraint in Equation (3.82), we can substitute Equation (3.88) into Equation (3.82) and solve for the Lagrange multipliers $\boldsymbol{\lambda}[n]$. Then we substitute $\boldsymbol{\lambda}[n]$ into the iteration equation in Equation (3.88) and arrive at:

$$\begin{aligned}\underline{\mathbf{w}}[n+1] = \underline{\mathbf{w}}[n] &- \mu\big(\boldsymbol{I} - \underline{\underline{\boldsymbol{C}}}(\underline{\underline{\boldsymbol{C}}}^H\underline{\underline{\boldsymbol{C}}})^{-1}\underline{\underline{\boldsymbol{C}}}^H\big)\boldsymbol{R}\underline{\mathbf{w}}[n] \\ &+ \underline{\underline{\boldsymbol{C}}}(\underline{\underline{\boldsymbol{C}}}^H\underline{\underline{\boldsymbol{C}}})^{-1}\big(\mathbf{f} - \underline{\underline{\boldsymbol{C}}}^H\underline{\mathbf{w}}[n]\big)\end{aligned} \tag{3.89}$$

Upon defining the shorthand $\boldsymbol{P} = \boldsymbol{I} - \underline{\underline{\boldsymbol{C}}}(\underline{\underline{\boldsymbol{C}}}^H\underline{\underline{\boldsymbol{C}}})^{-1}\underline{\underline{\boldsymbol{C}}}^H$, the algorithm in Equation (3.89) can be rewritten as:

$$\underline{\mathbf{w}}[n+1] = \underline{\mathbf{w}}_q + \boldsymbol{P}\big(\underline{\mathbf{w}}[n] - \mu\boldsymbol{R}\underline{\mathbf{w}}[n]\big) \tag{3.90}$$

Not knowing the true second order statistics in $\boldsymbol{R}$, the covariance matrix can be replaced by its simple approximation $\tilde{\boldsymbol{R}} = \underline{\underline{\boldsymbol{\Gamma}}}[n]^{\mathrm{H}}\boldsymbol{G}\underline{\underline{\boldsymbol{\Gamma}}}[n]$. This results in the minimization of the instantaneous squared output rather than the mean squared output, and leads to the following constrained stochastic gradient algorithm:

$$\underline{\mathbf{w}}[n+1] = \underline{\mathbf{w}}_q + \boldsymbol{P}\big(\underline{\mathbf{w}}[n] - \mu\underline{\underline{\boldsymbol{\Gamma}}}[n]^{\mathrm{H}}\underline{\mathbf{y}}[n]\big) \tag{3.91}$$

To increase the convergence speed of the algorithm, we can minimize a sum of the squared frequency-domain outputs in a similar way as the RLS adaptive algorithm (Benesty and Morgan, 2000), although with a much increased complexity.

The algorithm steps and their associated cost of Equation (3.91) in term of the number of real multiplications are detailed in Table 3.5, assuming real-valued array signals.

Table 3.5 Algorithm steps and computational cost for the constrained frequency-domain adaptive algorithm

Initialization:
$\underline{\mathbf{w}}_q = \underline{\underline{C}}(\underline{\underline{C}}^{\mathrm{H}}\underline{\underline{C}})^{-1}\mathbf{f}, \quad \boldsymbol{P} = \boldsymbol{I} - \underline{\underline{C}}(\underline{\underline{C}}^{H}\underline{\underline{C}})^{-1}\underline{\underline{C}}^{H}$
Iteration for every block of J input samples:
1: $\underline{\underline{\boldsymbol{\Gamma}}}_m[n] = \mathrm{diag}\{\boldsymbol{T}\ [\mathbf{x}_m^{\mathrm{T}}[n+J]\ \ \mathbf{x}_m^{\mathrm{T}}[n]]^{\mathrm{H}}\}$
Cost: $2MJ\log_2(2J)$
2: $\underline{\mathbf{y}}[n] = \boldsymbol{G}\ \underline{\underline{\boldsymbol{\Gamma}}}[n]\ \underline{\mathbf{w}}[n]$
Cost: $2MJ + 4J\log_2(2J)$
3: $\mu\underline{\underline{\boldsymbol{\Gamma}}}[n]^{\mathrm{H}}\underline{\mathbf{y}}[n]$
Cost: $2J + 2MJ$
4: $\boldsymbol{P}\big(\underline{\mathbf{w}}[n] - \mu\underline{\underline{\boldsymbol{\Gamma}}}[n]^{\mathrm{H}}\underline{\mathbf{y}}[n]\big)$
Cost: $4M^2J^2$

The complexity of this algorithm accrues to:

$$\begin{aligned} C_{\text{freq}} &= (2MJ\log_2(2J) + 2MJ + 4J\log_2(2J) \\ &\quad +2MJ + 2J + 4M^2J^2)/J \\ &= 2M(2MJ + \log_2 J + 3) + 4\log_2 J + 6 \end{aligned} \tag{3.92}$$

real multiplications per sampling period, whereby a factor of $1/J$ accounts for the updating which only occurs once per block. Compared to the $M^2J^2 + 2MJ + 1$ real multiplications per sampling period of the original Frost algorithm, the reduction of the computational complexity is:

$$C_{\text{reduction}} = M^2(J^2 - 4J) + 2M(J - \log_2 J - 3) - 4\log_2 J - 5 \tag{3.93}$$

which is significant when J is large.

3.6.3 Frequency-Domain GSC

In the time domain, the LCMV beamforming algorithm can be implemented alternatively by the GSC structure. Correspondingly, for the constrained frequency-domain adaptive beamforming algorithm, we can also find its unconstrained implementation using the GSC structure in a straightforward way.

The approach of the GSC is to separate the weight vector into two orthogonal components, $\underline{\mathbf{w}} = \underline{\mathbf{w}}_{\mathrm{q}} - \underline{\mathbf{v}}$, with a quiescent vector $\underline{\mathbf{w}}_{\mathrm{q}}$ representing a projection onto the constraints, and a projection away from the constraints, $-\underline{\mathbf{v}}$, given by:

$$\underline{\mathbf{v}} = \underline{\underline{\boldsymbol{B}}}\mathbf{w}_{\mathrm{a}} \tag{3.94}$$

utilizing the blocking matrix $\underline{\underline{\boldsymbol{B}}} = \mathrm{span}\{\underline{\underline{C}}^{\perp}\}$, which spans the nullspace of the constraint matrix $\underline{\underline{C}}$.

Therefore, with Equation (3.79), the beamformer output is given by:

$$\underline{\mathbf{y}}[n] = \boldsymbol{G}\,\underline{\underline{\boldsymbol{\Gamma}}}[n]\,\left(\underline{\mathbf{w}}_{\mathrm{q}} - \underline{\underline{\boldsymbol{B}}}\underline{\mathbf{w}}_{\mathrm{a}}\right) \tag{3.95}$$

Using the instantaneous squared error as a cost function $\xi = \underline{\mathbf{y}}^{\mathrm{H}}[n]\underline{\mathbf{y}}[n]$, we obtain a stochastic gradient:

$$\begin{aligned}\hat{\nabla}\xi &= \frac{\partial\xi}{\partial\underline{\mathbf{w}}_{\mathrm{a}}^{*}} \\ &= \frac{\partial\xi}{\partial}\left[\left(\underline{\mathbf{w}}_{\mathrm{q}} - \underline{\underline{\boldsymbol{B}}}\underline{\mathbf{w}}_{\mathrm{a}}\right)^{\mathrm{H}}\underline{\underline{\boldsymbol{\Gamma}}}^{\mathrm{H}}[n]\,\boldsymbol{G}^{\mathrm{H}}\boldsymbol{G}\underline{\underline{\boldsymbol{\Gamma}}}[n]\left(\underline{\mathbf{w}}_{\mathrm{q}} - \underline{\underline{\boldsymbol{B}}}\underline{\mathbf{w}}_{\mathrm{a}}\right)\right] \\ &= -\underline{\underline{\boldsymbol{B}}}^{\mathrm{H}}\;\underline{\underline{\boldsymbol{\Gamma}}}^{\mathrm{H}}[n]\;\underline{\mathbf{y}}[n]\end{aligned} \tag{3.96}$$

where $\boldsymbol{G}^{\mathrm{H}}\boldsymbol{G} = \boldsymbol{G}$ has been exploited.

The update equation for $\underline{\mathbf{w}}_{\mathrm{a}}$ can then be written as:

$$\underline{\mathbf{w}}_{\mathrm{a}}[n+1] = \underline{\mathbf{w}}_{\mathrm{a}}[n] + \mu\underline{\underline{\boldsymbol{B}}}^{\mathrm{H}}\underline{\underline{\boldsymbol{\Gamma}}}^{\mathrm{H}}[n]\underline{\mathbf{y}}[n] \tag{3.97}$$

The algorithm steps and their associated cost in term of the number of real multiplications are detailed in Table 3.6. The complexity of this algorithm accrues to:

$$\begin{aligned}C_{\mathrm{fgsc}} &= (2MJ(2MJ-r) + 2MJ\log_2(2J) \\ &\quad +2MJ + 4J\log_2(2J) + 2J + 2MJ + 2MJ(2MJ-r))/J \\ &= 2M(4MJ - 2r + \log_2 J + 3) + 4\log_2 J + 6\end{aligned} \tag{3.98}$$

multiplications per sampling period. Compared to the constrained algorithm in Equation (3.91), the frequency-domain GSC has a higher computational complexity and the extra

Table 3.6 Algorithm steps and computational cost for the frequency-domain GSC

Initialization:
$\underline{\mathbf{w}}_{q} = \underline{\underline{\boldsymbol{C}}}(\underline{\underline{\boldsymbol{C}}}^{\mathrm{H}}\underline{\underline{\boldsymbol{C}}})^{-1}\mathbf{f}, \quad \underline{\underline{\boldsymbol{B}}} = \underline{\underline{\boldsymbol{C}}}^{\perp}$

Iteration for every block of J input samples:

1: $\underline{\mathbf{w}}[n] = \underline{\mathbf{w}}_{\mathrm{q}} - \underline{\underline{\boldsymbol{B}}}\underline{\mathbf{w}}_{\mathrm{a}}[n]$
Cost: $2MJ(2MJ-r)$

2: $\underline{\underline{\boldsymbol{\Gamma}}}_{m}[n] = \mathrm{diag}\{\boldsymbol{T}\;[\mathbf{x}_{m}^{\mathrm{T}}[n+J]\;\;\mathbf{x}_{m}^{\mathrm{T}}[n]]^{\mathrm{H}}\}$
Cost: $2MJ\log_2(2J)$

3: $\underline{\mathbf{y}}[n] = \boldsymbol{G}\,\underline{\underline{\boldsymbol{\Gamma}}}[n]\,\underline{\mathbf{w}}[n]$
Cost: $2MJ + 4J\log_2(2J)$

4: $\mu\hat{\nabla}[n] = -\mu\underline{\underline{\boldsymbol{B}}}^{\mathrm{H}}\underline{\underline{\boldsymbol{\Gamma}}}[n]^{\mathrm{H}}\underline{\mathbf{y}}[n]$
Cost: $2J + 2MJ + 2MJ(2MJ-r)$

5: $\underline{\mathbf{w}}_{\mathrm{a}}[n+1] = \underline{\mathbf{w}}_{\mathrm{a}}[n] + \mu\hat{\nabla}[n]$
Cost: 0

cost is:

$$C_{\text{increase}} = 2M(2MJ - 2r) \tag{3.99}$$

Although this algorithm has a higher computational complexity, it provides more flexibility in the choice of $\underline{\underline{B}}$, i.e. the column dimension of the frequency-domain blocking matrix $\underline{\underline{B}}$ is not necessary to be $2MJ - r$, and can be any value smaller than $2MJ - r$, which leads to a partially adaptive frequency-domain GSC.

3.6.4 Simulations

The first set of simulations is based on a beamformer with $M = 5$ sensors and a TDL length $J = 32$. The signal of interest comes from the broadside and is corrupted by two interfering signals covering the frequency interval $[0.40\pi\ 0.90\pi]$ with DOA angles -30° and -60°, respectively, with an SIR of -24 dB. Additionally, the sensors record spatially and temporally white Gaussian noise at a 20 dB SNR. The frequency-domain constrained adaptive algorithm is compared with the Frost algorithm. Their step sizes are chosen empirically to achieve approximately the same steady-state values of the mean square residual errors, which are 5.9×10^{-6} for the frequency-domain method and 2×10^{-5} for the Frost algorithm.

Figure 3.41 shows the learning curves, where it is evident that the frequency-domain method converges slower than the time-domain one. Considering the computational complexity, the time-domain method needs 25921 real multiplications per sampling period, whereas the frequency-domain method only requires 3306 real multiplications, which is about 13% of the time-domain implementation. To improve the convergence speed of the frequency-domain algorithm, one possible choice is to replace the stepsize μ by

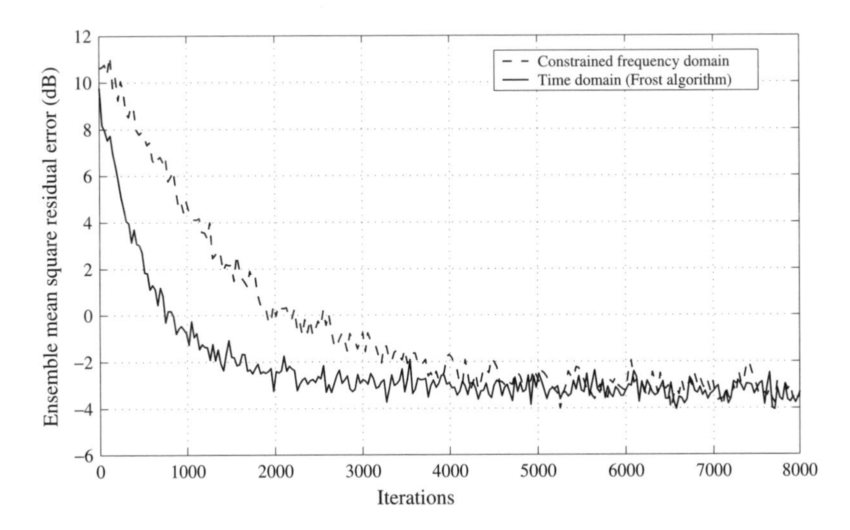

Figure 3.41 Learning curves for the first set of simulations

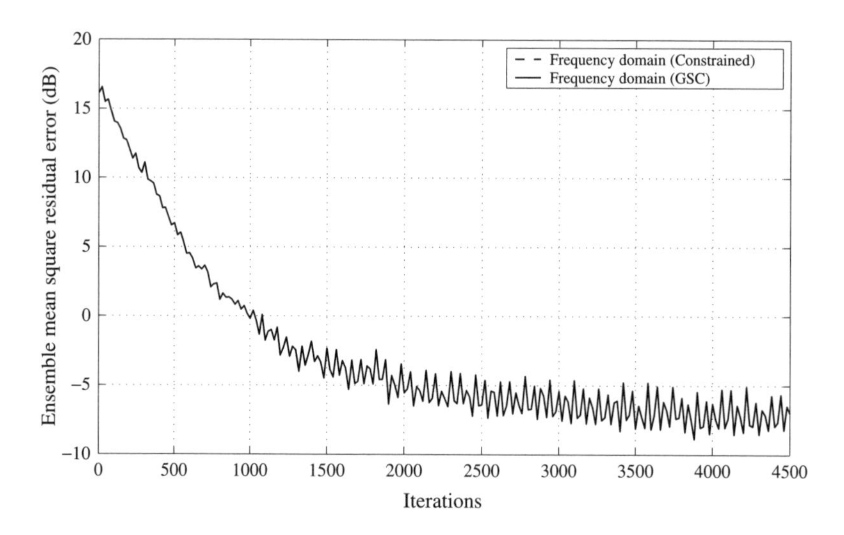

Figure 3.42 Learning curves for the second set of simulations

the inverse of the correlation matrix of the input frequency-domain signal (Benesty and Morgan, 2000).

Next, we compare the performances of the constrained frequency-domain algorithm and its alternative implementation using the GSC structure, where the blocking matrix $\underline{\underline{B}}$ is obtained by the SVD of the constraint matrix $\underline{\underline{C}}$. Extensive simulations show no difference in the learning rate of the two systems. Here we give an example based on a beamformer with $M = 3$ sensors, with one interfering signal coming from a DOA angle of $-30°$. All other conditions are identical to the first set of simulations. Figure 3.42 shows the two learning curves with the same step size of 10^{-5}, with no discernible difference between them. In term of their computational complexities, the constrained frequency-domain method needs 1226 real multiplications per sampling period, whereas its alternative implementation requires 1994 per sampling period, which is about 1.6 times that of the constrained frequency-domain method, but still much lower than the 9409 of the Frost algorithm.

3.7 Transform-Domain Adaptive Beamforming

In the GSC structure, the unconstrained adaptive part can be replaced by frequency-domain adaptive algorithms directly and the resultant frequency-domain adaptive GSC can be considered as a special case of the subband adaptive GSC studied in Section 3.4, with the DFT matrix replacing the analysis filter bank and the IDFT matrix replacing the synthesis filter bank. In such a frequency-domain GSC, the frequency-domain weight vector are updated every J signal samples received by the array. We can simplify this system by updating its coefficients each time we receive a new sample and removing the IDFT operation as a result, which leads to a structure called the transform-domain GSC (TGSC) (Chen and Fang, 1992), as shown in Figure 3.43.

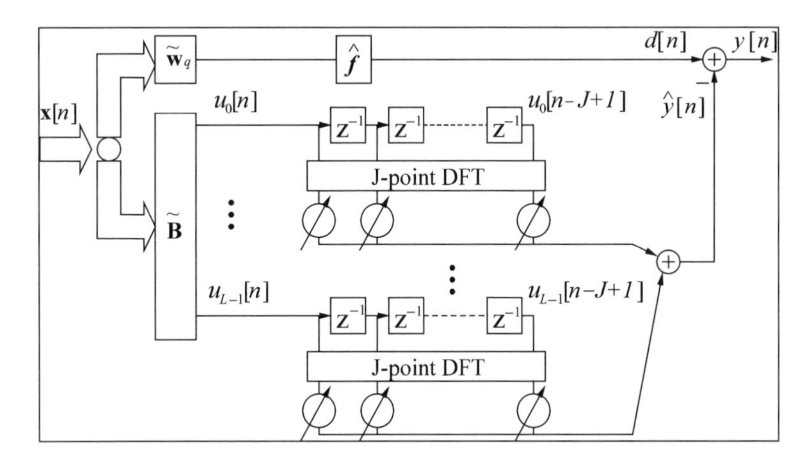

Figure 3.43 A transform-domain adaptive GSC structure

In this structure, a J-tap 1-D DFT is applied to each of the blocking matrix outputs; the frequency-bin outputs of the DFTs are then combined together adaptively to form the lower branch output $\hat{y}[n]$ and the adaptive coefficients are adjusted by minimizing the output variance of the beamformer output $y[n]$. In Chen and Fang (1992), an LMS algorithm with a self-orthogonalizing property is applied. Following this work, a 2-D transformation was introduced to the GSC in An and Champagne (1994), which improves the convergence rate further due to the approximate estimation of both the spatial and the temporal correlations. With the advantage of a higher convergence speed, the TGSC however poses the problem of a higher computational complexity due to the extra processing of the transformation matrix.

Actually the previously described temporally/spatially subband-selective blocking matrix can be combined with the transform-domain GSC to reduce its computational complexity, which leads to a subband-selective TGSC (STGSC) (Liu *et al.*, 2002a, 2003b). In this STGSC, when applying the DFT to the outputs of the blocking matrix with band selectivity, some of the frequency-bin outputs of the DFT will be approximately zero and can be omitted from the following adaptive processing. Because of the relatively high sidelobe level of the DFT (Oppenheim and Schafer, 1975), it is advantageous to apply a window function to the blocking matrix outputs before performing the DFT.

In this section, we will first review the transform-domain GSC proposed in Chen and Fang (1992) and then study the subband-selective transform-domain GSC in Liu *et al.* (2002a, 2003b).

3.7.1 Transform-Domain GSC

The structure of a transform-domain GSC is shown in Figure 3.43, where the blocking matrix output $\mathbf{u}[n] = \left[u_0[n]\, u_1[n] \ldots u_{L-1}[n]\right]^T$ is obtained by $\mathbf{u}[n] = \tilde{\boldsymbol{B}}^H \mathbf{x}[n]$ and $L = M - S$ for a fully adaptive GSC. A J-point DFT is applied to each of the tapped delay-line vectors $\mathbf{u}_l[n]$, $l = 0, 1, \ldots, L-1$, where:

$$\mathbf{u}_l[n] = [u_l[n]\ u_l[n-1]\ \cdots\ u_l[n-J+1]]^T \tag{3.100}$$

The output of the lth DFT block is:

$$\mathbf{v}_l[n] = \mathrm{DFT}\{\mathbf{u}_l[n]\} \tag{3.101}$$

where $\mathbf{v}_l[n] = \left[v_{l,0}[n]\ v_{l,1}[n]\ \cdots\ v_{l,J-1}[n]\right]^T$.

Stacking the DFT outputs as:

$$\mathbf{v}[n] = \left[\mathbf{v}_0[n]^T\ \ \mathbf{v}_1[n]^T\ \ \cdots\ \ \mathbf{v}_{L-1}[n]^T\right]^T \tag{3.102}$$

we can formulate $\hat{y}[n] = \tilde{\mathbf{w}}^H\mathbf{v}[n]$, where $\tilde{\mathbf{w}}$ is the weight vector including all the corresponding weights in the transform-domain LMS algorithm. These weights $\tilde{\mathbf{w}}$ are updated continuously to minimize the variance of the difference signal $y[n] = d[n] - \hat{y}[n]$ by a self-orthogonalizing LMS algorithm (Gitlin and Magee, 1977):

$$\tilde{\mathbf{w}}[n+1] = \tilde{\mathbf{w}}[n] + 2\gamma y^*[n]\boldsymbol{R}_{vv}^{-1}\mathbf{v}[n] \tag{3.103}$$

where:

$$\boldsymbol{R}_{vv} = E\{\mathbf{v}[n]\mathbf{v}^H[n]\} \tag{3.104}$$

and $0 < \gamma < 1/(LJ)$ to ensure convergence of the algorithm. The role of $\boldsymbol{R}_{vv}^{-1}$ is to reduce the eigenvalue spread of the matrix governing the adaptation process.

Note that $\boldsymbol{R}_{vv}$ is unknown in practice and we here use the following approach to approximate it by a diagonal matrix $\tilde{\boldsymbol{R}}_{vv}$ (An and Champagne, 1994; Lee and Un, 1986):

$$\tilde{\boldsymbol{R}}_{vv} = \mathrm{diag}\left[r_{0,0}, \cdots, r_{0,J-1}, \cdots, r_{L-1,0}, \cdots, r_{L-1,J-1}\right] \tag{3.105}$$

where:

$$r_{l,j} = E\{|v_{l,j}[n]|^2\}, \quad l = 0, 1, \ldots, L-1, \quad j = 0, 1, \ldots, J-1 \tag{3.106}$$

is the power of the corresponding frequency bin output of the DFT. The diagonal elements $r_{l,j}$ in turn can be recursively estimated at time instance n through the following equation:

$$\tilde{r}_{l,j}[n] = \beta\tilde{r}_{l,j}[n-1] + (1-\beta)|v_{l,j}[n]|^2 \tag{3.107}$$

where $0 \leq \beta \leq 1$ is a forgetting factor. Then the estimate $\tilde{\boldsymbol{R}}_{vv}^{-1}$ of $\boldsymbol{R}_{vv}^{-1}$ is given by:

$$\tilde{\boldsymbol{R}}_{vv}^{-1} = \mathrm{diag}\left[\tilde{r}_{0,0}^{-1}, \cdots, \tilde{r}_{0,J-1}^{-1}, \cdots, \tilde{r}_{L-1,0}^{-1}, \cdots, \tilde{r}_{L-1,J-1}^{-1}\right] \tag{3.108}$$

and we obtain the new update equation:

$$\tilde{\mathbf{w}}[n+1] = \tilde{\mathbf{w}}[n] + 2\gamma y^*[n]\tilde{\boldsymbol{R}}_{vv}^{-1}\mathbf{v}[n] \tag{3.109}$$

Although the TGSC accelerates the convergence speed, it also increases the computational complexity of the system. In the next section, we sacrifice some DOFs of the system by introducing the previously proposed subband-selective blocking matrix in order to achieve a lower computational complexity.

Table 3.7 Computational complexities for the subband-selective TGSC and the original TGSC

GSC realizations	Complex multiplications per cycle (LMS)
TGSC	$(M - S)J\log_2 J + 3.5(M - S)J$
STGSC	$LJ\log_2 J + 1.75LJ$

3.7.2 Subband-Selective Transform-Domain GSC

The introduction of the subband-selective TGSC (STGSC) is straightforward. The standard blocking matrix in Figure 3.43 is simply replaced by the subband-selective blocking matrix. As noted before, its outputs $u_l[n]$, $l = 0, \ldots, L-1$ contain signals with tighter and tighter highpass spectra, as the index l increases. If we apply a DFT to the output signal $\mathbf{u}_l[n]$, some of the frequency bins will possess negligible energy and can be omitted from the following adaptive process. In order to best exploit this property, we need to select a suitable window function with good frequency selectivity, which will be multiplied with the time-domain signals prior to applying the DFT.

Now we analyse the computational complexity of the system. Since for a fully adaptive GSC the output dimension of the blocking matrix is $L = M - S$, the total number of weights in a partially adaptive system is reduced by $L/(M - S)$. Concerning the DFT and the adaptive part under ideal conditions, i.e. if sufficiently selective column vectors $\mathbf{b}_l$ and a good window function can be designed, the last DFT output $\mathbf{v}_{L-1}$ will have only approximately two non-zero frequency bins for real-valued signals (or one for complex-valued signals) and $\mathbf{v}_{L-2}$ has four (or two for complex-valued signals), and so on. Finally, only $\mathbf{v}_0$ does not have any negligible frequency bins. Thus, under ideal conditions, the total number of weights to be updated will be further halved. Considering the overall subband-selective TGSC, its computational complexity is summarized in Table 3.7, which also provides a comparison with the TGSC proposed in Chen and Fang (1992).

3.7.3 Simulations

Simulations are performed to compare the performance of the GSC, TGSC and STGSC, which are based on a setup with $M = 17$ sensors and a TDL length of $J = 32$. The signal of interest comes from the broadside at an SIR of -24 dB and SNR of 20 dB. There are two interfering signals, which cover the frequency intervals $[0.15\pi\ 0.45\pi]$ and $[0.55\pi, 0.85\pi]$, with DOA angles of 20° and -60°, respectively.

A 32-point DFT with a Hamming window is applied in the STGSC, whereby the frequency response of the window function is shown in Figure 3.44. The dimension of the blocking matrix is 17×16 ($L = 16$ and $S = 1$), which is obtained by the transformation method discussed in Section 3.5.3. The frequency responses of this transformation matrix $\boldsymbol{T}$ are shown in Figure 3.45. As $L = M - S$, the STGSC is a fully adaptive beamformer.

The frequency bins discarded for the STGSC are shown in Table 3.8, where the elements with zero value mean that the corresponding frequency bin outputs are discarded, while those having a value of 1 are retained. We compare the performance of the STGSC with the TGSC and the general GSC with $\tilde{\boldsymbol{B}}$ obtained using the CCD method. The corresponding

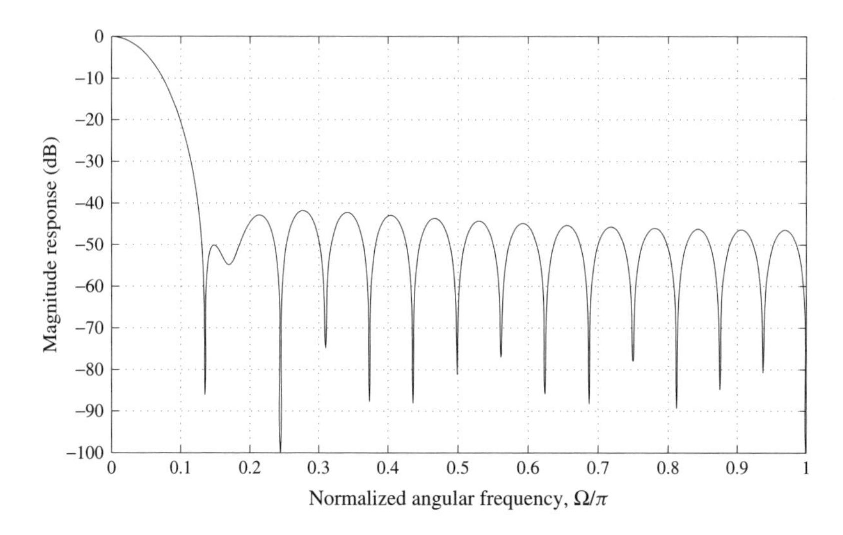

Figure 3.44 Frequency response of a 32-tap hamming window

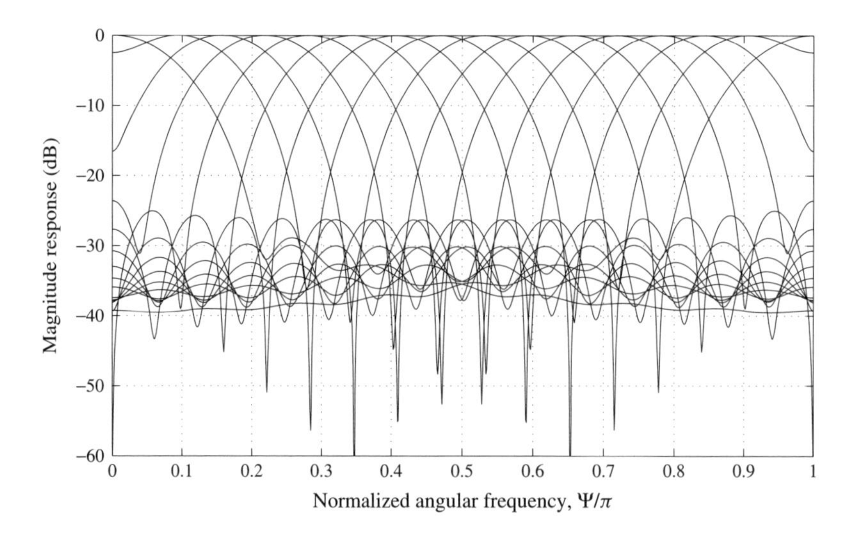

Figure 3.45 Frequency responses of the row vectors of the 16×16 transformation matrix

step size parameters γ used for the STGSC, TGSC, GSC are respectively 6.42×10^{-4}, 4.88×10^{-4} and 6.18×10^{-4}, which have been chosen empirically to achieve similar steady-state mean square residual error values.

From the simulation results shown in Figure 3.46, we can see that the TGSC converges faster than the time-domain GSC because of the temporal decorrelation effect of the DFT, whereas the STGSC is slightly faster than the TGSC due to its combined spatial/temporal decorrelation effect. In addition, although the computational complexity of the STGSC

Table 3.8 Frequency bin outputs discarded in the STGSC in the simulation of Section 3.7.3

$\mathbf{v}_l$	$\mathbf{v}_0 \rightarrow \mathbf{v}_4$	$\mathbf{v}_5$	$\mathbf{v}_6$	$\mathbf{v}_7$	$\mathbf{v}_8$	$\mathbf{v}_9$	$\mathbf{v}_{10}$	$\mathbf{v}_{11}$	$\mathbf{v}_{12}$	$\mathbf{v}_{13}$	$\mathbf{v}_{14}$	$\mathbf{v}_{15}$
$\mathbf{v}_{l,0}$	1	0	0	0	0	0	0	0	0	0	0	0
$\mathbf{v}_{l,1}$	1	1	0	0	0	0	0	0	0	0	0	0
$\mathbf{v}_{l,2}$	1	1	1	0	0	0	0	0	0	0	0	0
$\mathbf{v}_{l,3}$	1	1	1	1	0	0	0	0	0	0	0	0
$\mathbf{v}_{l,4}$	1	1	1	1	1	0	0	0	0	0	0	0
$\mathbf{v}_{l,5}$	1	1	1	1	1	1	0	0	0	0	0	0
$\mathbf{v}_{l,6}$	1	1	1	1	1	1	1	0	0	0	0	0
$\mathbf{v}_{l,7}$	1	1	1	1	1	1	1	1	0	0	0	0
$\mathbf{v}_{l,8}$	1	1	1	1	1	1	1	1	1	1	0	0
$\mathbf{v}_{l,9}$	1	1	1	1	1	1	1	1	1	1	1	0
$\mathbf{v}_{l,10}$	1	1	1	1	1	1	1	1	1	1	1	1
$\vdots$	$\vdots$	$\vdots$	$\vdots$	$\vdots$	$\vdots$	$\vdots$	$\vdots$	$\vdots$	$\vdots$	$\vdots$	$\vdots$	$\vdots$
$\mathbf{v}_{l,22}$	1	1	1	1	1	1	1	1	1	1	1	1
$\mathbf{v}_{l,23}$	1	1	1	1	1	1	1	1	1	1	1	0
$\mathbf{v}_{l,24}$	1	1	1	1	1	1	1	1	1	1	0	0
$\mathbf{v}_{l,25}$	1	1	1	1	1	1	1	1	0	0	0	0
$\mathbf{v}_{l,26}$	1	1	1	1	1	1	1	0	0	0	0	0
$\mathbf{v}_{l,27}$	1	1	1	1	1	1	0	0	0	0	0	0
$\mathbf{v}_{l,28}$	1	1	1	1	1	0	0	0	0	0	0	0
$\mathbf{v}_{l,29}$	1	1	1	1	0	0	0	0	0	0	0	0
$\mathbf{v}_{l,30}$	1	1	1	0	0	0	0	0	0	0	0	0
$\mathbf{v}_{l,31}$	1	1	0	0	0	0	0	0	0	0	0	0

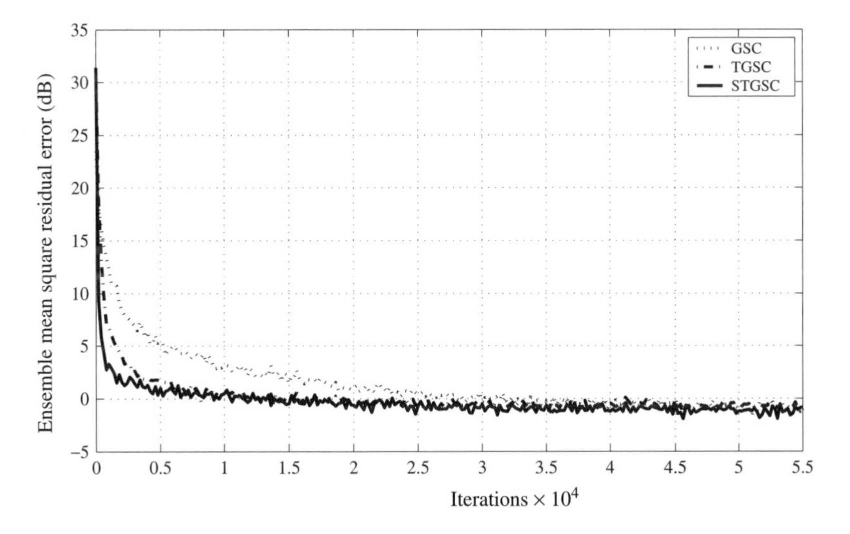

Figure 3.46 Learning curves for the GSC, TGSC and STGSC

is about 3.7 times that of the time-domain GSC, it is only about 90% of the complexity required for the TGSC in the considered example.

3.8 Summary

In this chapter, we have studied various subband techniques and structures for adaptive wideband beamforming, which can achieve a higher convergence rate and a lower computational complexity in most cases.

Since subband decomposition is realized by filter banks, a review of this technique and its application to adaptive filtering were provided first. Thereafter, a general subband adaptive beamforming structure was introduced with the reference signal based beamformer and the GSC as two implementation examples. Replacing the unconstrained adaptive part of the simplified TDL-based GSC by subband adaptation directly, we obtained a subband adaptive GSC structure.

Further improvement to the performance of the subband adaptive GSC was achieved by employing a temporally/spatially subband-selective blocking/transformation matrix. A special property of this matrix is that it can select the received array signals based on not only their DOA angles, but also their frequencies. Combined with the normal subband decomposition operation, some of the lower subbands in the following multi-channel adaptation can be discarded without affecting the beamformer's performance in the ideal case. More importantly, the discarded subbands are determined a priori, independent of the specific signal environments. This idea was then extended to the general subband adaptive beamforming structure by introducing such a transformation matrix between the received array signals and the following subband decomposition.

Since the DFT and IDFT pair can be considered as a maximally decimated filter banks system, frequency-domain adaptation techniques were also studied in this chapter. In particular, a special class of frequency-domain adaptive algorithms applied to the original LCMV beamformer and the GSC was presented, which can solve the LCMV problem with significantly reduced computational complexity, although with a slower convergence speed compared to the time-domain approach.

A class of transform-domain based beamformers emerges when the IDFT operation is omitted in the frequency-domain adaptation structure. The cost is that we cannot reduce the sampling rate of the system any more and the overall computational complexity is therefore higher than the fullband beamformer. A reduction of its computational complexity was achieved by employing the previously introduced temporally/spatially subband-selective blocking matrix in the structure.

4

Design of Fixed Wideband Beamformers

We have studied different classes of adaptive wideband beamformers in the last two chapters. As mentioned in Section 1.3, Chapter 1, the weight coefficients of a wideband beamformer can be fixed irrespective of the signal environments. As a result, the beamformer will maintain a fixed response independent of the signal/interference scenarios. Although such a beamformer may not be able to achieve a high output SINR as in the adaptive case, it has a lower computational complexity and can be implemented easily in practice. Moreover, for some very complicated situations, such as the multipath or reverberant case, the adaptive beamformer may not work well and a fixed beamformer could be the only viable choice when the main direction of the desired source signal is known.

The fixed beamformer design problem is also called an array pattern synthesis problem and there are mainly two classes of design approaches for a fixed wideband beamformer. The first one is the iterative optimization approach, where many iterative optimization methods can be employed directly; the second one is the analytical approach, which includes the classical least squares formulation and the eigenfilter based solutions.

Without loss of generality, all of the design examples are based on the uniformly spaced linear arrays and it is straightforward to extend the discussed design to other array geometries in most of the cases where the general form of a beamformer's response $P(\Omega, \theta) = \mathbf{w}^H \mathbf{d}(\Omega, \theta)$ is used.

4.1 Iterative Optimization

4.1.1 Traditional Methods

Given the desired beam pattern $P_d(\Omega, \theta)$, the design of a wideband beamformer meeting the desired response can be considered as a general optimization problem and solved by employing all kinds of iterative optimization methods.

For example, the design can be formulated as a weighted Chebyshev approximation problem or a minmax problem (Kamp and Thiran, 1975) as given in Nordebo *et al.*, (1994) and Nordholm *et al.* (1998):

$$\min_{\mathbf{w}}\{\max_{\Omega,\theta} v(\Omega, \theta)|\mathbf{w}^H \mathbf{d}(\Omega, \theta) - P_d(\Omega, \theta)|\} \tag{4.1}$$

where $v(\Omega, \theta)$ is a weighting function with real positive values applied to the difference between the desired response $P_d(\Omega, \theta)$ and the designed response $P(\Omega, \theta) = \mathbf{w}^H\mathbf{d}(\Omega, \theta)$.

The cost function is evaluated on all values of Ω and θ within the frequency range of interest and the direction range of the impinging signals. Although it may not be necessary for some algorithms (Lau *et al.*, 1999; Nordholm *et al.*, 1998), in practice, we may choose to discretize Ω and θ and evaluate the cost function on a finite number of grid points as an approximation and the design problem is then changed to:

$$\min_{\mathbf{w}}\{\max_{\forall i,j} v(\Omega_i, \theta_j)|\mathbf{w}^H\mathbf{d}(\Omega_i, \theta_j) - P_d(\Omega_i, \theta_j)|\} \tag{4.2}$$

where the frequency range of interest is discretized into I_Ω points, $\Omega_i, i = 0, 1, \ldots, I_\Omega - 1$ and the direction range into J_θ points, θ_j, $j = 0, 1, \ldots, J_\theta - 1$.

The weight coefficients $\mathbf{w}$ can be obtained by a sequential quadratic programming method (Fletcher, 2000) or those proposed in Lau *et al.* (1999) and Nordholm *et al.* (1998).

Another example is to ignore the phase part of the beam response and formulate the design into the following minimization problem (Kajala and Hämäläinen, 1999):

$$\min_{\mathbf{w}} \Phi(\mathbf{w}) \tag{4.3}$$

where:

$$\Phi(\mathbf{w}) = \int_\Omega \int_\theta v(\Omega, \theta)[|\mathbf{w}^H\mathbf{d}(\Omega, \theta)| - |P_d(\Omega, \theta)|]^2 \mathrm{d}\Omega \mathrm{d}\theta \tag{4.4}$$

In this formulation, the coefficients vector $\mathbf{w}$ cannot be completely extracted from the double integral and for each iteration with a new $\mathbf{w}$, the double integral needs to be evaluated numerically, which leads to a very high computational complexity for the iterative optimization process. To reduce the complexity, a modification to $\Phi(\mathbf{w})$ was introduced as follows in Doclo and Moonen (2003b):

$$\Phi(\mathbf{w}) = \int_\Omega \int_\theta v(\Omega, \theta)[|\mathbf{w}^H\mathbf{d}(\Omega, \theta)|^2 - |P_d(\Omega, \theta)|^2]^2 \mathrm{d}\Omega \mathrm{d}\theta \tag{4.5}$$

This problem can be solved readily using existing optimization methods such as the quasi-Newton method (Fletcher, 2000).

4.1.2 Convex Optimization

Recently, with the development of convex optimization techniques, especially the interior-point methods (Boyd and Vandenberghe, 2004; Nesterov and Nemirovskii, 1994), convex optimization has become a popular and efficient tool for solving the wideband array pattern synthesis problem (Duan *et al.*, 2008; Lebret and Boyd, 1997; Scholnik and Coleman, 2000a,b, 2001; Yan and Ma, 2005; Yan, 2006; Yan *et al.*, 2007; Zhao *et al.*, 2008) and other beamforming problems, such as robust adaptive beamforming (El-Keyi *et al.*, 2005; Lorenz and Boyd, 2005; Rübsamen and Gershman, 2008; Vorobyov *et al.*, 2003; Yu *et al.*, 2008a,b, 2009b).

An optimization problem is considered to be convex when both its objective function and its constraint functions are convex, expressed in the following general form:

$$\min_{\hat{\mathbf{w}}} f_0(\hat{\mathbf{w}})$$
$$\text{subject to } f_i(\hat{\mathbf{w}}) \leq b_i$$
$$i = 1, \ldots, m \tag{4.6}$$

where the vector $\hat{\mathbf{w}}$ represents a set of real-valued variables in this context (if the variables are complex-valued, as in the narrowband beamformer case, they can be represented by a real-valued vector with their imaginary and real parts listed separately in the vector), $f_i(\hat{\mathbf{w}})$, $i = 0, 1, \ldots, m$, are convex functions, and b_i is a constant giving the upper bound for the corresponding constraint function. A function $f_i(\hat{\mathbf{w}})$ is said to be convex if it satisfies:

$$f_i(\alpha\hat{\mathbf{w}}_1 + (1-\alpha)\hat{\mathbf{w}}_2) \leq \alpha f_i(\hat{\mathbf{w}}_1) + (1-\alpha) f_i(\hat{\mathbf{w}}_2) \tag{4.7}$$

for all real-valued α and real-valued vectors $\hat{\mathbf{w}}_1$ and $\hat{\mathbf{w}}_2$, which lie in the same space as $\hat{\mathbf{w}}$, i.e. $\hat{\mathbf{w}}_1$ and $\hat{\mathbf{w}}_2$ are all of the possible values of $\hat{\mathbf{w}}$. Examples of convex functions include the norms $|\hat{\mathbf{w}}|$ and $|\hat{\mathbf{w}}|^2$ of the vector $\hat{\mathbf{w}}$, and the quadratic vector function $\hat{\mathbf{w}}^T \boldsymbol{R}\hat{\mathbf{w}}$, where $\boldsymbol{R}$ is a symmetric positive semi-definite matrix.

As discussed in Lebret and Boyd (1997) and other wideband array pattern synthesis papers with convex optimization, most of the pattern synthesis problems can be reformulated into the convex form in Equation (4.6) and therefore can be solved efficiently employing the interior-point methods or other appropriate convex optimization algorithms.

As an example, consider the design of a wideband linear array. Suppose the frequency range of interest is represented by $\Omega_{pb} = [\Omega_{\min}\ \Omega_{\max}]$ and the sidelobe area of the beamformer is denoted by Θ_{sl}. The look direction of the beamformer is θ_0 and we discretize the sidelobe area Θ_{sl} into $J_\theta - 1$ points (θ_j, $j = 1, \ldots, J_\theta - 1$) and the frequency range Ω_{pb} into I_Ω points (Ω_i, $i = 0, 1, \ldots, I_\Omega - 1$). We minimize the maximum value of the beamformer response at the sidelobe area Θ_{sl} within the frequency range Ω_{pb} subject to the constraints that it has a distortion-less response at the look direction θ_0 over the whole frequency range Ω_{pb}, i.e. a pure delay of T_0 as given in Equation (2.56) of Section 2.3.2, shown below:

$$\begin{bmatrix} \mathbf{x}_c^T \\ \mathbf{x}_c^T \end{bmatrix} \mathbf{w} = \begin{bmatrix} \cos(\Omega_i T_0/T_s) \\ \sin(\Omega_i T_0/T_s) \end{bmatrix} \tag{4.8}$$

For convenience, we rewrite the above equation as:

$$\boldsymbol{C}(\Omega, \theta_0)^T \mathbf{w} = \mathbf{f}(\Omega) \tag{4.9}$$

Note that in this case the magnitude response of the beamformer with respect to Ω and θ is simply $|\boldsymbol{C}(\Omega, \theta)^T \mathbf{w}|$.

Moreover, with a known direction θ_k of the possible interfering signals, we also want to constrain the response of the beamformer at the direction θ_k to be smaller than a very small constant δ_k. Then this design problem can be formulated as:

$$\min_{\mathbf{w}}\{\max_{\substack{i=0,\ldots,I_\Omega-1\\ j=1,\ldots,J_\theta-1}} |\boldsymbol{C}(\Omega_i,\theta_j)^T\mathbf{w}|\}$$

$$\text{subject to } \boldsymbol{C}(\Omega_i,\theta_0)^T\mathbf{w}=\mathbf{f}(\Omega_i),\ i=0,\ldots,I_\Omega-1$$

$$|\boldsymbol{C}(\Omega_i,\theta_k)^T\mathbf{w}|<\delta_k,\ i=0,\ldots,I_\Omega-1 \tag{4.10}$$

It can be solved conveniently using existing convex optimization toolboxes (Lofberg, 2004; Sturm, 1999)

A design example is shown in Figure 4.1 for a wideband beamformer with $M = 10$ sensors and $J = 20$ taps following each sensor. The desired direction is the broadside, i.e. $\theta_0 = 0°$, and the value of T_0/T_s in $\mathbf{f}(\Omega_i)$ is 10. The sidelobe area is from $-90°$ to $-20°$ and $20°$ to $90°$ and discretized into $(J_\theta - 1) = 100$ points. The frequency range of interest is $\Omega_{pb} = [0.3\pi\ \pi]$ and discretized into $I_\Omega = 20$ points. For the constrained direction θ_k, we set $\theta_k = 50°$ and $\delta_k = 0.01$. From the figure we can see clearly the null at 50° and due to the minmax formulation, we have an approximately equal-ripple response at the sidelobe area with an attenuation about -20 dB. A 3-D version of this response is shown in Figure 4.2.

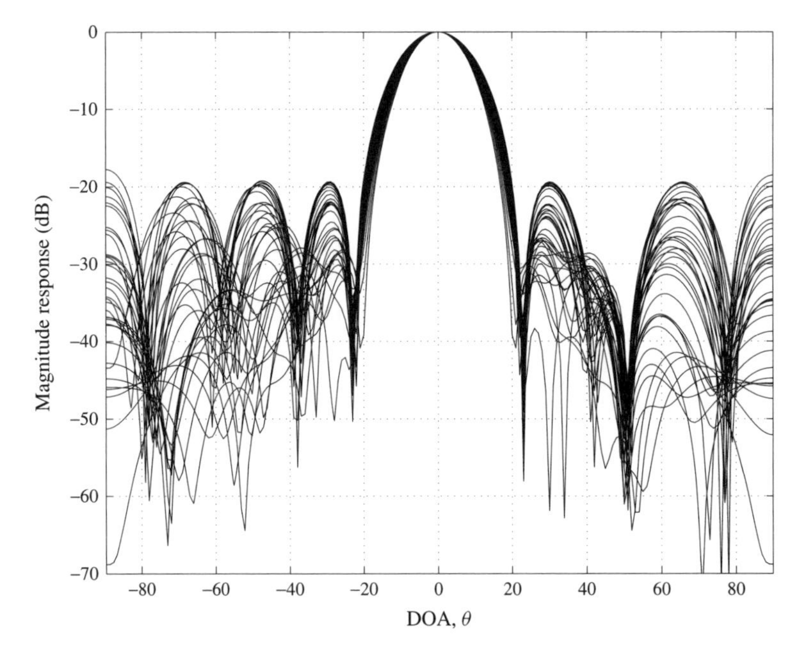

Figure 4.1 A design example using the convex optimization method in Equation (4.10) based on a wideband beamformer with $M = 10$ sensors and $J = 20$ taps following each sensor, for the frequency range $\Omega_{pb} \in [0.3\pi\ \pi]$

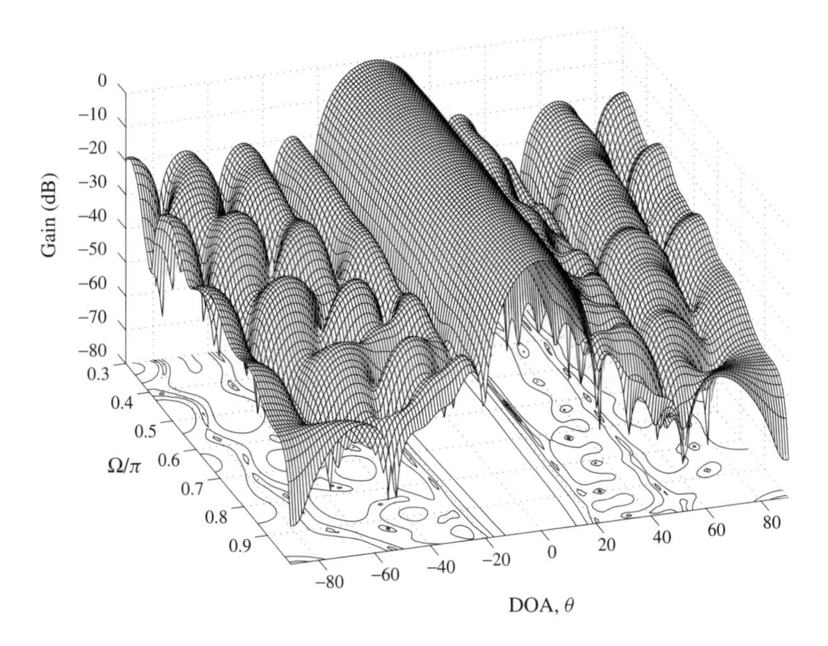

Figure 4.2 A 3-D version of the design example shown in Figure 4.1

We can also minimize the sidelobe response in the least square sense and formulate the problem into the following form accordingly:

$$\min_{\mathbf{w}}\{\sum_{i=0}^{I_\Omega-1}\sum_{j=1}^{J_\theta-1}|\boldsymbol{C}(\Omega_i,\theta_j)^T\mathbf{w}|^2=\mathbf{w}^H\boldsymbol{C}_{ls}\mathbf{w}\}$$

$$\text{subject to } \boldsymbol{C}(\Omega_i,\theta_0)^T\mathbf{w}=\mathbf{f}(\Omega_i),\ i=0,\ldots,I_\Omega-1$$

$$|\boldsymbol{C}(\Omega_i,\theta_k)^T\mathbf{w}|<\delta_k,\ i=0,\ldots,I_\Omega-1 \tag{4.11}$$

where:

$$\boldsymbol{C}_{ls}=\sum_{i=0}^{I_\Omega-1}\sum_{j=1}^{J_\theta-1}\boldsymbol{C}(\Omega_i,\theta_j)\boldsymbol{C}(\Omega_i,\theta_j)^T \tag{4.12}$$

The symmetric matrix $\boldsymbol{C}_{ls}$ can be decomposed into the following form:

$$\boldsymbol{C}_{ls}=\boldsymbol{V}_{ls}\boldsymbol{U}_{ls}\boldsymbol{V}_{ls}^H \tag{4.13}$$

where $\boldsymbol{U}_{ls}$ is a diagonal matrix with its diagonal elements being the eigenvalues of $\boldsymbol{C}_{ls}$ and $\boldsymbol{V}_{ls}$ a full-rank matrix with its columns being the corresponding eigenvectors. Then the cost function changes to a convex form:

$$\mathbf{w}^H\boldsymbol{C}_{ls}\mathbf{w}=\mathbf{w}^H\boldsymbol{V}_{ls}\boldsymbol{U}_{ls}\boldsymbol{V}_{ls}^H\mathbf{w}=|\boldsymbol{U}_{ls}^{\frac{1}{2}}\boldsymbol{V}_{ls}^H\mathbf{w}|^2 \tag{4.14}$$

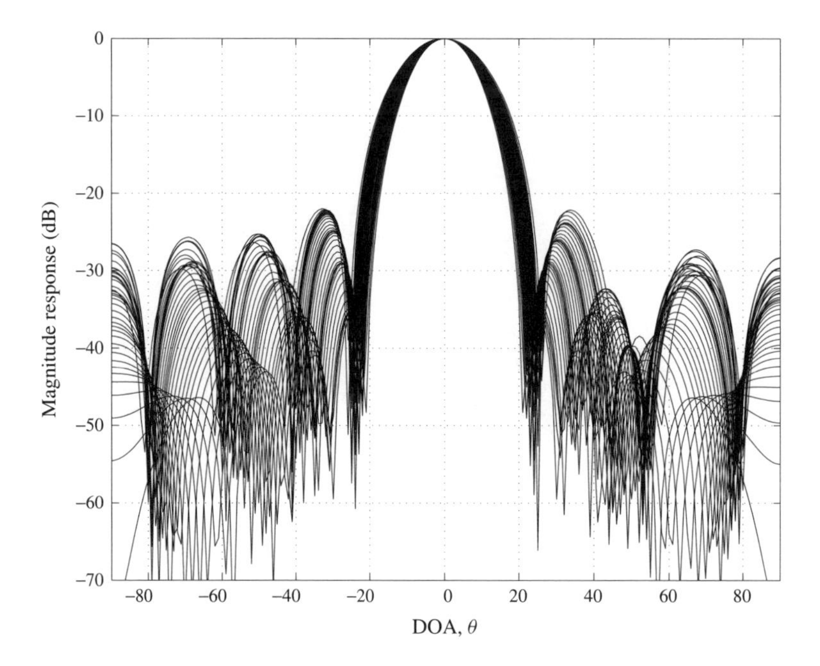

Figure 4.3 A design example using the convex optimization method based on the least square criterion in Equation (4.11)

A design example based on this new criterion with the same parameters as in Figure 4.1 is shown in Figure 4.3. A change of the response from the minmax form into the least square form can be seen clearly at the sidelobe area.

If we want to increase the width of the main beam to design a beam response with a flat top, for example, over the angle range $\theta \in [-5^\circ\ 5^\circ]$, we can add the following additional constraints to either Equation (4.10) or Equation (4.11):

$$\boldsymbol{C}(\Omega_i, \theta_l)^T \mathbf{w} = \mathbf{f}(\Omega_i),\ i = 0, \ldots, I_\Omega - 1 \tag{4.15}$$

for $l = 1, \ldots, L_\theta$ and $\theta_l \in [-5^\circ\ 5^\circ]$, where we have split the angle range $[-5^\circ\ 5^\circ]$ into L_θ discrete points. However, these equality constraints may take up too many of the freedoms in the design and a better way is to limit the error between the response at the point (Ω_i, θ_l) and the look direction (Ω_i, θ_0) to a very small value δ_L as follows:

$$\begin{aligned} &|\boldsymbol{C}(\Omega_i, \theta_l)^T \mathbf{w} - \boldsymbol{C}(\Omega_i, \theta_0)^T \mathbf{w}| \\ &= |(\boldsymbol{C}(\Omega_i, \theta_l) - \boldsymbol{C}(\Omega_i, \theta_0))^T \mathbf{w}| < \delta_L,\quad i = 0, \ldots, I_\Omega - 1 \end{aligned} \tag{4.16}$$

Figure 4.4 shows a design result based on the least square criterion using the constraints of Equation (4.16) with $L_\theta = 10$, $\delta_L = 0.2$ and a frequency range of interest $\Omega_{pb} = [0.5\pi\ \pi]$. All of the other parameters are the same as before. Although we have achieved a relative flat top for the main beam as desired, the response at the transition angle range has been out of control. A remedy is to constrain the magnitude response at those

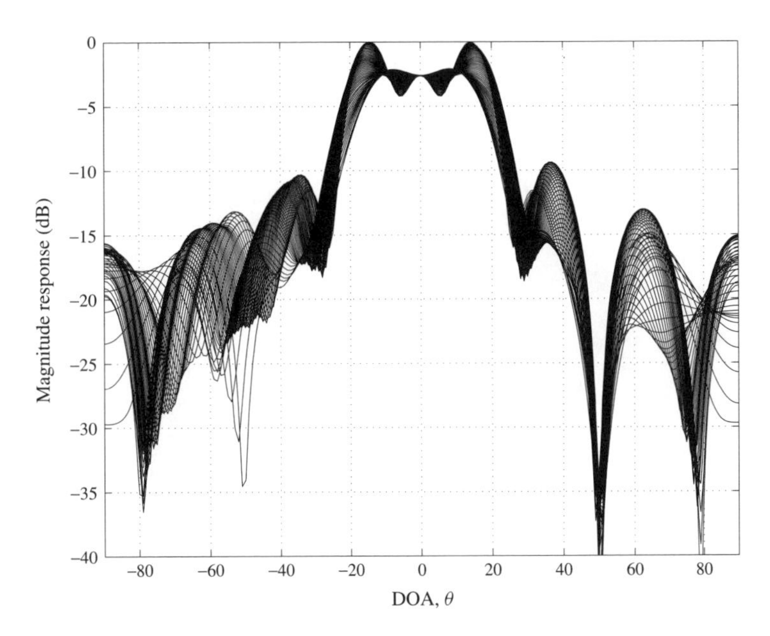

Figure 4.4 A design example with a flat top over the frequency range of interest $\Omega_{pb} = [0.5\pi\ \pi]$ using the formulation in Equation (4.16)

directions from -20° to -5° and from 5° to 20° to be smaller than unity, which is the desired response at the look direction θ_0. This further constraint can be expressed as:

$$|\boldsymbol{C}(\Omega_i, \theta)^T \mathbf{w}| < 1 \text{ for } \theta \in \Theta_{ts}, \quad i = 0, \ldots, I_\Omega - 1 \tag{4.17}$$

where Θ_{ts} denotes the transition area.

Figure 4.5 shows a design result based on this modified formulation with the additional constraints imposed on the transition part of the DOA angle. The undesired much higher magnitude response at the transition area has been suppressed effectively with a satisfactory flat top at the main beam direction.

In all of the formulations for convex optimization discussed so far, we have used the real-valued response given in Equation (4.8). However, we can use the original complex-valued formulation given by $P(\Omega, \theta) = \mathbf{w}^H \mathbf{d}(\Omega, \theta)$ directly as many existing convex optimization toolboxes can deal with complex-valued problems. Otherwise, we would need to separate the real part and the complex part and then combine them together into an equivalent form using real-valued parameters before we can solve it using those toolboxes.

Note that the formulations discussed here are only some representative examples and there are many different variations to them depending on the specific requirements. For example, we can limit the noise gain of the beamformer in the design by adding the following constraints as suggested in Scholnik and Coleman (2000a,b):

$$\mathbf{w}^H \mathbf{w} < \delta_{norm} \tag{4.18}$$

where δ_{norm} is the upper bound of the norm.

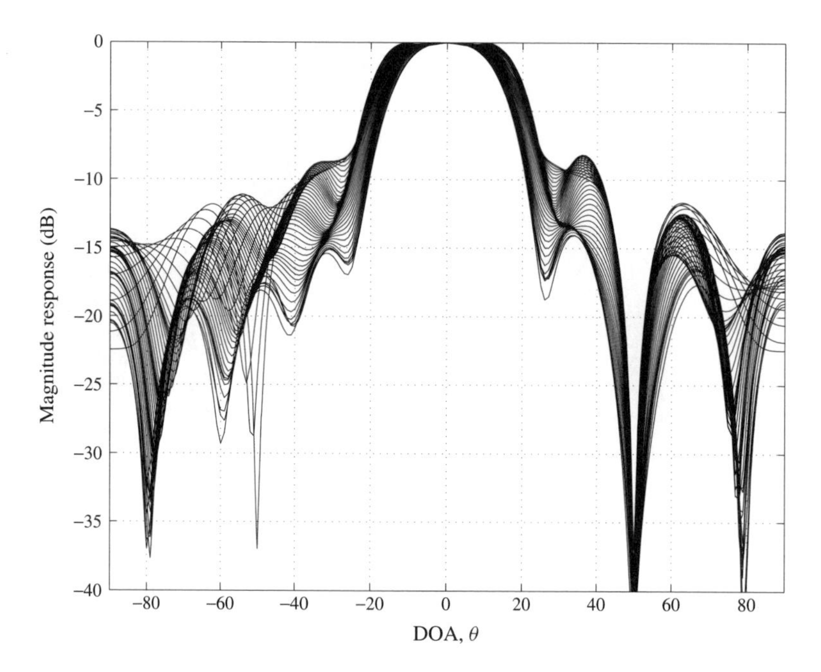

Figure 4.5 A design example with a flat top over the frequency range of interest $\Omega_{pb} = [0.5\pi \ \pi]$ with the additional constraints introduced into the transition area, as given in Equation (4.17)

In addition to the methods discussed so far, it is also possible to design a wideband beamformer employing adaptive array techniques. The basic idea is to simulate an environment with many interfering signals from different directions and then the optimum array coefficients with the desired pattern can be obtained after array adaptation to form a beam in the desired direction and low sidelobes at the interfering directions. For details, please refer to C and Keeping (1982), Olen and Compton (1990) and Zhou and Ingram (1999).

4.2 The Least Squares Approach

Although the fixed beamformer design problem can be solved using the iterative optimization approaches, they become less efficient for the case with a very large number of coefficients and cannot provide a closed form solution to the problem. In the remaining part of this chapter, we will discuss two classes of analytical approaches which can provide such a closed-form solution: the least squares approach and the eigenfilter approach. We first discuss the least squares approach in this section.

4.2.1 Standard Formulation

The least squares problem is a traditional subject and has been well-studied in the past (Björck, 1996; Lawson and Hanson, 1974). Given the desired beam pattern $P_d(\Omega, \theta)$, such a problem arises naturally by minimizing the sum of the squares of the error between $P_d(\Omega,\theta)$ and the designed response $P(\Omega,\theta)$ over the frequency range of interest Ω_{pb} and the range of signal arrival angle Θ:

$$\min_{\mathbf{w}} \int_{\Omega_{pb}} \int_{\Theta} |P(\Omega,\theta) - P_d(\Omega,\theta)|^2 \mathrm{d}\Omega \mathrm{d}\theta \tag{4.19}$$

We can add the weighting function $v(\Omega,\theta)$ as before to form a weighted least squares problem (Doclo and Moonen, 2003b):

$$\min_{\mathbf{w}} \int_{\Omega_{pb}} \int_{\Theta} v(\Omega,\theta)|P(\Omega,\theta) - P_d(\Omega,\theta)|^2 \mathrm{d}\Omega \mathrm{d}\theta \tag{4.20}$$

The cost function can be expanded into the following form:

$$\begin{aligned}
J_{ls}(\mathbf{w}) &= \int_{\Omega_{pb}} \int_{\Theta} v(\Omega,\theta)|P(\Omega,\theta) - P_d(\Omega,\theta)|^2 \mathrm{d}\Omega \mathrm{d}\theta \\
&= \int_{\Omega_{pb}} \int_{\Theta} v(\Omega,\theta)(P(\Omega,\theta) - P_d(\Omega,\theta))(P(\Omega,\theta) - P_d(\Omega,\theta))^H \mathrm{d}\Omega \mathrm{d}\theta \\
&= \int_{\Omega_{pb}} \int_{\Theta} v(\Omega,\theta)(|P(\Omega,\theta)|^2 + |P_d(\Omega,\theta)|^2 - 2Re\{P(\Omega,\theta)P_d^*(\Omega,\theta)\} \mathrm{d}\Omega \mathrm{d}\theta \\
&= \mathbf{w}^H \boldsymbol{G}_{ls}\mathbf{w} - \mathbf{w}^H \bar{\mathbf{g}}_{ls} - \bar{\mathbf{g}}_{ls}^H \mathbf{w} + g_{ls} \\
&= \mathbf{w}^H \boldsymbol{G}_{ls}\mathbf{w} - 2\mathbf{w}^H \mathbf{g}_{ls} + g_{ls} \quad \text{(for real-valued} \quad \mathbf{w})
\end{aligned} \tag{4.21}$$

where:

$$\begin{aligned}
\boldsymbol{G}_{ls} &= \int_{\Omega_{pb}} \int_{\Theta} v(\Omega,\theta)(\mathbf{d}(\Omega,\theta)\mathbf{d}^H(\Omega,\theta)) \mathrm{d}\Omega \mathrm{d}\theta \\
&= \int_{\Omega_{pb}} \int_{\Theta} v(\Omega,\theta)\boldsymbol{D}(\Omega,\theta) \mathrm{d}\Omega \mathrm{d}\theta \\
\bar{\mathbf{g}}_{ls} &= \int_{\Omega_{pb}} \int_{\Theta} v(\Omega,\theta)(\mathbf{d}(\Omega,\theta)P_d^*(\Omega,\theta)) \mathrm{d}\Omega \mathrm{d}\theta \\
\mathbf{g}_{ls} &= \int_{\Omega_{pb}} \int_{\Theta} v(\Omega,\theta)(\mathbf{d}_R(\Omega,\theta)P_{d,R}(\Omega,\theta) + \mathbf{d}_I(\Omega,\theta)P_{d,I}(\Omega,\theta)) \mathrm{d}\Omega \mathrm{d}\theta \\
g_{ls} &= \int_{\Omega_{pb}} \int_{\Theta} v(\Omega,\theta)|P_d(\Omega,\theta)|^2 \mathrm{d}\Omega \mathrm{d}\theta
\end{aligned} \tag{4.22}$$

$\mathbf{d}_R(\Omega,\theta)$ and $P_{d,R}(\Omega,\theta)$ denote the real parts of $\mathbf{d}(\Omega,\theta)$ and $P_d(\Omega,\theta)$, and $\mathbf{d}_I(\Omega,\theta)$ and $P_{d,I}(\Omega,\theta)$ are their imaginary parts.

With $\boldsymbol{D}(\Omega,\theta) = \mathbf{d}(\Omega,\theta)\mathbf{d}^H(\Omega,\theta)$, it can be decomposed into two parts: the real part $\boldsymbol{D}_R(\Omega,\theta)$ and the imaginary part $\boldsymbol{D}_I(\Omega,\theta)$, namely:

$$\boldsymbol{D}(\Omega,\theta) = \boldsymbol{D}_R(\Omega,\theta) + j\boldsymbol{D}_I(\Omega,\theta) \tag{4.23}$$

Since $\mathbf{w}^H \boldsymbol{D}\mathbf{w}$ is real-valued, for real-valued $\mathbf{w}$, we have:

$$\begin{aligned}\mathbf{w}^H \boldsymbol{D}\mathbf{w} &= \frac{1}{2}(\mathbf{w}^H \boldsymbol{D}\mathbf{w} + (\mathbf{w}^H \boldsymbol{D}\mathbf{w})^*)\\ &= \frac{1}{2}(\mathbf{w}^H(\boldsymbol{D} + \boldsymbol{D}^*)\mathbf{w})\\ &= \mathbf{w}^H \boldsymbol{D}_R\mathbf{w}\end{aligned} \tag{4.24}$$

Then $\boldsymbol{G}_{ls}$ changes to:

$$\boldsymbol{G}_{ls} = \int_{\Omega_{pb}}\int_{\Theta} v(\Omega,\theta)\boldsymbol{D}_R(\Omega,\theta)\mathrm{d}\Omega\mathrm{d}\theta \tag{4.25}$$

Minimizing the cost function J_{ls} with respect to the coefficients vector $\mathbf{w}$ gives the standard least squares solution:

$$\mathbf{w}_{opt} = \boldsymbol{G}_{ls}^{-1}\mathbf{g}_{ls} \tag{4.26}$$

For the frequency range of interest Ω_{pb}, if the desired response $P_d(\Omega,\theta)$ is $\mathrm{e}^{-j(T_0/T_s)\Omega}$ for the mainlobe area Θ_{ml}, zero for the sidelobe area Θ_{sl}, and the weighting function is α for the mainlobe and $(1-\alpha)$ for the sidelobe, then the cost function changes to:

$$\begin{aligned}\boldsymbol{G}_{ls} &= \alpha\int_{\Omega_{pb}}\int_{\Theta_{ml}} \boldsymbol{D}_R(\Omega,\theta)\mathrm{d}\Omega\mathrm{d}\theta + (1-\alpha)\int_{\Omega_{pb}}\int_{\Theta_{sl}} \boldsymbol{D}_R(\Omega,\theta)\mathrm{d}\Omega\mathrm{d}\theta\\ \mathbf{g}_{ls} &= \alpha\int_{\Omega_{pb}}\int_{\Theta_{ml}} \left(\mathbf{d}_R(\Omega,\theta)\cos\left(\frac{T_0}{T_s}\Omega\right) - \mathbf{d}_I(\Omega,\theta)\sin\left(\frac{T_0}{T_s}\Omega\right)\right)\mathrm{d}\Omega\mathrm{d}\theta\\ g_{ls} &= \alpha\int_{\Omega_{pb}}\int_{\Theta_{ml}} 1\mathrm{d}\Omega\mathrm{d}\theta\end{aligned} \tag{4.27}$$

A design example is shown in Figure 4.6 for a wideband beamformer with $M = 10$ sensors and $J = 20$ taps following each sensor. The main lobe direction (look direction) is chosen to be the broadside, i.e. $\theta_0 = 0°$ and Θ_{ml} only includes one single direction θ_0, at which the desired response is $\mathrm{e}^{-j10\Omega}$. The sidelobe area is from $-90°$ to $-20°$ and $20°$ to $90°$ and discretized into $(J_\theta - 1) = 100$ points, 50 points for each side. The frequency range of interest is $\Omega_{pb} = [0.3\pi\ \pi]$ and discretized into $I_\Omega = 20$ points. The discretization of both frequency and angle is used for calculating the integrations in Equation (4.27). For the weighting function, $\alpha = 0.6$. A two-dimensional version is shown in Figure 4.7.

4.2.2 Constrained Least Squares

In the least squares formulation, it is possible to add some additional constraints to constrain the response of the beamformer at some specific directions or frequencies. These constraints can be of either equality or inequality and for the linear equality constraints, a closed-form solution can be found by reducing it to a low-dimensional unconstrained least squares problem using different methods (Björck, 1996; Lawson and Hanson, 1974), such as the transformation employed in translating the LCMV problem into the unconstrained GSC.

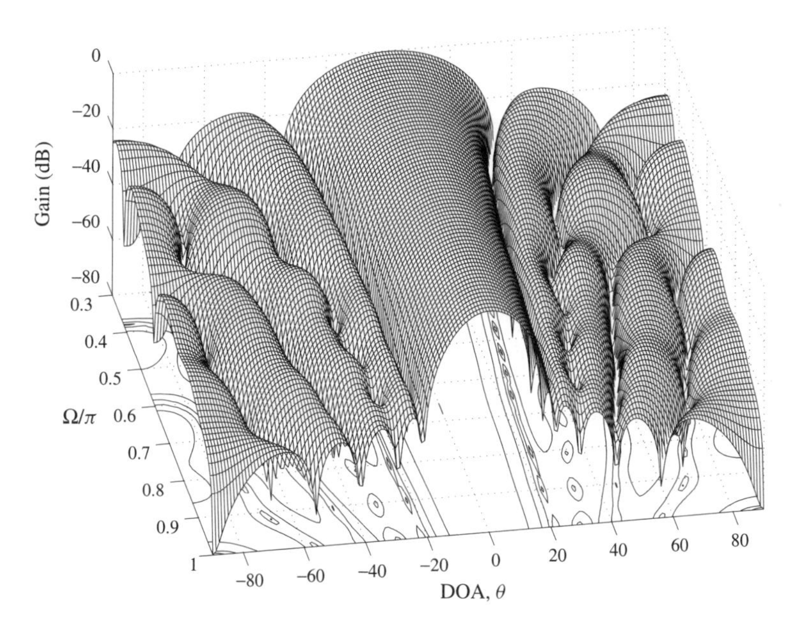

Figure 4.6 The resulting 3-D beam response for the design example using the standard least squares approach as formulated in Equation (4.27)

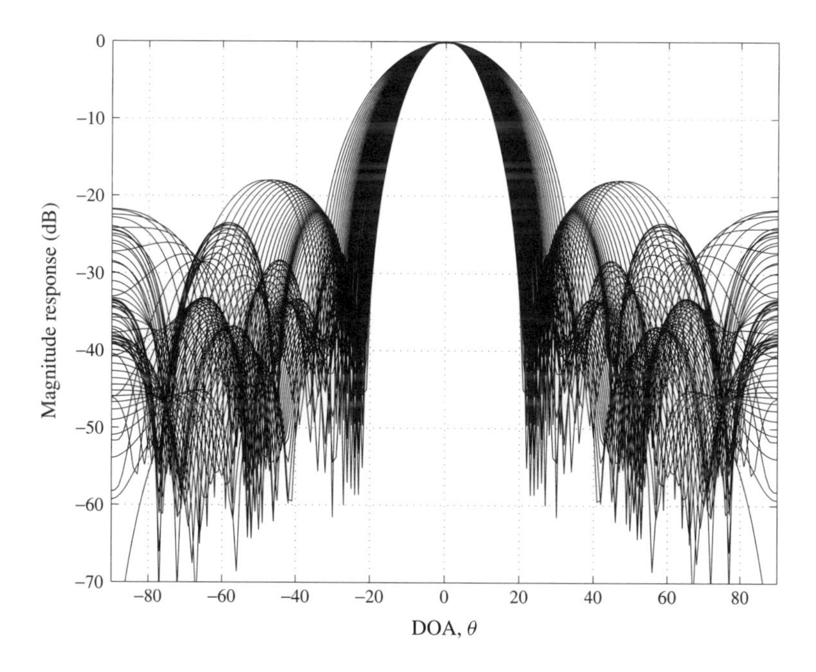

Figure 4.7 The resulting 2-D beam response for the design example shown in Figure 4.6

Now consider the constraints in the following general form:

$$\boldsymbol{C}^H \mathbf{w} = \mathbf{f} \tag{4.28}$$

Then the constrained least squares problem can be formulated as:

$$\min_{\mathbf{w}} J_{ls}(\mathbf{w}) \qquad \text{subject to } \boldsymbol{C}^H \mathbf{w} = \mathbf{f} \tag{4.29}$$

We here derive its solution by the method of Lagrange multipliers in a similar way as in Section 2.2.2.

We first form the new Lagrangian cost function by the objective function $J_{ls}(\mathbf{w})$, plus the real part of the constraint function of $\boldsymbol{C}^H\mathbf{w} - \mathbf{f}$, weighted element wise by the vector of undetermined Lagrange multipliers $\boldsymbol{\lambda}$, given by:

$$\frac{1}{2}(J_{ls}(\mathbf{w}) + \boldsymbol{\lambda}^H(\boldsymbol{C}^H\mathbf{w} - \mathbf{f}) + \boldsymbol{\lambda}^T(\boldsymbol{C}^T\mathbf{w}^* - \mathbf{f}^*)) \tag{4.30}$$

Differentiating it with respect to $\mathbf{w}^*$ and setting the result equal to zero, we have:

$$\boldsymbol{G}_{ls}\mathbf{w} - \mathbf{g}_{ls} + \boldsymbol{C}\boldsymbol{\lambda} = 0 \tag{4.31}$$

Then the optimal weight vector $\mathbf{w}_{opt}$ can be expressed as:

$$\mathbf{w}_{opt} = \boldsymbol{G}_{ls}^{-1}(\mathbf{g}_{ls} - \boldsymbol{C}\boldsymbol{\lambda}) \tag{4.32}$$

Since $\mathbf{w}_{opt}$ must satisfy the constraints, we have:

$$\boldsymbol{C}^H\boldsymbol{G}_{ls}^{-1}(\mathbf{g}_{ls} - \boldsymbol{C}\boldsymbol{\lambda}) = \mathbf{f} \tag{4.33}$$

Solving this equation for $\boldsymbol{\lambda}$, we have:

$$\boldsymbol{\lambda} = (\boldsymbol{C}^H\boldsymbol{G}_{ls}^{-1}\boldsymbol{C})^{-1}(\boldsymbol{C}^H\boldsymbol{G}_{ls}^{-1}\mathbf{g}_{ls} - \mathbf{f}) \tag{4.34}$$

Substituting $\boldsymbol{\lambda}$ into Equation (4.32) yields:

$$\mathbf{w}_{opt} = \boldsymbol{G}_{ls}^{-1}\mathbf{g}_{ls} - \boldsymbol{G}_{ls}^{-1}\boldsymbol{C}(\boldsymbol{C}^H\boldsymbol{G}_{ls}^{-1}\boldsymbol{C})^{-1}(\boldsymbol{C}^H\boldsymbol{G}_{ls}^{-1}\mathbf{g}_{ls} - \mathbf{f}) \tag{4.35}$$

Now if we express the desired response at the passband ($\Omega \in \Omega_{pb}$ and $\theta \in \Theta_{ml}$) into the form of constraints, then we only need to minimize the error in the sidelobe region and therefore can ignore the mainlobe part in the cost function $J_{ls}(\mathbf{w})$. In this case, Equation (4.27) is reduced to:

$$\begin{aligned}
\boldsymbol{G}_{ls} &= (1-\alpha)\int_{\Omega_{pb}}\int_{\Theta_{sl}} \boldsymbol{D}_R(\Omega,\theta)\mathrm{d}\Omega\mathrm{d}\theta \\
\mathbf{g}_{ls} &= 0 \\
g_{ls} &= 0
\end{aligned} \tag{4.36}$$

and the optimum solution is given by:

$$\mathbf{w}_{opt} = \boldsymbol{G}_{ls}^{-1}\boldsymbol{C}(\boldsymbol{C}^H\boldsymbol{G}_{ls}^{-1}\boldsymbol{C})^{-1}\mathbf{f} \tag{4.37}$$

Note the value of α does not matter any more in this scenario.

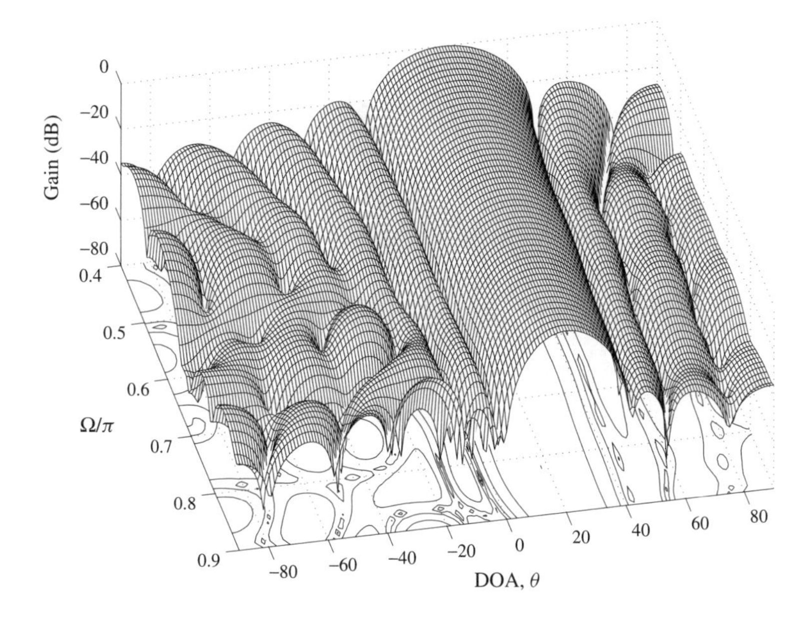

Figure 4.8 The resulting 3-D beam response for the design example using the constrained least squares approach as formulated in Equation (4.37)

As an example, consider the design of a wideband beamformer with $M = 10$ sensors and $J = 20$ taps following each sensor. The look direction of the main beam is $\theta_0 = 20^\circ$ and the desired response at this direction is $\mathrm{e}^{-j10\Omega}$. Then the constraint equation can be expressed as:

$$\boldsymbol{C}(\Omega_i, \theta_0)^T \mathbf{w} = \mathbf{f}(\Omega_i) = \begin{bmatrix} \cos(10\Omega_i) \\ \sin(10\Omega_i) \end{bmatrix}, \quad i = 0, \ldots, I_\Omega - 1 \tag{4.38}$$

where the frequency range of interest $\Omega_i \in \Omega_{pb} = [0.4\pi \ 0.9\pi]$ has been discretized into $I_\Omega = 20$ points. The sidelobe region is from -90° to 0° and 40° to 90° and discretized uniformly into 100 points in total. The 3-D response of the design result is shown in Figure 4.8 and its 2-D version is shown in Figure 4.9.

4.3 The Eigenfilter Approach

Another approach which can provide a closed-form solution to the fixed wideband beamformer design problem is the eigenfilter approach. Unlike the least squares one, no matrix inversion is required for this class of methods.

The term ‘eigenfilter’ is referred to as a filter with its coefficients being the elements of an eigenvector (Makhoul, 1981). The eigenfilter approach was first proposed for the design of digital filters (Chen, 1993; Pei and Shyu, 1990; Pei and Tseng, 2001; Tkacenko *et al.*, 2003; Vaidyanathan and Nguyen, 1987) and then extended to the design of spatial filters or beamformers (Chen, 1993; Doclo and Moonen, 2003a,b; Korompis *et al.*, 1994; Tkacenko *et al.*, 2003). In addition to the traditional/standard eigenfilter approach, there

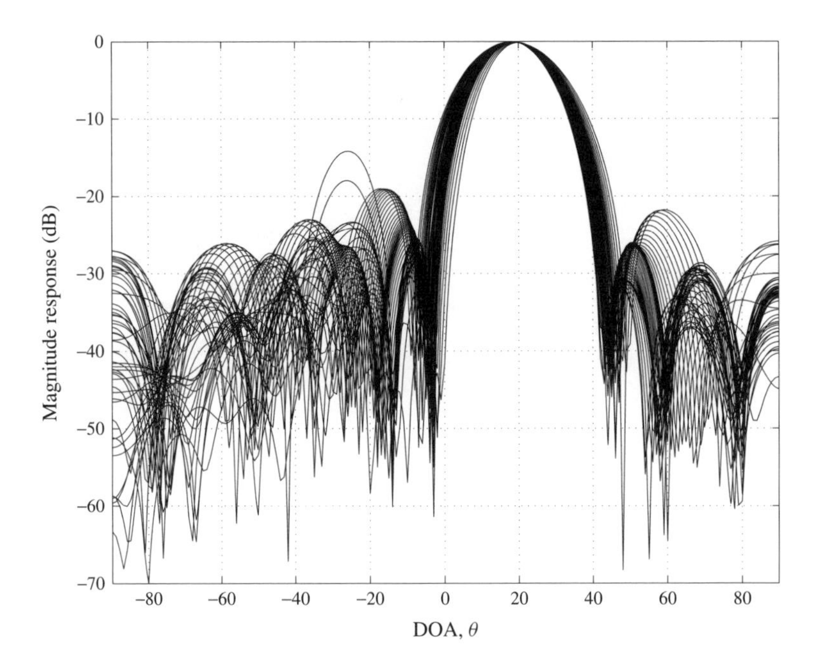

Figure 4.9 The resulting 2-D beam response for the design example shown in Figure 4.8

are also many different variations of it, in which finding the generalized eigenvector of two matrices is normally involved. In this part, we will focus on three of them, namely the standard approach, the maximum energy approach and the total least squares approach.

4.3.1 Standard Approach

The standard eigenfilter approach is based on the Rayleigh–Ritz theorem, which is named after two British physicists (Horn and Johnson, 1985). It states that for a Hermitian matrix $\boldsymbol{R}$, its Rayleigh–Ritz ratio

$$\frac{\mathbf{w}^H \boldsymbol{R}\mathbf{w}}{\mathbf{w}^H \mathbf{w}} \tag{4.39}$$

reaches its maximum when $\mathbf{w}$ is the eigenvector corresponding to the maximum eigenvalue of $\boldsymbol{R}$ and reaches its minimum when $\mathbf{w}$ is the eigenvector corresponding to the minimum eigenvalue of $\boldsymbol{R}$. The maximum and minimum values of this ratio are respectively the maximum and minimum eigenvalues.

In the design of FIR filters with an arbitrary desired response, in order to formulate the problem into such a form, a reference frequency point is used and the resulting Hermitian matrix $\boldsymbol{R}$ is positive definite (Nguyen, 1993; Pei and Shyu, 1993) with its eigenvector corresponding to the minimum eigenvalue being the desired solution. This method can be easily extended to the design of wideband beamformers (Chen, 1993; Doclo and Moonen, 2003a,b; Tkacenko *et al.*, 2003).

Assume $P_d(\Omega, \theta)$ is the desired response, then the cost function for this design problem is formulated as:

$$J_{ef}(\mathbf{w}) = \int_{\Omega_{pb}} \int_{\Theta} v(\Omega, \theta) \left| P(\Omega, \theta) - P(\Omega_r, \theta_r) \frac{P_d(\Omega, \theta)}{P_d(\Omega_r, \theta_r)} \right|^2 \mathrm{d}\Omega \mathrm{d}\theta \tag{4.40}$$

where $v(\Omega, \theta)$ is the weighting function and (Ω_r, θ_r) is the reference point, which, for a good design result, should be chosen in such a way so that $P_d(\Omega_r, \theta_r)$ has the largest magnitude response. This formulation leads to a zero value for the cost function at the reference point.

With $P(\Omega, \theta) = \mathbf{w}^H \mathbf{d}(\Omega, \theta)$, $J_{ef}(\mathbf{w})$ is reduced to a quadratic form:

$$J_{ef}(\mathbf{w}) = \mathbf{w}^H \boldsymbol{G}_{ef}(\Omega, \theta) \mathbf{w} \tag{4.41}$$

where:

$$\boldsymbol{G}_{ef} = \int_{\Omega_{pb}} \int_{\Theta} v(\Omega, \theta) \left(\mathbf{d}(\Omega, \theta) - \mathbf{d}(\Omega_r, \theta_r) \frac{P_d(\Omega, \theta)}{P_d(\Omega_r, \theta_r)} \right) \times \left(\mathbf{d}(\Omega, \theta) - \mathbf{d}(\Omega_r, \theta_r) \frac{P_d(\Omega, \theta)}{P_d(\Omega_r, \theta_r)} \right)^H \mathrm{d}\Omega \mathrm{d}\theta \tag{4.42}$$

In order to avoid the trivial solution $\mathbf{w} = \mathbf{0}$, a constraint on the norm of $\mathbf{w}$ is added to the problem. A complete formulation of the eigenfilter design problem is then given by:

$$\min_{\mathbf{w}} \mathbf{w}^H \boldsymbol{G}_{ef}(\Omega, \theta) \mathbf{w} \text{ subject to } \mathbf{w}^H \mathbf{w} = 1 \tag{4.43}$$

or alternatively:

$$\min_{\mathbf{w}} \frac{\mathbf{w}^H \boldsymbol{G}_{ef}(\Omega, \theta) \mathbf{w}}{\mathbf{w}^H \mathbf{w}} \tag{4.44}$$

The minimum value is equal to the minimum eigenvalue of $\boldsymbol{G}_{ef}(\Omega, \theta)$ and the corresponding eigenvector is the optimum weight vector $\mathbf{w}_{opt}$.

Instead of having the norm constraint, it is possible to add some linear constraints in the form of $\boldsymbol{C}^H \mathbf{w} = \mathbf{f}$ to the design, which leads to the following minimization problem:

$$\min_{\mathbf{w}} \mathbf{w}^H \boldsymbol{G}_{ef}(\Omega, \theta) \mathbf{w} \text{ subject to } \boldsymbol{C}^H \mathbf{w} = \mathbf{f} \tag{4.45}$$

This is a special form of Equation (4.29) and a similar solution has been given by Equations (4.37) and (2.36) and in this case it is:

$$\mathbf{w}_{opt} = \boldsymbol{G}_{ef}^{-1} \boldsymbol{C} (\boldsymbol{C}^H \boldsymbol{G}_{ef}^{-1} \boldsymbol{C})^{-1} \mathbf{f} \tag{4.46}$$

We can also add the linear constraints to the original eigenfilter problem directly as follows:

$$\min_{\mathbf{w}} \frac{\mathbf{w}^H \boldsymbol{G}_{ef}(\Omega, \theta) \mathbf{w}}{\mathbf{w}^H \mathbf{w}} \text{ subject to } \boldsymbol{C}^H \mathbf{w} = \mathbf{f} \tag{4.47}$$

Its solution can be obtained by first transforming the constraint into the form (Pei *et al.*, 1998):

$$\hat{\mathbf{C}}^H \mathbf{w} = \mathbf{0} \tag{4.48}$$

with:

$$\hat{\mathbf{C}} = \mathbf{C} - \frac{\mathbf{d}(\Omega_r, \theta_r)}{P(\Omega_r, \theta_r)} \mathbf{f}^H \tag{4.49}$$

To meet this constraint equation, $\mathbf{w}$ must lie in the null space of $\hat{\mathbf{C}}$. Suppose $\tilde{\mathbf{C}}$ is a unitary matrix with its columns being the bases of the null space. Then we have $\mathbf{w} = \tilde{\mathbf{C}}\tilde{\mathbf{w}}$ and then the problem is reduced to finding the new unknown vector $\tilde{\mathbf{w}}$ in the following minimizing problem:

$$\min_{\tilde{\mathbf{w}}} \frac{\tilde{\mathbf{w}}^H \tilde{\mathbf{C}}^H \mathbf{G}_{ef} \tilde{\mathbf{C}} \tilde{\mathbf{w}}}{\tilde{\mathbf{w}}^H \tilde{\mathbf{C}}^H \tilde{\mathbf{C}} \tilde{\mathbf{w}}} = \frac{\tilde{\mathbf{w}}^H \tilde{\mathbf{C}}^H \mathbf{G}_{ef} \tilde{\mathbf{C}} \tilde{\mathbf{w}}}{\tilde{\mathbf{w}}^H \tilde{\mathbf{w}}} \tag{4.50}$$

This is again a standard eigenfilter problem and the optimum $\tilde{\mathbf{w}}$ is the eigenvector corresponding to the smallest eigenvalue of the matrix $\tilde{\mathbf{C}}^H \mathbf{G}_{ef} \tilde{\mathbf{C}}$.

To illustrate the performance of the standard eigenfilter approach, we give a design example based on the formulation in Equation (4.44) for a wideband beamformer with $M = 10$ sensors and $J = 10$ taps following each sensor. The look direction is $\theta_0 = 10^\circ$ with a desired response given by $\mathrm{e}^{-j10\Omega}$. In the mainlobe we only consider the error at the single direction θ_0 and the frequency range of interest is $\Omega_{pb} = [0.4\pi \ 0.9\pi]$ with $\Omega_r = 0.65\pi$ and $\theta_r = 10^\circ$ as the reference point. The weighting function is $\alpha = 0.6$ at the look direction and $(1 - \alpha) = 0.4$ at the sidelobe region from -90° to -10° and 30° to 90°, where the desired response is zero. Then $\mathbf{G}_{ef}$ becomes:

$$\begin{aligned}\mathbf{G}_{ef} = \alpha \int_{\Omega_{pb}} &(\mathbf{d}(\Omega, \theta_0) - \mathrm{e}^{-j10(\Omega-\Omega_r)}\mathbf{d}(\Omega_r, \theta_r))(\mathbf{d}(\Omega, \theta_0) - \mathrm{e}^{-j10(\Omega-\Omega_r)}\mathbf{d}(\Omega_r, \theta_r))^H \mathrm{d}\Omega \\ &+ (1-\alpha) \int_{\Omega_{pb}} \int_{\Theta_{sl}} \mathbf{d}(\Omega, \theta)\mathbf{d}(\Omega, \theta)^H \mathrm{d}\Omega \mathrm{d}\theta \end{aligned} \tag{4.51}$$

Since $\mathbf{w}$ is real-valued, $\mathbf{G}_{ef}$ can be changed to the following form without affecting the final result:

$$\begin{aligned}\mathbf{G}_{ef} = \alpha \int_{\Omega_{pb}} &\mathrm{Re}\{(\mathbf{d}(\Omega, \theta_0) - \mathrm{e}^{-j10(\Omega-\Omega_r)}\mathbf{d}(\Omega_r, \theta_r)) \\ &(\mathbf{d}(\Omega, \theta_0) - \mathrm{e}^{-j10(\Omega-\Omega_r)}\mathbf{d}(\Omega_r, \theta_r))^H\} \mathrm{d}\Omega \\ &+ (1-\alpha) \int_{\Omega_{pb}} \int_{\Theta_{sl}} \mathbf{D}_R(\Omega, \theta) \mathrm{d}\Omega \mathrm{d}\theta \end{aligned} \tag{4.52}$$

The result with a 3-D beam response is shown in Figure 4.10 and the two-dimensional one in Figure 4.11. The sidelobe area is discretized into 100 points and the frequency range

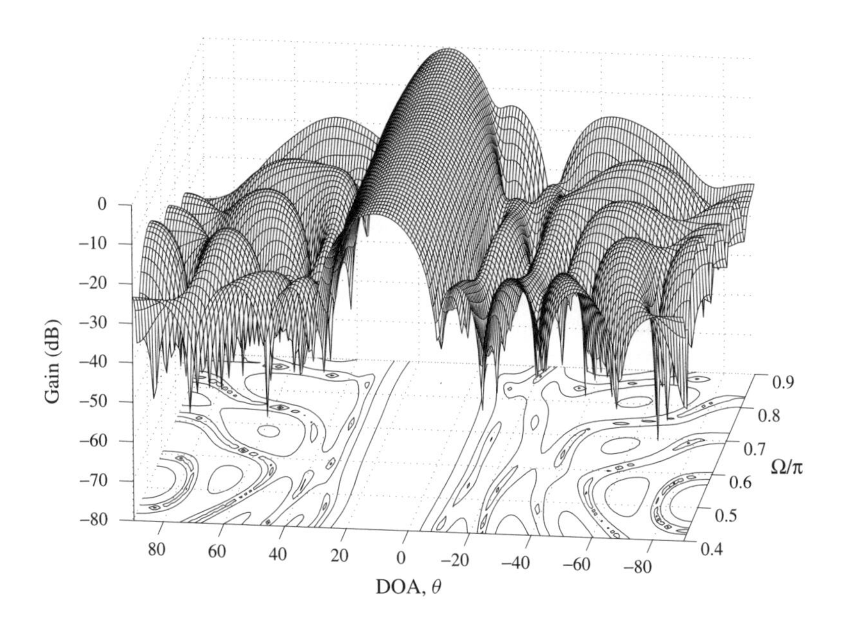

Figure 4.10 The resulting 3-D beam response for the design example using the standard eigenfilter approach as formulated in Equation (4.44)

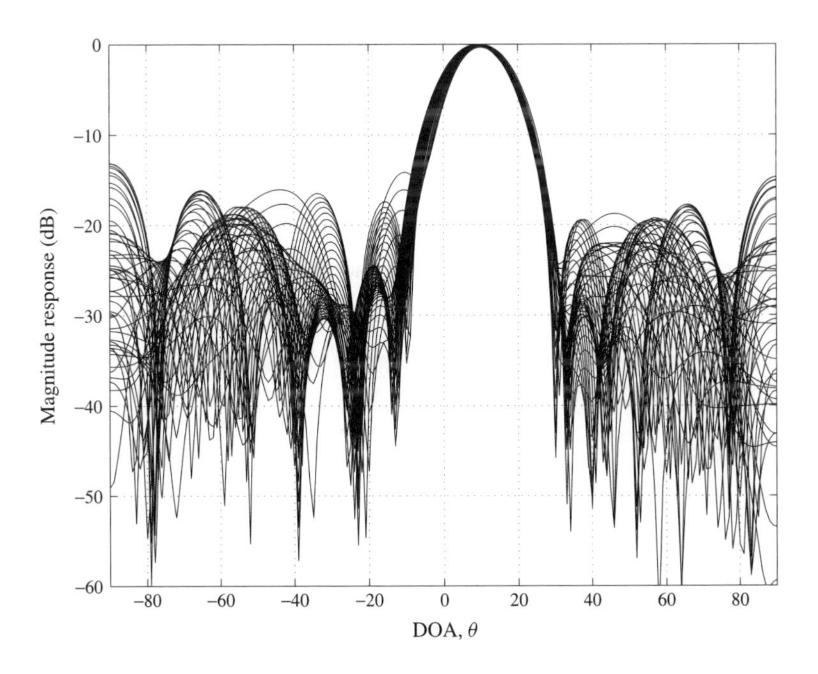

Figure 4.11 The resulting 2-D beam response for the design example shown in Figure 4.10

of interest Ω_{pb} discretized into $I_\Omega = 20$ points, to calculate the integration numerically, although it is possible to obtain the result without discretization, as suggested in Doclo and Moonen (2003b).

However, we find that the design is not consistent for different look direction θ_0 and different number of sensors M and TDL length J. For some cases the result is not acceptable. This can be seen by checking $\boldsymbol{G}_{ef}$ of the cost function in Equation (4.52), where, for $\theta_r = \theta_0$, the first part of $\boldsymbol{G}_{ef}$ is actually used for measuring the error between the response of the beamformer at the look direction and that at the reference point $(\Omega_r, \theta_r) = (\Omega_r, \theta_0)$. The term $e^{-j10(\Omega-\Omega_r)}$ is a compensation for the different phase shifts of the response at different frequencies. By this arrangement, the relative variation of the beamformer's response over the look direction for different frequencies is minimized. However, there is no mechanism to control the absolute response of the beamformer and even if the response at the look direction is zero, it is still possible for the cost function to reach a minimum. A remedy to this problem is to add a linear constraint for the beamformer to have a unity response at the reference point. However, this will lead to a solution in the form of Equation (4.46) and matrix inversion will be required.

Since there is no physical meaning for the norm of the weight vector $\mathbf{w}$ in the context of wideband beamforming, a new quadratic energy constraint was introduced to replace the norm constraint in Doclo and Moonen (2003b), given by:

$$\int_\Omega \int_\Theta |\mathbf{w}^H \mathbf{d}(\Omega, \theta)|^2 \mathrm{d}\Omega \mathrm{d}\theta = \mathbf{w}^H \boldsymbol{G}_c \mathbf{w} = 1 \tag{4.53}$$

with:

$$\boldsymbol{G}_c = \int_\Omega \int_\Theta \mathbf{d}(\Omega, \theta)\mathbf{d}^H(\Omega, \theta) \mathrm{d}\Omega \mathrm{d}\theta \tag{4.54}$$

where the integration is over the whole area of Ω and θ. For real-valued $\mathbf{w}$, similar to $\boldsymbol{G}_{ef}$, $\boldsymbol{G}_c$ can also be changed to:

$$\boldsymbol{G}_c = \int_\Omega \int_\Theta \boldsymbol{D}_R(\Omega, \theta) \mathrm{d}\Omega \mathrm{d}\theta \tag{4.55}$$

Then the eigenfilter design with this quadratic energy constraint becomes:

$$\min_{\mathbf{w}} \mathbf{w}^H \boldsymbol{G}_{ef} \mathbf{w} \text{ subject to } \mathbf{w}^H \boldsymbol{G}_c \mathbf{w} = 1 \tag{4.56}$$

or:

$$\min_{\mathbf{w}} \frac{\mathbf{w}^H \boldsymbol{G}_{ef} \mathbf{w}}{\mathbf{w}^H \boldsymbol{G}_c \mathbf{w}} \tag{4.57}$$

This is actually a generalized eigenvector problem (Golub and Van Loan, 1996) and the solution is the generalized eigenvector corresponding to the smallest generalized eigenvalue for the matrix pair $\boldsymbol{G}_{ef}$ and $\boldsymbol{G}_c$. We will discuss this class of methods based on the generalized eigenvector problem in the next section.

4.3.2 Maximum Energy

The maximum energy approach to the design of fixed wideband beamformers is based on the generalized eigenvector problem. Given the Hermitian matrix $\boldsymbol{R}$ and a positive definite matrix $\boldsymbol{B}$, the generalized eigenvector problem is described as finding a vector $\mathbf{w}$ which satisfies (Golub and Van Loan, 1996):

$$\boldsymbol{R}\mathbf{w} = \lambda \boldsymbol{B}\mathbf{w} \tag{4.58}$$

where λ is the corresponding generalized eigenvalue. As a result, the following generalised Rayleigh–Ritz ratio:

$$\frac{\mathbf{w}^H \boldsymbol{R}\mathbf{w}}{\mathbf{w}^H \boldsymbol{B}\mathbf{w}} \tag{4.59}$$

reaches its maximum when $\mathbf{w}$ is the generalized eigenvector corresponding to the maximum generalized eigenvalue of the matrix pair $\boldsymbol{R}$ and $\boldsymbol{B}$; it reaches its minimum when $\mathbf{w}$ is the generalized eigenvector corresponding to its minimum generalized eigenvalue.

Based on this formulation, the design of a fixed wideband beamformer can be achieved by maximizing the ratio of the beamformer's energy at the main lobe and the total energy of the beamformer over the whole area, as follows:

$$\max_{\mathbf{w}} \frac{\int_{\Omega_{pb}} \int_{\Theta_{ml}} |\mathbf{w}^H \mathbf{d}(\Omega, \theta)|^2 \mathrm{d}\Omega \mathrm{d}\theta}{\int_{\Omega} \int_{\Theta} |\mathbf{w}^H \mathbf{d}(\Omega, \theta)|^2 \mathrm{d}\Omega \mathrm{d}\theta} \tag{4.60}$$

where the numerator represents the energy of the beamformer at the mainlobe area over the frequency range of interest and the denominator is the total energy of the beamformer over the whole area given also by Equation (4.53). A modification could be made to the integration range Ω by replacing it with Ω_{pb} and we can also add the weighting function $v(\Omega, \theta)$ when we calculate the main lobe energy.

Using the previous notation $\boldsymbol{D}(\Omega, \theta) = \mathbf{d}(\Omega, \theta)\mathbf{d}^H(\Omega, \theta)$, Equation (4.60) changes to:

$$\max_{\mathbf{w}} \frac{\int_{\Omega_{pb}} \int_{\Theta_{ml}} \mathbf{w}^H \boldsymbol{D}(\Omega, \theta)\mathbf{w} \mathrm{d}\Omega \mathrm{d}\theta}{\int_{\Omega} \int_{\Theta} |\mathbf{w}^H \mathbf{w}^H \boldsymbol{D}(\Omega, \theta)\mathbf{w} \mathrm{d}\Omega \mathrm{d}\theta} = \frac{\mathbf{w}^H \boldsymbol{G}_{ml}\mathbf{w}}{\mathbf{w}^H \boldsymbol{G}_c \mathbf{w}} \tag{4.61}$$

where:

$$\boldsymbol{G}_{ml} = \int_{\Omega} \int_{\Theta_{ml}} \boldsymbol{D}(\Omega, \theta) \mathrm{d}\Omega \mathrm{d}\theta \tag{4.62}$$

$\boldsymbol{D}(\Omega, \theta)$ can be replaced by its real part $\boldsymbol{D}_R(\Omega, \theta)$ for a real-valued $\mathbf{w}$.

The solution is obtained by finding the generalized eigenvector corresponding to the maximum eigenvalue of $\boldsymbol{G}_{ml}$ and $\boldsymbol{G}_c$, which leads to a maximum energy ratio, hence the name maximum energy.

A variation to this approach is to maximize the ratio of the energy of the beamformer at the mainlobe and that at the sidelobe area, given by:

$$\max_{\mathbf{w}} \frac{\mathbf{w}^H \boldsymbol{G}_{ml} \mathbf{w}}{\int_{\Omega} \int_{\Theta_{sl}} \mathbf{w}^H \boldsymbol{D}(\Omega, \theta) \mathbf{w} \mathrm{d}\Omega \mathrm{d}\theta} = \frac{\mathbf{w}^H \boldsymbol{G}_{ml} \mathbf{w}}{\mathbf{w}^H \boldsymbol{G}_{sl} \mathbf{w}} \tag{4.63}$$

with:

$$\boldsymbol{G}_{sl} = \int_{\Omega} \int_{\Theta_{sl}} \boldsymbol{D}(\Omega, \theta) \mathrm{d}\Omega \mathrm{d}\theta \tag{4.64}$$

or:

$$\boldsymbol{G}_{sl} = \int_{\Omega} \int_{\Theta_{sl}} \boldsymbol{D}_R(\Omega, \theta) \mathrm{d}\Omega \mathrm{d}\theta \tag{4.65}$$

for a real-valued $\mathbf{w}$.

However, a problem with this approach is that there is no control on the exact value of the beamformer response over the main lobe at different frequencies. A maximum energy ratio does not guarantee a smooth consistent response over the main lobe for different frequencies. This can be seen by a design example shown in Figure 4.12, with the same parameters as in Figure 4.6 except for the main lobe area Θ_{ml} which is from $-5°$ to $5°$ for calculating $\boldsymbol{G}_{ml}$.

This problem can be solved by adding linear constraints in the form of $\boldsymbol{C}^H \mathbf{w} = \mathbf{f}$ to the design so that a smooth consistent response can be achieved at the main lobe. The

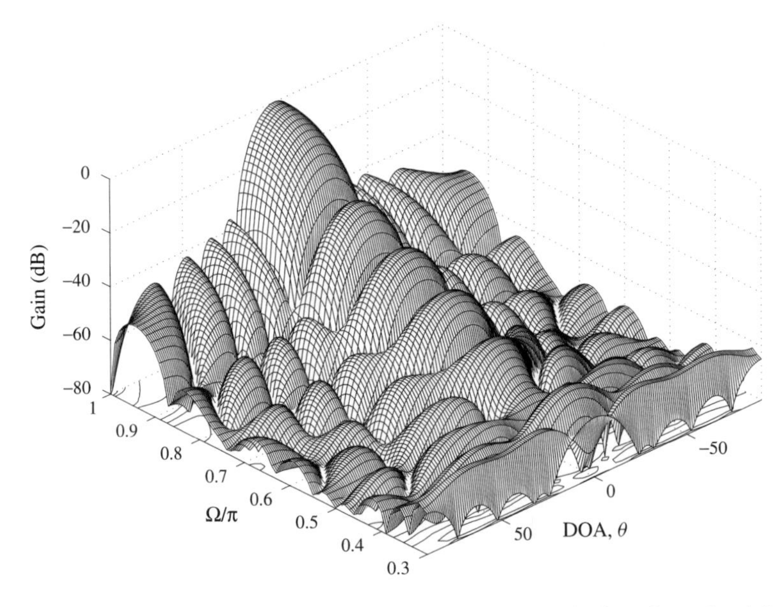

Figure 4.12 A design example using the maximum energy method as formulated in Equation (4.63)

modified problem is:

$$\max_{\mathbf{w}} \frac{\mathbf{w}^H \boldsymbol{G}_{ml} \mathbf{w}}{\mathbf{w}^H \boldsymbol{G}_{sl} \mathbf{w}} \quad \text{subject to} \quad \boldsymbol{C}^H \mathbf{w} = \mathbf{f} \tag{4.66}$$

which can be transformed into the following form (Doclo and Moonen, 2003b):

$$\max_{\hat{\mathbf{w}}} \frac{\hat{\mathbf{w}}^H \hat{\boldsymbol{G}}_{ml} \hat{\mathbf{w}}}{\hat{\mathbf{w}}^H \hat{\boldsymbol{G}}_{sl} \hat{\mathbf{w}}} \quad \text{subject to} \quad \hat{\boldsymbol{C}}^H \hat{\mathbf{w}} = \mathbf{0} \tag{4.67}$$

with:

$$\begin{aligned} \hat{\mathbf{w}} &= [\mathbf{w}^H, -1]^H, \hat{\boldsymbol{G}}_{ml} = \begin{bmatrix} \boldsymbol{G}_{ml} & \mathbf{0} \\ \mathbf{0} & 0 \end{bmatrix} \\ \hat{\boldsymbol{G}}_{sl} &= \begin{bmatrix} \boldsymbol{G}_{sl} & \mathbf{0} \\ \mathbf{0} & 0 \end{bmatrix}, \hat{\boldsymbol{C}} = [\boldsymbol{C}^H \ \mathbf{f}]^H \end{aligned} \tag{4.68}$$

Similar to Equation (4.50), Equation (4.67) can be transformed into an unconstrained one by introducing the matrix $\tilde{\boldsymbol{C}}$, whose columns are the bases of the null space of $\hat{\boldsymbol{C}}$. Then we have $\hat{\mathbf{w}} = \tilde{\boldsymbol{C}}\tilde{\mathbf{w}}$, which leads to the following standard generalized eigenvector problem without constraints:

$$\max_{\tilde{\mathbf{w}}} \frac{\tilde{\mathbf{w}}^H \tilde{\boldsymbol{C}}^H \hat{\boldsymbol{G}}_{ml} \tilde{\boldsymbol{C}} \tilde{\mathbf{w}}}{\tilde{\mathbf{w}}^H \tilde{\boldsymbol{C}}^H \hat{\boldsymbol{G}}_{sl} \tilde{\boldsymbol{C}} \tilde{\mathbf{w}}} \tag{4.69}$$

The optimum $\tilde{\mathbf{w}}$ is the generalized eigenvector corresponding to the largest generalized eigenvalue of the matrix pair $\tilde{\boldsymbol{C}}^H \hat{\boldsymbol{G}}_{ml} \tilde{\boldsymbol{C}}$ and $\tilde{\boldsymbol{C}}^H \hat{\boldsymbol{G}}_{sl} \tilde{\boldsymbol{C}}$. Note the last element of $\hat{\mathbf{w}}$ obtained in this way needs to be scaled to -1 before we obtain $\mathbf{w}$ by taking the first MJ elements of $\hat{\mathbf{w}}$.

As an example, a broadside constraint is imposed on the design shown in Figure 4.12 with the desired response at $\theta_0 = 0^\circ$ given by $e^{-j10\Omega}$. The result is shown in Figure 4.13 with the two-dimensional one in Figure 4.14. Compared to Figure 4.12, the design result has been improved significantly.

4.3.3 *Total Least Squares*

The traditional least squares approach involves matrix inversion as indicated in Equation (4.26). However, it is possible to solve this least squares problem based on the generalized eigenvector idea by reformulating it using the total least squares method as follows (Doclo and Moonen, 2003b; Pei and Tseng, 2001):

$$\min_{\mathbf{w}} \frac{\int_{\Omega_{pb}} \int_{\Theta} v(\Omega, \theta) |P(\Omega, \theta) - P_d(\Omega, \theta)|^2 \mathrm{d}\Omega \mathrm{d}\theta}{\mathbf{w}^H \mathbf{w} + 1} \tag{4.70}$$

where the numerator is the standard least squares cost function as discussed in Section 4.2. A modification can be made by replacing $\mathbf{w}^H \mathbf{w}$ in the denominator by the total energy

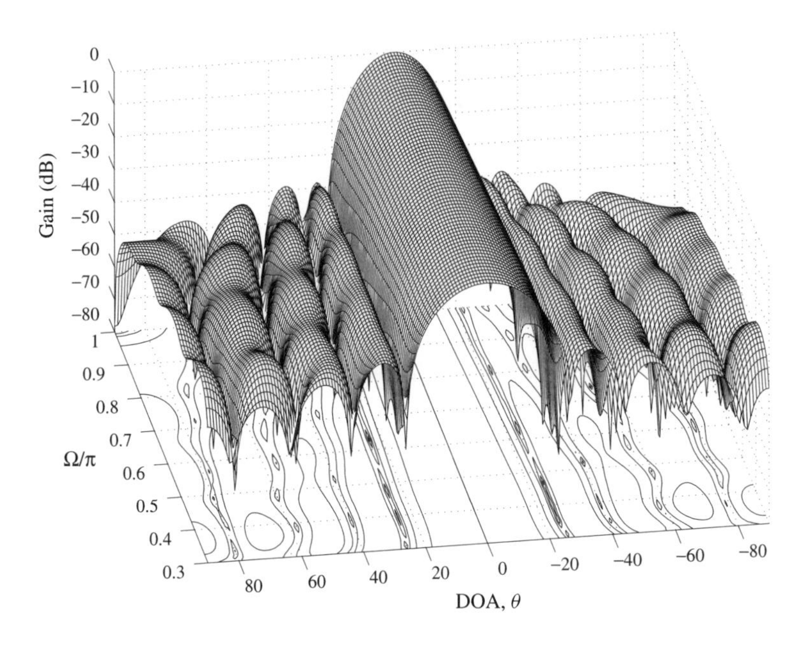

Figure 4.13 A design example using the maximum energy method with an additional broadside constraint as formulated in Equation (4.66)

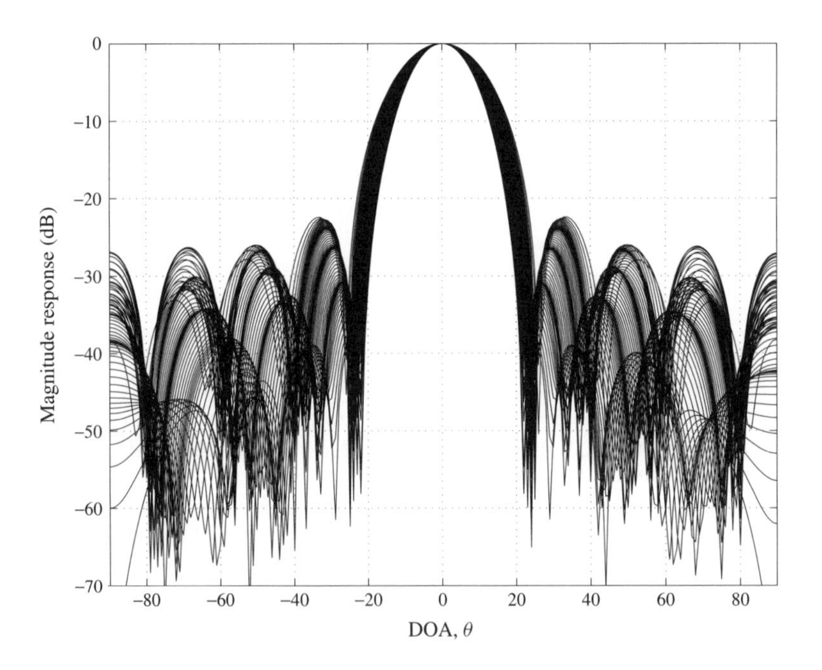

Figure 4.14 The two-dimensional version of Figure 4.13

$\mathbf{w}^H \boldsymbol{G}_c \mathbf{w}$ in Equation (4.53), as suggested in Doclo and Moonen (2003b). With this modification and the result in Equation (4.21), we have the following new problem:

$$\min_{\mathbf{w}} \quad \frac{\int_{\Omega_{pb}} \int_{\Theta} v(\Omega, \theta)|P(\Omega, \theta) - P_d(\Omega, \theta)|^2 \mathrm{d}\Omega \mathrm{d}\theta}{\mathbf{w}^H \boldsymbol{G}_c \mathbf{w} + 1}$$
$$= \frac{\mathbf{w}^H \boldsymbol{G}_{ls} \mathbf{w} - \mathbf{w}^H \bar{\mathbf{g}}_{ls} - \bar{\mathbf{g}}_{ls}^H \mathbf{w} + g_{ls}}{\mathbf{w}^H \boldsymbol{G}_c \mathbf{w} + 1} \tag{4.71}$$

With:

$$\hat{\mathbf{w}} = [\mathbf{w}^H, -1]^H \ , \ \boldsymbol{G}_{tls} = \begin{bmatrix} \boldsymbol{G}_{ls} & \bar{\mathbf{g}}_{ls} \\ \bar{\mathbf{g}}_{ls}^H & g_{ls} \end{bmatrix}$$
$$\hat{\boldsymbol{G}}_c = \begin{bmatrix} \boldsymbol{G}_c & \mathbf{0} \\ \mathbf{0} & 1 \end{bmatrix} \tag{4.72}$$

we have:

$$\min_{\hat{\mathbf{w}}} \frac{\hat{\mathbf{w}}^H \boldsymbol{G}_{tls} \hat{\mathbf{w}}}{\hat{\mathbf{w}}^H \hat{\boldsymbol{G}}_c \hat{\mathbf{w}}} \tag{4.73}$$

The optimum $\hat{\mathbf{w}}$ is the generalized eigenvector corresponding to the smallest generalized eigenvalue of the matrix pair $\boldsymbol{G}_{tls}$ and $\hat{\boldsymbol{G}}_c$. Similar to the maximum energy case with linear constraints, the last element of $\hat{\mathbf{w}}$ needs to be scaled to -1 before exporting the first MJ elements to form $\mathbf{w}$. A design example is shown in Figure 4.15 with the same parameters as in Figure 4.6.

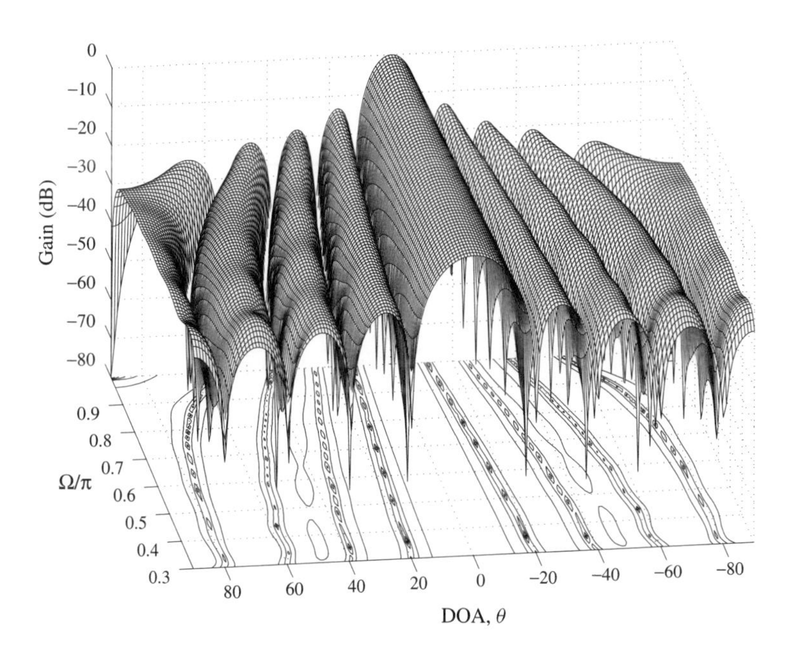

Figure 4.15 A design example based on the total least squares approach with the same parameters as in Figure 4.6

4.4 Summary

In this chapter the fixed wideband beamformer design problem has been studied. Three classes of methods were reviewed, including the iterative optimization method, the least squares method and the eigenfilter method.

The last two methods can provide a close-form solution to the design problem with low computational complexity. However, the first one is more robust and can deal with some design problems which the other two cannot, such as the minimax design problem. There exist many variations to each of the three methods and they have been introduced in the corresponding sections. A number of design examples were also provided to illustrate the performance of the corresponding design methods.

5

Frequency Invariant Beamforming

In Chapter 4, we have studied the design of fixed wideband beamformers. From the design results we can see that the beamwidth of the resulting beamformer normally increases with the decrease of signal frequency, which is due to the fact that for a fixed aperture, the spatial resolution of a beamformer is proportional to frequency.

To achieve a response independent of frequency, therefore with a constant beamwidth, a new class of wideband beamformers was proposed and then studied extensively in recent years. Since their responses are independent of frequency, they are often called frequency independent beamformers or frequency invariant beamformers (FIBs). Sometimes to emphasize its constant beamwidth property, they are also called constant beamwidth beamformers. In this book we will use the term frequency invariant beamformers to refer to this class of fixed wideband beamformers.

5.1 Introduction

In order to achieve a frequency independent response, many methods have been proposed in the past. Harmonic nesting is a widely used method in order to step towards frequency invariance, whereby for a number of frequency bands, different subarrays with appropriate apertures and sensor spacings are operated (Hixson and Au, 1970; Smith, 1970; Van Trees, 2002; Weiss *et al.*, 2002;). This method can be based on frequency bin processing (Hixson and Au, 1970; Smith, 1970; Van Trees, 2002) or a decomposition into octave bands by means of filter banks (Weiss *et al.*, 2002). Subsequently, each octave band or group of frequency bins lying within one octave draw their inputs from one specific subarray. While the resulting beampattern is octave-independent, the spatial resolution within each octave band is still dependent on frequency.

To attain invariance within each octave, Chou (1995) combines harmonic nesting and filter-and-sum beamforming together, whereby each element of a subarray is followed by an FIR filter whose response is determined by the desired beam pattern. Frequency bin dependent windowing of the array elements can lead to a constant beamwidth for the main beam (Van Trees, 2002), whereby the same method can be applied in the time domain with approPariately designed lowpass filters following each sensor (Goodwin and Elko, 1993). In both approaches (Goodwin and Elko, 1993; Van Trees, 2002), sensor elements close to the array's end positions are disemphasized at higher frequencies, thus yielding a constant beam width and near frequency independent beam pattern.

Wideband Beamforming Wei Liu and Stephan Weiss

Different from the above methods, an approach employing the asymptotic theory of unequally spaced arrays has been suggested in Doles and Benedict (1988), where the relationship between the beam pattern properties and array properties is derived and exploited for the broadband linear array design. Most of the above work is focused on linear arrays, and recently the design of frequency invariant beamformer for circular arrays has also been studied (Chan and Pun, 2002) and then extended to the case of concentric circular arrays (Chan and Chen, 2005, 2007; Chen *et al.*, 2007) and concentric spherical arrays (Chan and Chen, 2006; Chen and Chan, 2007).

Instead of focusing on one basic type of arrays, a systematic method has been proposed in Ward *et al.* (1995), which can be applied to one-dimensional, two-dimensional and three-dimensional arrays. In this new method, each element of the array is followed by its own primary filter and the outputs of these primary filters share a common secondary filter to form the final output. Although the design for a 1-D array is relatively simple because of the dilation property of the primary filters, for higher-dimensional arrays this property is not guaranteed, which makes the general design case too complicated and no design examples for a 2-D array were provided there. In the following references (Liu and Mandic, 2005; Liu and Weiss, 2004a,b, 2005, 2008b, 2009b; Liu *et al.*, 2005, 2007c; Sekiguchi and Karasawa, 2000), the design is achieved based on simple multi-dimensional inverse Fourier transforms by exploiting the relationship between the array's spatial and temporal parameters and its beam pattern. This approach can be applied to 1-D, 2-D and 3-D arrays, for both continuous and discrete sensors and signals.

In Trucco *et al.* (2006), the case when the array aperture is shorter than the involved signal wavelength was studied, while in Parra (2006) a decoupling approach to separate the frequency response of the array from its spatial response was proposed for an arbitrary array geometry. Based on the design methods discussed in Chapter 4, we may optimize the response of the array directly with respect to its coefficients (Duan *et al.*, 2008; Yan *et al.*, 2007; Zhao *et al.*, 2008, 2009b,c,d). The key is to introduce an element in the design to control the frequency invariance property during the optimization.

In the following sections, we will first review the design based on multi-dimensional inverse Fourier transforms in Section 5.2, with its subband implementation in Section 5.3; the design based on the phase-mode transformation for circular arrays is presented in Section 5.4 and the direct optimization approach using the methods of Chapter 4 is introduced in Section 5.5; an application of the frequency invariant beamformers to adaptive beamforming is studied in Section 5.6.

5.2 Design Based on Multi-Dimensional Inverse Fourier Transform

We first address the case of continuous sensors and signals in Section 5.2.1 and then extend the approach to the discrete case in Section 5.2.2. Design examples are provided in Section 5.2.3 and a further generalization to the design is presented in Section 5.2.4.

5.2.1 Continuous Sensor and Signals

5.2.1.1 One-Dimensional Array

Figure 5.1 shows a one-dimensional continuous sensor array aligned with the x-axis, where a plane wave with angular frequency ω illuminates the array from an angle θ

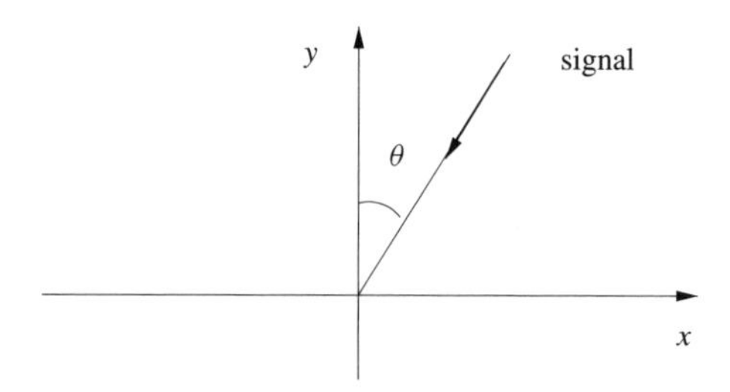

Figure 5.1 A continuous sensor array aligned with the x-axis

measured from the broadside. The response of this linear array with respect to frequency and angle of arrival is given by:

$$P(\omega, \theta) = \int_{-\infty}^{\infty} \mathrm{e}^{-j\frac{\omega \sin\theta}{c}x} D(x, \omega)\mathrm{d}x \tag{5.1}$$

where c is the propagation speed and $D(x, \omega)$ is the frequency response of the sensor at location x with respect to the angular frequency ω. Then in general $P(\omega, \theta)$ is a function of both ω and θ, while for a frequency invariant beamformer the response $P(\omega, \theta)$ is required to be independent of ω.

5.2.1.2 Design Idea

Suppose that the frequency response $D(x, \omega)$ is derived from the Fourier transform of a function $d(x, t)$, namely:

$$D(x, \omega) = \int_{-\infty}^{+\infty} d(x, t)\mathrm{e}^{-j\omega t}\mathrm{d}t \tag{5.2}$$

then for $P(\omega, \theta)$ we have:

$$P(\omega, \theta) = \iint_{-\infty}^{+\infty} d(x, t)\mathrm{e}^{-j\frac{\omega \sin\theta}{c}x}\mathrm{e}^{-j\omega t}\mathrm{d}x\mathrm{d}t \tag{5.3}$$

Therefore, given a desired response $P(\omega, \theta)$, the required time domain response $d(x, t)$ for the sensor located at the position x can be calculated by an inverse transformation corresponding to Equation (5.3). Note that the resulting $d(x, t)$ could in general be non-causal, although a causal approximate response can be attained by suitably translating and truncating $d(x, t)$.

In order to simplify the inversion from $P(\omega, \theta)$ to $d(x, t)$ and later to achieve simple rules for frequency invariance, we want to express $P(\omega, \theta)$ as a two-dimensional Fourier transform of $d(x, t)$. Note that by introducing the substitutions $\omega_1 = (\omega \sin\theta)/c$ and $\omega_2 = \omega$ into Equation (5.3), we obtain:

$$P(\omega_1, \omega_2) = \iint_{-\infty}^{+\infty} d(x, t)\mathrm{e}^{-j\omega_1 x}\mathrm{e}^{-j\omega_2 t}\mathrm{d}x\mathrm{d}t \tag{5.4}$$

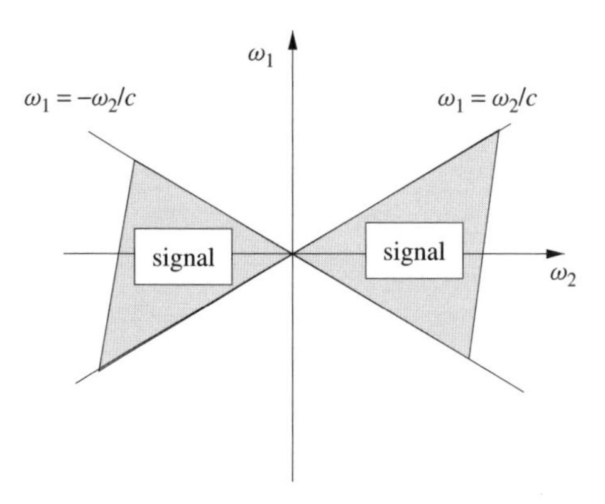

Figure 5.2 The possible location of the spatio-temporal spectrum of the impinging signal on the (ω_1, ω_2) plane

Thus from Equation (5.4) it is clear that the beam pattern of a continuous linear array can be obtained by first applying a 2-D Fourier transform to $d(x, t)$ and then re-substituting $\omega_1 = (\omega \sin\theta)/c$ and $\omega_2 = \omega$.

The spatio-temporal spectrum of a signal impinging from an angle θ is located on the line $\omega_1 = (\omega_2 \sin\theta)/c$ of the (ω_1, ω_2) plane. Due to the range limitation $\sin\theta \in [-1\ 1]$, the relation $|\omega_1/\omega_2| \leq 1/c$ emerges. Thus, Figure 5.2 shows the possible location of the spatio-temporal spectrum of any impinging signal. With respect to the division by ω_2 above, note that the case of $\omega_2 = \omega = 0$ can be excluded from our analysis, since for DC signals no spatial discrimination can be achieved by a sensor array.

For the beam response to be frequency invariant, the 2-D Fourier transform $P(\omega_1, \omega_2)$ must be a function of only θ, or more precisely $\sin\theta$. Let $F(\sin\theta)$ be such a frequency invariant beam response. In order to match $F(\sin\theta)$, the function $P(\omega_1, \omega_2)$ must, after re-substituting $\omega_1 = (\omega \sin\theta)/c$ and $\omega_2 = \omega$, be identical to $F(\sin\theta)$. In order to achieve this, the variables ω_1 and ω_2 must obey a specific dependency in the expression of $P(\omega_1, \omega_2)$ for ω to disappear. We note that if $P(\omega_1, \omega_2)$ depends on ω_1 and ω_2, such that it can be written as $P(c(\omega_1)/\omega_2)$, then after the re-substitutions, $c(\omega_1/\omega_2)$ will change to:

$$c\frac{\omega_1}{\omega_2} = c\frac{\omega \sin\theta}{c\omega} = \sin\theta \tag{5.5}$$

thus eliminating any dependency on ω. Therefore, we can set $P(\omega_1, \omega_2) = F((c\omega_1)/\omega_2)$ for a desired frequency invariant response $F(\sin\theta)$. Due to the specific relationship between ω_1 and ω_2, the resulting $P(\omega_1, \omega_2)$ will represent the 2-D Fourier transform of a frequency invariant beamformer's coefficients.

5.2.1.3 Design Procedure

With the above analysis, given a desired frequency invariant response $F(\sin\theta)$, as shown in Figure 5.3 as an example, which may be obtained by a narrowband beamformer design method, we proceed to a response $P(\omega_1, \omega_2)$ by the above substitutions. Thereafter, an

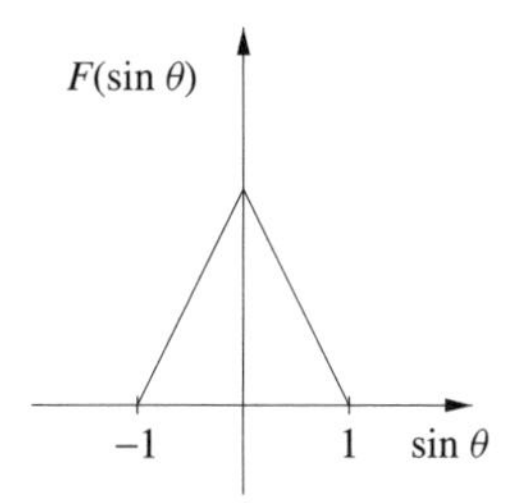

Figure 5.3 An example for a frequency invariant beam response

inverse 2-D Fourier transform yields the time domain parameters of the desired beamformer. The steps taken in this design are outlined below.

Step 1. Given a desired frequency invariant beam response $F(\sin\theta)$ and the substitution $\sin\theta = c(\omega_1)/\omega_2$, a 2-D response function $\hat{P}(\omega_1, \omega_2)$ is obtained:

$$\hat{P}(\omega_1, \omega_2) = \begin{cases} F(\omega_1 c/\omega_2) & \text{for } |c\omega_1|/|\omega_2| \leq 1 \\ a(\omega_1, \omega_2) & \text{for } |c\omega_1|/|\omega_2| > 1 \\ a(0, 0) & \text{for } \omega_2 = \omega_1 = 0 \end{cases} \tag{5.6}$$

where $a(\omega_1, \omega_2)$ is an arbitrary function to define values for $\hat{P}(\omega_1, \omega_2)$ where a beam pattern does not exist according to Figure 5.2. The choice of $a(\omega_1, \omega_2)$ therefore does not influence the desired beam pattern, but is likely to affect the design result near the boundary $|c(\omega_1)/\omega_2| = 1$.

Step 2. Suppose the maximum frequency of any signal recorded by the array is limited, i.e. $\omega \leq \omega_{1-D}$. Then the spatio-temporal spectrum of the impinging signal on the (ω_1, ω_2) plane changes to Figure 5.4. We apply the following modification to

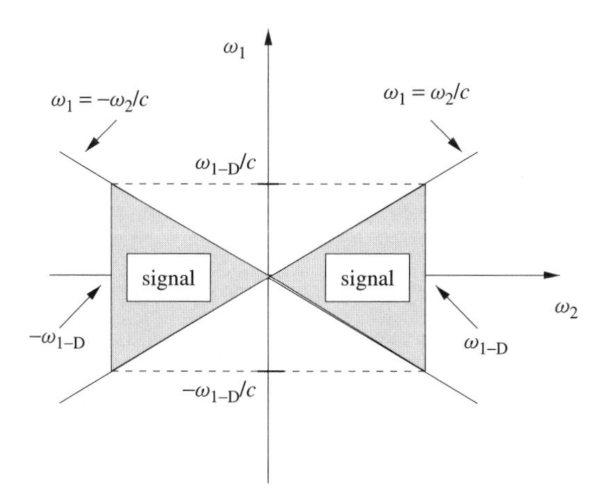

Figure 5.4 The spatio-temporal spectrum of the impinging signal on the (ω_1, ω_2) plane with the limit $\omega \leq \omega_{1-D}$

$\hat{P}(\omega_1, \omega_2)$ to yield $P(\omega_1, \omega_2)$:

$$P(\omega_1, \omega_2) = \begin{cases} 0 & \text{for } |\omega_2| > \omega_{1-\mathrm{D}} \ \vee \ |\omega_1| > \omega_{1-\mathrm{D}}/c \\ \hat{P}(\omega_1, \omega_2) & \text{otherwise} \end{cases} \tag{5.7}$$

Since $\lim_{\omega_1, \omega_2 \to \infty} |P(\omega_1, \omega_2)| = 0$, $P(\omega_1, \omega_2)$ is absolutely integrable and its inverse Fourier transform exists. Figure 5.5 shows the response of $P(\omega_1, \omega_2)$ for the example of Figure 5.3 with $a(\omega_1, \omega_2) = 0 \ \forall \omega_1, \omega_2$.

Step 3. The desired response $d(x, t)$ is obtained by the 2-D inverse Fourier transform of $P(\omega_1, \omega_2)$, which may be analytically difficult. Hence numerical methods may have to be employed to perform the inverse transform.

The approach outlined above can be applied to any desired frequency invariant beam response $F(\sin\theta)$.

5.2.1.4 Off-Broadside Case

If the main beam should point towards a direction θ_0 off the broadside, i.e. $\theta_0 \neq 0$, then this can be taken into account by a suitably defined $F(\sin\theta)$ and following the steps 1–3 above.

Alternatively, the off-broadside case can be based on a template $F(\sin\theta)$ pointing towards broadside, in combination with either electronic pre-steering or a modification of the design approach, with the latter one outlined below.

Let $F(\sin\theta)$ be the desired response with a broadside main beam. Note that the response of the beamformer $F(\sin\theta)$ to a signal from the broadside is $F(0)$. To design a beam

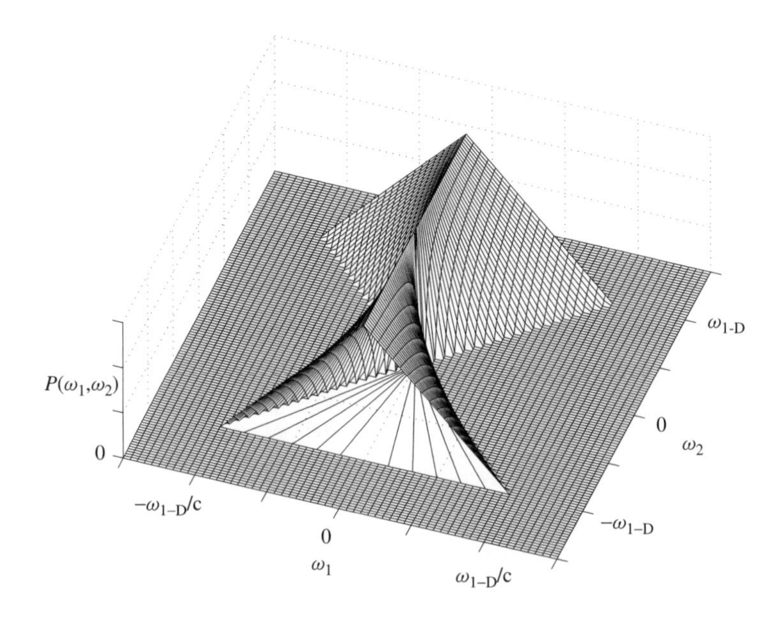

Figure 5.5 The response of $P(\omega_1, \omega_2)$ for the example of Figure 5.3

pointing to the direction $\theta_0 \neq 0$, we use the substitution $\sin\theta = (c\omega_1)/\omega_2 - \sin\theta_0$ in Step 1. For a signal from direction θ_0, we have $(c\omega_1)/\omega_2 = \sin\theta_0$. Thus:

$$\hat{P}(\omega_1, \omega_2) = F\left(\frac{c\omega_1}{\omega_2} - \sin\theta_0\right) \quad \text{for } \omega \neq 0 \tag{5.8}$$

will assume the response value $F(0)$ for the signal from the direction θ_0. However a difference to the previously introduced method arises, since the range of $(c\omega_1)/\omega_2 - \sin\theta_0$ is shifted by $\sin\theta_0$ for $-\pi/2 \leq \theta_0 \leq \pi/2$. To keep $\hat{P}(\omega_1, \omega_2)$ valid, we need to consider $F(\hat{\omega})$ as a periodic function with a period of 2, where $\hat{\omega}$ is a general variable replacing $\sin\theta$ in $F(\sin\theta)$.

5.2.1.5 Two-Dimensional Array

Figure 5.6 shows a 2-D continuous sensor array on the (x, y) plane. The response of this continuous array to a signal with angular frequency ω from the direction (θ, ϕ) is given by:

$$P(\omega, \theta, \phi) = \iint_{-\infty}^{+\infty} e^{-j\frac{\omega}{c}(x\sin\theta\cos\phi + y\sin\theta\sin\phi)} D(x, y, \omega) \mathrm{d}x\mathrm{d}y \tag{5.9}$$

where $D(x, y, \omega)$ is the frequency response of the sensor at the point (x, y). Similarly, using the inverse Fourier transform $d(x, y, t)$ of $D(x, y, \omega)$, we have:

$$P(\omega, \theta, \phi) = \iiint_{-\infty}^{+\infty} d(x, y, t) e^{-j\left(\frac{\omega\sin\theta\cos\phi}{c}\right)x} e^{-j\left(\frac{\omega\sin\theta\sin\phi}{c}\right)y} e^{-j\omega t} \mathrm{d}x\mathrm{d}y\mathrm{d}t \tag{5.10}$$

The substitutions $\omega_1 = (\omega\sin\theta\cos\phi)/c$, $\omega_2 = (\omega\sin\theta\sin\phi)/c$ and $\omega_3 = \omega$ into Equation (5.10) result in:

$$P(\omega_1, \omega_2, \omega_3) = \iiint_{-\infty}^{+\infty} d(x, y, t) e^{-j\omega_1 x} e^{-j\omega_2 y} e^{-j\omega_3 t} \mathrm{d}x\mathrm{d}y\mathrm{d}t \tag{5.11}$$

i.e. $P(\omega_1, \omega_2, \omega_3)$ is the 3-D Fourier transform of $d(x, y, t)$. In this case, the spatio-temporal spectrum of the impinging signal is located on the lines $(\omega_1 c)/\omega_3 = \sin\theta\cos\phi$

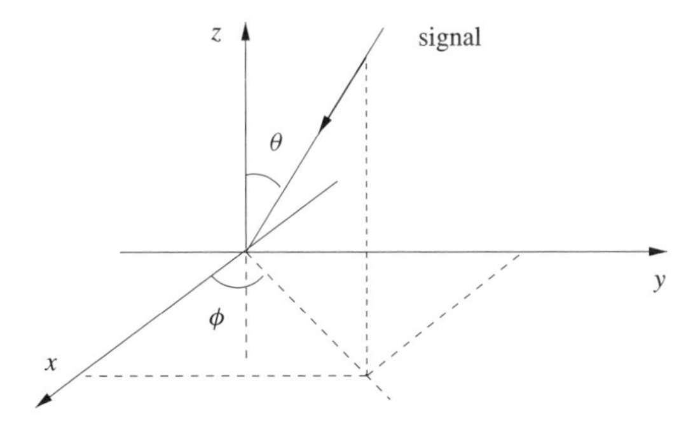

Figure 5.6 A continuous 2-D sensor array, where the signal impinges from the direction (θ, ϕ)

and $(\omega_2 c)/\omega_3 = \sin\theta \sin\phi$. If this Fourier transform can be expressed as $P(\omega_1, \omega_2, \omega_3) = F((\omega_1 c)/\omega_3, (\omega_2 c)/\omega_3)$, then the resulting function $F(\sin\theta\cos\phi, \sin\theta\sin\phi)$ will represent a frequency invariant response.

Now suppose $F(\sin\theta\cos\phi, \sin\theta\sin\phi)$ is the desired frequency invariant beam response. Analogous to the 1-D design in Section 5.2.1, the 2-D frequency invariant beamformer can be designed as follows.

Step 1. With the substitutions $\sin\theta\cos\phi = (\omega_1 c)/\omega_3$ and $\sin\theta\sin\phi = (\omega_2 c)/\omega_3$, we have:

$$\hat{P}(\omega_1, \omega_2, \omega_3) = \begin{cases} F\left(c\omega_1/\omega_3, c\omega_2/\omega_3\right), & \text{for } c\sqrt{\omega_1^2 + \omega_2^2}/|\omega_3| \le 1 \\ a(\omega_1, \omega_2, \omega_3) & \text{for } c\sqrt{\omega_1^2 + \omega_2^2}/|\omega_3| > 1 \\ a(0, 0, 0) & \text{for } \omega_1 = \omega_2 = \omega_3 = 0 \end{cases} \tag{5.12}$$

For the case $\omega = \omega_1 = \omega_2 = \omega_3 = 0$, or $(c\sqrt{\omega_1^2 + \omega_2^2})/|\omega_3| > 1$, $\hat{P}(\omega_1, \omega_2, \omega_3)$ is treated in the same way as in Step 1 in Section 5.2.1 by assigning an arbitrary function $a(\omega_1, \omega_2, \omega_3)$ which is outside the domain of a beam pattern and can be used to alleviate the design at the beam pattern's margins.

Step 2. Suppose the maximum frequency of the interested signal is limited by $\omega_{2-\mathrm{D}}$. We define:

$$P(\omega_1, \omega_2, \omega_3) = \begin{cases} 0 & \text{for } |\omega_3| > \omega_{2-\mathrm{D}} \vee |c\omega_2| > \omega_{2-\mathrm{D}} \vee |c\omega_1| > \omega_{2-\mathrm{D}} \\ \hat{P}(\omega_1, \omega_2, \omega_3) & \text{otherwise} \end{cases} \tag{5.13}$$

This modification to $\hat{P}(\omega_1, \omega_2, \omega_3)$ has the same effect as in the 1-D case, guaranteeing the existence of the inverse Fourier transform.

Step 3. To obtain the desired response $d(x, y, t)$, we apply a 3-D inverse Fourier transform to $P(\omega_1, \omega_2, \omega_3)$. As in the 1-D case, we may also need to translate the result along the t axis and truncate it in order to obtain a causal approximation.

For a beam pattern with its main beam pointing towards an arbitrary direction (θ_0, ϕ_0), in addition to the above method, we can use a similar alternative approach as in the 1-D case, and the substitutions $\sin\theta\cos\phi = (c\omega_1/\omega_3 - \sin\theta_0\cos\phi_0)$ and $\sin\theta\sin\phi = (c\omega_2/\omega_3 - \sin\theta_0\sin\phi_0)$ can be applied to a desired response $F(\sin\theta\cos\phi, \sin\theta\sin\phi)$ with a broadside main beam in step 1 of the above design method, while all the other design parameters remain unchanged.

5.2.1.6 Three-Dimensional Array

The response of a 3-D continuous sensor array, which in addition to Figure 5.6 also extends in the z-direction, is given by:

$$P(\omega, \theta, \phi) = \iiiint_{-\infty}^{+\infty} d(x, y, z, t) e^{-j(\frac{\omega\sin\theta\cos\phi}{c})x} e^{-j(\frac{\omega\sin\theta\sin\phi}{c})y} e^{-j(\frac{\omega\cos\theta}{c})z} e^{-j\omega t} \mathrm{d}x\mathrm{d}y\mathrm{d}z\mathrm{d}t \tag{5.14}$$

Here, the substitutions $\omega_1 = (\omega \sin\theta \cos\phi)/c$, $\omega_2 = (\omega \sin\theta \sin\phi)/c$, $\omega_3 = (\omega \cos\theta)/c$ and $\omega_4 = \omega$ lead to:

$$P(\omega_1, \omega_2, \omega_3, \omega_4) = \iiiint_{-\infty}^{+\infty} d(x, y, z, t)\, \mathrm{e}^{-j\omega_1 x}\mathrm{e}^{-j\omega_2 y}\mathrm{e}^{-j\omega_3 z}\mathrm{e}^{-j\omega_4 t}\mathrm{d}x\mathrm{d}y\mathrm{d}z\mathrm{d}t \quad (5.15)$$

By these substitutions, the spatio-temporal spectrum of the impinging signal falls onto the lines $c\omega_1/\omega_4 = \sin\theta\cos\phi$, $c\omega_2/\omega_4 = \sin\theta\sin\phi$ and $c\omega_3/\omega_4 = \cos\theta$, respectively. The design steps for a frequency invariant 3-D broadband array are analogous to the previous methods in Sections 5.2.1.3 and 5.2.1.5, and omitted here.

5.2.2 *Discrete Sensors and Signals*

Having established the theory for the continuous sensor and signals, the aim in this section is to extend the design to the discrete case. In general, this can be regarded as an approximation to the ideal response obtained for the continuous case, i.e. the array responses $d(x, t)$, $d(x, y, t)$ or $d(x, y, z, t)$ resulting from the proposed method in Section 5.2.1 need to be sampled in space and time, and truncated according to the sensor pattern and the temporal processing structure.

However, for a special class of sensor geometries, where all the sensors are positioned on lines parallel to the x-axis (1-D array), x- and y-axes (2-D array), or x-, y- and z-axes (3-D array) and uniformly spaced in each of the three directions, a corresponding theory can be developed based on the Fourier transform of a discrete series rather than on the sampling of a continuous function.

5.2.2.1 One-Dimensional Array

A uniformly spaced linear array with a sensor spacing of d_x is shown in Figure 5.7. The received signal by the mth sensor is sampled with a sampling period of T_s and then processed by a tapped delay-line with coefficients $d[m, n]$, $n = -\infty, \ldots, -1, 0, 1, \ldots, +\infty$. Corresponding to Equation (5.3) in the continuous case, we obtain its beam response $P(\omega, \theta)$ given by:

$$P(\omega, \theta) = \sum_{m,n=-\infty}^{+\infty} d[m, n]\mathrm{e}^{-jm\mu\Omega\sin\theta}\mathrm{e}^{-jn\Omega} \quad (5.16)$$

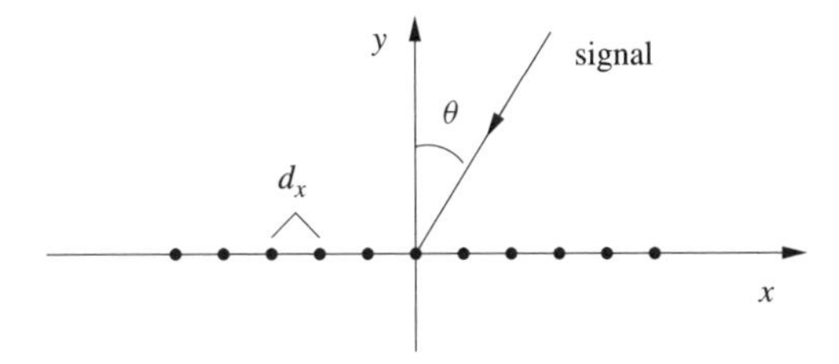

Figure 5.7 A uniformly spaced linear array with a sensor spacing of d_x, where the signal impinges from the direction θ

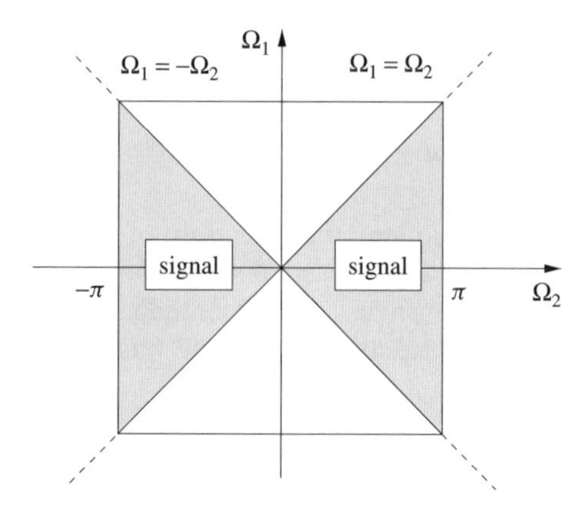

Figure 5.8 The possible location of the spatio-temporal spectrum of the impinging signal on the (Ω_1, Ω_2) plane

where $\mu = d_x/(cT_s)$, and $\Omega = \omega T_s$ is the normalized angular frequency. To avoid aliasing in both the spatial and temporal domains, we can choose $d_x = (\lambda_{\min})/2$ and $T_s = c/(2\lambda_{\min})$, where $\lambda_{\min}$ is the wavelength corresponding to the maximum frequency of interest. As a result, we have $\mu = 1$. With the substitutions $\Omega_1 = \mu\Omega \sin\theta$ and $\Omega_2 = \Omega$ into Equation (5.16), we have:

$$P(\Omega_1, \Omega_2) = \sum_{m,n=-\infty}^{+\infty} d[m, n]\, \mathrm{e}^{-jm\Omega_1}\, \mathrm{e}^{-jn\Omega_2} \tag{5.17}$$

With $\mu = 1$, the spatio-temporal spectrum of the impinging signal lies on the line $\Omega_1 = \Omega_2 \sin\theta$, which for a variable angle of arrival θ covers the area between the two lines $\Omega_1 = \Omega_2$ and $\Omega_1 = -\Omega_2$ as shown in Figure 5.8. Again a method can be developed to obtain a frequency invariant beam response, comprising the steps below.

Step 1. Let the desired response be $F(\sin\theta)$. With the substitution $\sin\theta = \Omega_1/\Omega_2$, for $(\Omega_1, \Omega_2) \in [-\pi\ \pi)$ we have:

$$P(\Omega_1, \Omega_2) = \begin{cases} F(\Omega_1/\Omega_2) & \text{for } |\Omega_1|/|\Omega_2| \leq 1 \\ A(\Omega_1, \Omega_2) & \text{for } |\Omega_1|/|\Omega_2| > 1 \\ A(0, 0) & \text{for } \Omega_1 = \Omega_2 = 0 \end{cases} \tag{5.18}$$

The function $A(\Omega_1, \Omega_2)$ can be selected in a similar way as $a(\omega_1, \omega_2)$ in Step 1 of Section 5.2.1 and it will not affect the beam pattern because no signal exists in this area according to Figure 5.8. Note that $P(\Omega_1, \Omega_2)$ is a function with a period of 2π. The response of $P(\Omega_1, \Omega_2)$ for the example of Figure 5.3 is shown in Figure 5.9 with $A(\Omega_1, \Omega_2) = 0$.

Step 2. Applying a 2-D inverse Fourier transform to $P(\Omega_1, \Omega_2)$ results in a $d[m, n]$ with infinite support. As in the continuous case, it is difficult to obtain a result analytically; therefore we can apply the 2-D inverse discrete Fourier transform

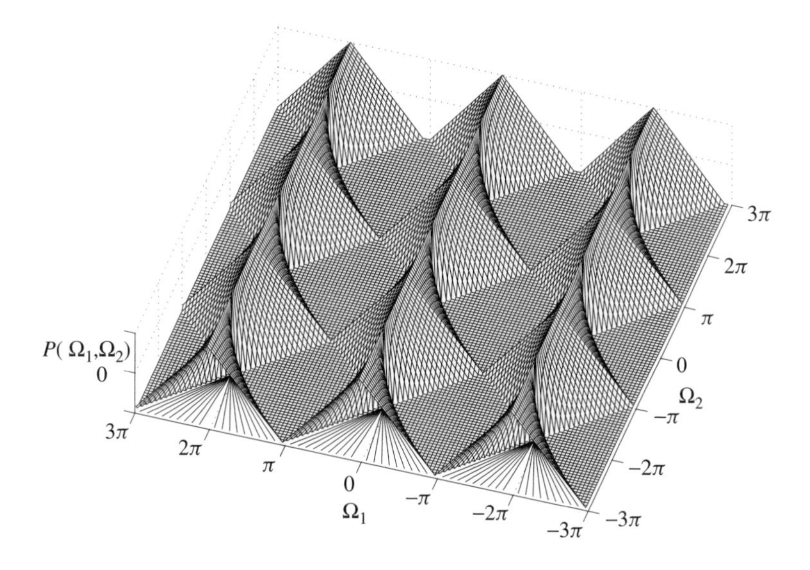

Figure 5.9 The response of $P(\Omega_1, \Omega_2)$ for the example of Figure 5.3

(DFT) as an approximation by sampling $P(\Omega_1, \Omega_2)$. After applying an inverse DFT, the resulting $d[m, n]$ may need to be delayed along the discrete time index n for reasons of causality and be truncated according to the number of sensors and the TDL length.

As in the continuous sensor and signal case, in this design the array's main beam in the desired response template $F(\sin\theta)$ can be oriented in any direction.

Alternatively, for a main beam pointing to an arbitrary direction θ_0, we can also design the beam pattern based on a template with the main beam towards the broadside first, and then either electronically steer the main beam towards the desired direction θ_0, or follow a slightly modified design approach.

For the latter, the substitution $\sin\theta = (\Omega_1/\Omega_2 - \sin\theta_0)$ is employed analogous to Section 5.2.1. Again, the template $F(\hat{\Omega})$ is required to be periodic with a period of 2, where $\hat{\Omega}$ is replacing $\sin\theta$ in $F(\sin\theta)$ as a general variable in order to cover the exceeded range in Equation (5.18).

5.2.2.2 Two-Dimensional Array

Figure 5.10 shows the structure of a planar array with sensor spacings of d_x and d_y respectively. With spatial indices l and m and the time index n, the 2-D array response is given by:

$$P(\omega, \theta, \phi) = \sum_{l,m,n=-\infty}^{+\infty} d[l, m, n]\, \mathrm{e}^{-jl\frac{\omega}{c}d_x \sin\theta\cos\phi} \mathrm{e}^{-jm\frac{\omega}{c}d_y \sin\theta\sin\phi} \mathrm{e}^{-jn\omega T_s} \tag{5.19}$$

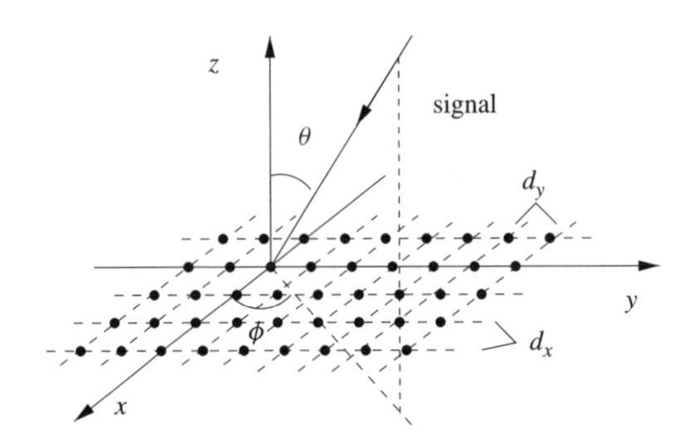

Figure 5.10 A uniformly spaced planar array with sensor spacings of d_x and d_y, respectively, where the signal impinges from the direction (θ, ϕ)

Assuming the same setup for spatial and temporal sampling as in Section 5.2.2 such that $d_x = d_y = (\lambda_{\min})/2 = cT_s$, we have:

$$P(\omega, \theta, \phi) = \sum_{l,m,n=-\infty}^{+\infty} d[l, m, n]\, \mathrm{e}^{-jl\Omega \sin\theta \cos\phi} \mathrm{e}^{-jm\Omega \sin\theta \sin\phi} \mathrm{e}^{-jn\Omega} \tag{5.20}$$

Substituting $\Omega_1 = \Omega \sin\theta \cos\phi$, $\Omega_2 = \Omega \sin\theta \sin\phi$ and $\Omega_3 = \Omega$ into Equation (5.20) gives:

$$P(\Omega_1, \Omega_2, \Omega_3) = \sum_{l,m,n=-\infty}^{+\infty} d[l, m, n] \mathrm{e}^{-jl\Omega_1} \mathrm{e}^{-jm\Omega_2} \mathrm{e}^{-jn\Omega_3} \tag{5.21}$$

The spatio-temporal spectrum of the impinging signal in the 2-D case lies on the lines $\Omega_1/\Omega_3 = \sin\theta \cos\phi$ and $\Omega_2/\Omega_3 = \sin\theta \sin\phi$, respectively. Based on the design of the 2-D continuous array, we can perform the 2-D discrete design as follows:

Step 1. Suppose $F(\sin\theta \cos\phi, \sin\theta \sin\phi)$ is the desired frequency invariant beam response. With the substitutions $\sin\theta \cos\phi = \Omega_1/\Omega_3$ and $\sin\theta \sin\phi = \Omega_2/\Omega_3$, we obtain $P(\Omega_1, \Omega_2, \Omega_3)$ defined over an interval of one period $\Omega_1, \Omega_2, \Omega_3 \in [-\pi; \pi)$ as:

$$P(\Omega_1, \Omega_2, \Omega_3) = \begin{cases} F((\Omega_1/\Omega_3), (\Omega_2/\Omega_3)) & \text{for } \sqrt{\Omega_1^2 + \Omega_2^2}/\Omega_3 \leq 1 \\ A(\Omega_1, \Omega_2, \Omega_3) & \text{for } \sqrt{\Omega_1^2 + \Omega_2^2}/\Omega_3 > 1 \\ A(0, 0, 0) & \text{for } \Omega_1 = \Omega_2 = \Omega_3 = 0 \end{cases} \tag{5.22}$$

For the case $\Omega_1 = \Omega_2 = \Omega_3 = 0$, or $(\sqrt{\Omega_1^2 + \Omega_2^2})/\Omega_3 > 1$, provisions similar to the previous designs, such as step 1 of the 1-D design case in Section 5.2.2, have to be made.

Step 2. Applying a 3-D inverse Fourier transform (or DFT as an approximation) to $P(\Omega_1, \Omega_2, \Omega_3)$ returns the desired response $d[l, m, n]$. For a causal and practical result, a truncation in the spatial l and m domains and the temporal n domain is necessary with a possible shift in time index n prior to truncation in order to ensure a good approximation with a causal response.

As in the 1-D case, this method can be applied to a desired beam pattern template with its main beam pointing towards any direction (θ_0, ϕ_0).

Alternatively, for an off-broadside main beam design, step 1 of the above algorithm can be modified by substituting the arguments of $F(\sin\theta\cos\phi, \sin\theta\sin\phi)$ in the broadside design by $(\Omega_1/\Omega_3 - \sin\theta_0\cos\phi_0)$ and $(\Omega_2/\Omega_3 - \sin\theta_0\sin\phi_0)$, respectively.

5.2.2.3 Three-Dimensional Array

With spatial indices k, l, m, and the TDL index n, the response of a 3-D array is given by

$$P(\omega, \theta, \phi) = \sum_{k,l,m,n=-\infty}^{+\infty} d[k, l, m, n]\, \mathrm{e}^{-jk\frac{\omega\sin\theta\cos\phi}{c}d_x} \mathrm{e}^{-jl\frac{\omega\sin\theta\sin\phi}{c}d_y} \mathrm{e}^{-ml\frac{\omega\cos\theta}{c}d_z} \mathrm{e}^{-jn\omega T_s} \tag{5.23}$$

By selecting the array parameters as in the 1-D and 2-D cases, we obtain $d_x = d_y = d_z = cT_s$ and $\Omega = \omega T_s$. The substitutions $\Omega_1 = \Omega\sin\theta\cos\phi$, $\Omega_2 = \Omega\sin\theta\sin\phi$, $\Omega_3 = \Omega\cos\theta$ and $\Omega_4 = \Omega$ applied to Equation (5.23) yield:

$$P(\Omega_1, \Omega_2, \Omega_3, \Omega_4) = \sum_{k,l,m,n=-\infty}^{+\infty} d[k, l, m, n]\, \mathrm{e}^{-jk\Omega_1}\mathrm{e}^{-jl\Omega_2}\mathrm{e}^{-jm\Omega_3}\mathrm{e}^{-jn\Omega_4} \tag{5.24}$$

Based on the above substitutions, a design approach for a 3-D array with frequency invariant response can be developed analogously to the previous designs, for which we omit the details here.

However, there are some other factors to be considered in the design. For a 3-D array, it is very likely that the sensors within the array will not be able to receive the signals properly from some specific directions as the view of those sensors has been blocked by the sensors at the outer layer. Similarly reflection and diffraction can be a serious problem in practical designs for a 3-D array. Simply ignoring these problems in the design may cause unaccounted degradation in the performance. But, when the size of each sensor is very small compared to their spacings, the 3-D design result for the ideal situation may provide a good approximation to the real problem or at least some useful guidance when we design a 3-D array.

5.2.3 Design Examples

5.2.3.1 n-D Design Examples for Broadside Look Direction

To show the effectiveness of the methods, we provide two simple examples. The first design is for a linear array with 16 sensors and a TDL length of 20; the second one is a planar array with 24×24 sensors and a TDL length of 24. Both arrays comprise of uniformly spaced sensors with $d_x = d_y = cT_s = (\lambda_{\min})/2$ and have their main beams directed towards the broadside. The desired frequency invariant responses are given by:

$$F_{1D}(\sin\theta) = \frac{1}{5}\sum_{m=-2}^{2} \mathrm{e}^{-jm\pi\sin\theta} \tag{5.25}$$

and:

$$F_{2D}(\sin\theta\cos\phi, \sin\theta\sin\phi) = \frac{1}{49}\sum_{l=-3}^{3}\sum_{m=-3}^{3} \mathrm{e}^{-jl\pi\sin\theta\cos\phi}\mathrm{e}^{-jm\pi\sin\theta\sin\phi} \tag{5.26}$$

respectively.

It is obvious that $F_{1D}(\sin\theta)$ is the response of a narrowband uniformly spaced linear array with a uniform weighting and $F_{2D}(\sin\theta\cos\phi, \sin\theta\sin\phi)$ is the response of a narrowband uniformly spaced planar array with a uniform weighting.

In the linear array design, we employ a 32×32 point 2-D inverse DFT on the resultant periodic function $P(\Omega_1, \Omega_2)$ and set $A(\Omega_1, \Omega_2) = 0$ for the areas outside the definition of the beam pattern in Equation (5.18). The array coefficients obtained by an inverse DFT are also of dimensions 32×32, which are truncated directly to the final result with dimensions 16×20. The value distribution of the resulting coefficients $d[m, n]$ is shown in Figure 5.11.

For the planar array, according to Equation (5.22), we first obtain the response $P(\Omega_1, \Omega_2, \Omega_3)$ and then set $A(\Omega_1, \Omega_2, \Omega_3) = 0$ for the area outside the beam pattern's domain, where no array signal exists. A $32 \times 32 \times 32$ 3-D inverse DFT is employed and the

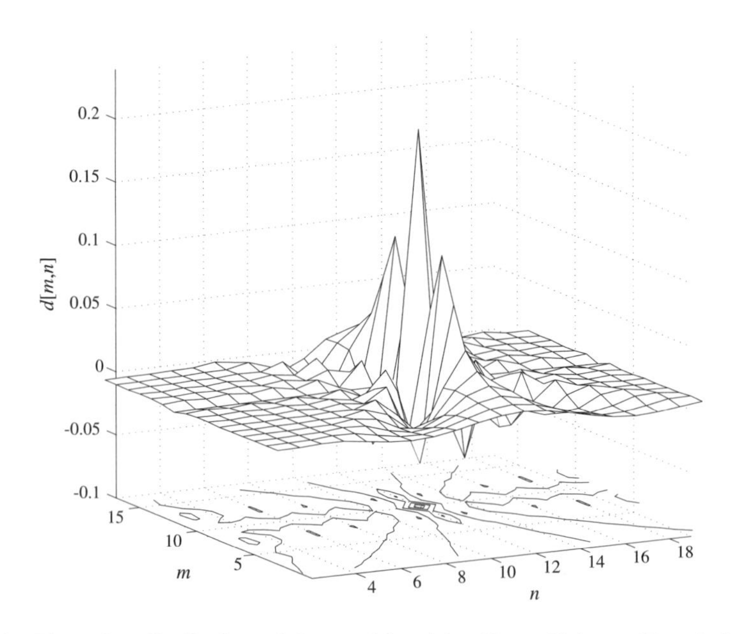

Figure 5.11 The value distribution of the resulting 16×20 coefficients $d[m, n]$ for the linear array design example with a broadside main beam

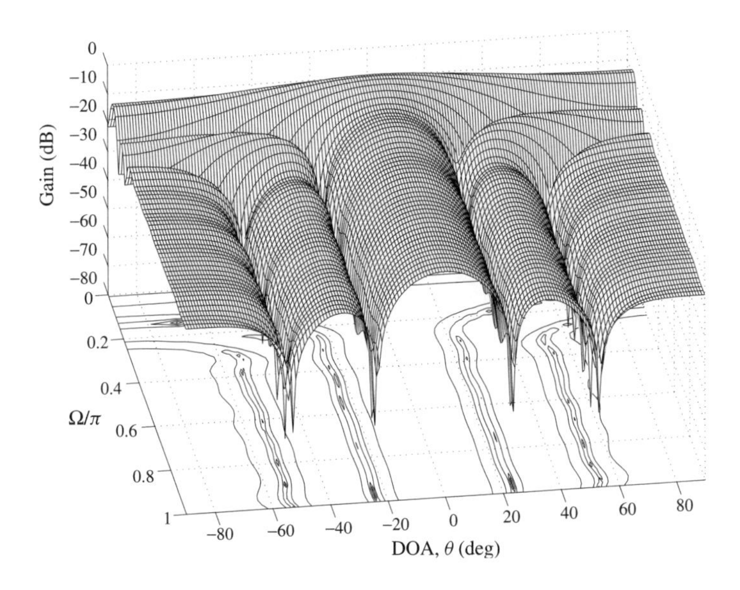

Figure 5.12 The resultant beam pattern for the linear array with a broadside main beam

$32 \times 32 \times 32$ temporal coefficients are subsequently truncated to the required array dimensions $24 \times 24 \times 24$.

The resultant beam pattern for the linear array is shown in Figure 5.12, which exhibits good frequency invariance for $\Omega > 0.3\pi$. For the four-dimensional beam pattern of the planar array, some exemplary snapshots are given below.

Figures 5.13 and 5.14 are the planar array's response to the frequencies $\Omega = 0.4\pi$ and $\Omega = 0.9\pi$, respectively. The frequency invariant property can be shown by slices of its

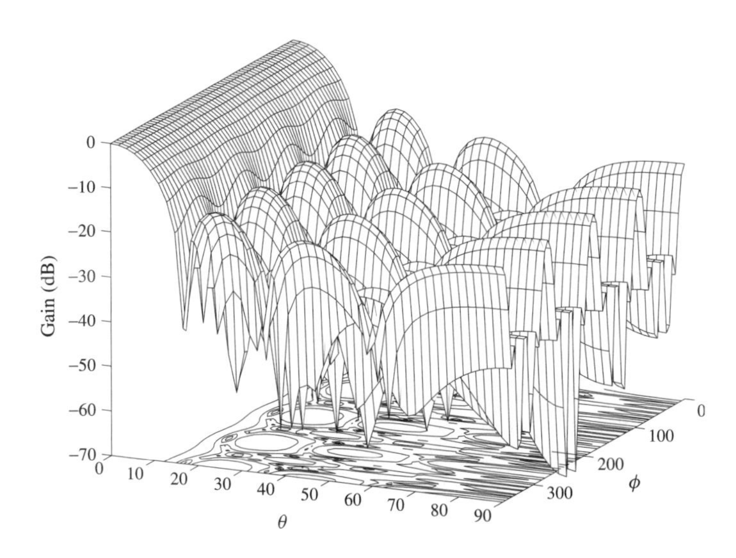

Figure 5.13 The resultant beam pattern of the planar array at $\Omega = 0.4\pi$

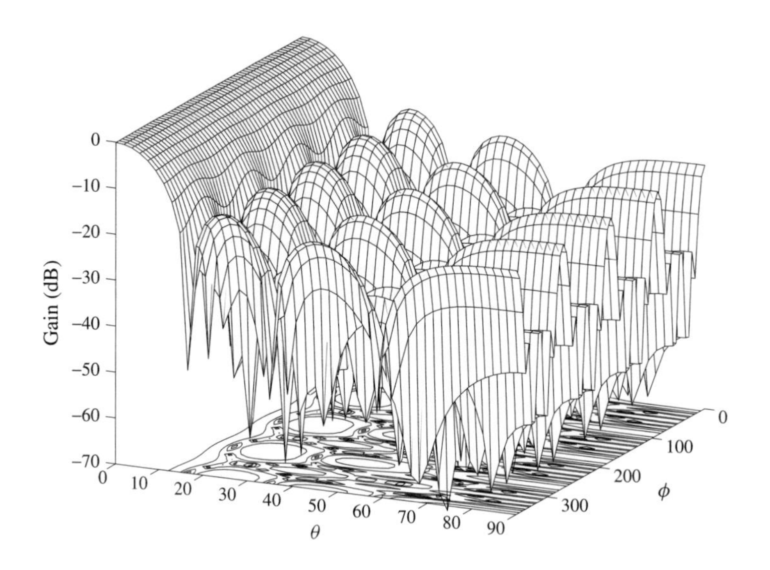

Figure 5.14 The resultant beam pattern of the planar array at $\Omega = 0.9\pi$

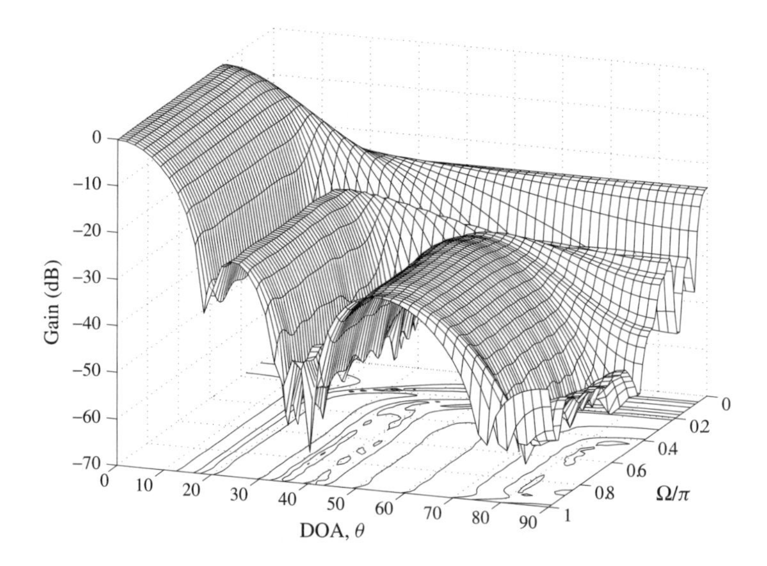

Figure 5.15 A slice of the beam pattern at $\phi = 60^{\circ}$

beam pattern at different values of ϕ, i.e. for each value of ϕ, we draw its response to signals varying over the parameters Ω and θ. Three representative slices are given in Figures 5.15, 5.16 and 5.17 with $\phi = 60^{\circ}$, 120°, and 180°, respectively, where for $\Omega > 0.3\pi$ the beam pattern is nearly frequency invariant.

From the beam patterns shown in Figures 5.12 to 5.17, it is noted that this method cannot achieve a good frequency invariance property for the lower frequencies. The reason for this phenomenon lies in the fact that the array aperture in terms of the signal wavelength

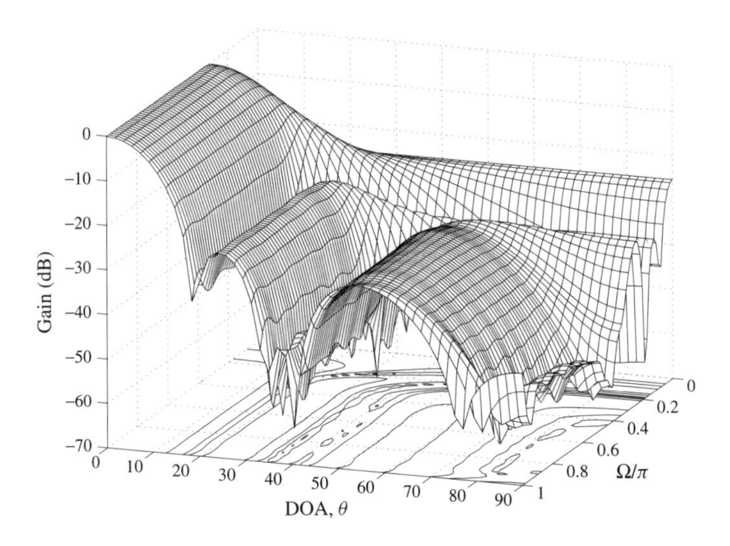

Figure 5.16 A slice of the beam pattern at $\phi = 120^\circ$

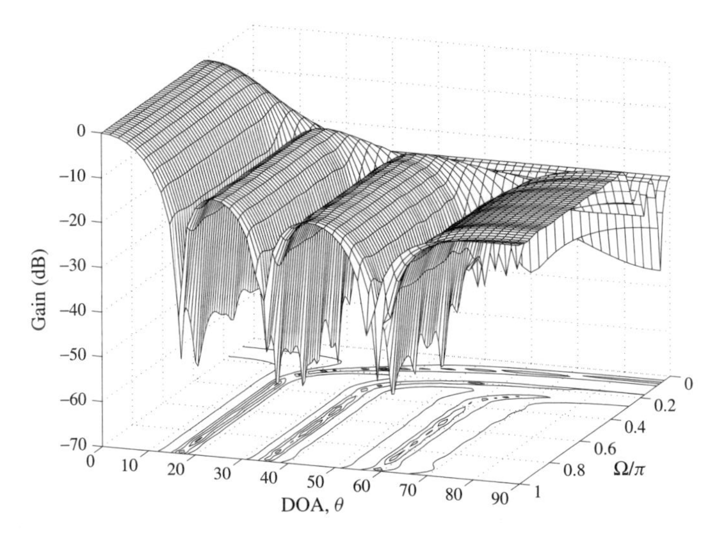

Figure 5.17 A slice of the beam pattern at $\phi = 180^\circ$

becomes smaller as the frequency is reduced. Mathematically, the density of the spatio-temporal spectrum of the input signal is much higher at small values of Ω_2 for the linear array, and Ω_3 for the planar array. The low frequency region cannot be represented as sufficiently as the higher frequency part when performing the inverse DFT. To show this, we consider the linear array as an example.

According to Figure 5.8, the incoming signals for a fixed angle θ lie on one of the dashed lines in Figure 5.18. When the inverse DFT is executed in order to calculate the temporal coefficients, we sample the (Ω_1, Ω_2) plane on the grid points given in

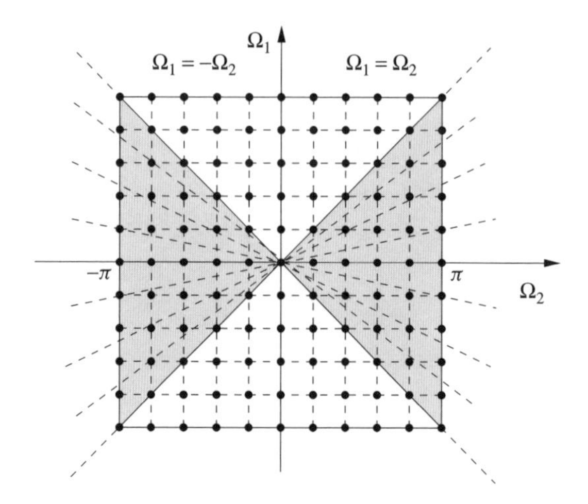

Figure 5.18 Sampling of the spatio-temporal spectrum of the signal for a linear array

Figure 5.18. Clearly, in the whole shaded area, where signals may exist, for smaller values of Ω_2, the number of acquired samples is much lower than that for larger values of Ω_2, which means the sampling resolution decreases with decreasing Ω_2. Therefore, with respect to frequency invariance, at low frequencies the array resulting from the introduced design cannot perform as well as at high frequencies. The DC component corresponds to the single point $(0, 0)$ on the (Ω_1, Ω_2) plane and hence the sampling density for the DC component is always the same and cannot be increased; that is why we cannot achieve a meaningful beam at DC, no matter how large the array is.

5.2.3.2 Example with Off-broadside Main Beam and its Problems

Based on the above design method, we can design an FIB with its main beam in an arbitrary direction. However, although the design result for a broadside main beam is very good, for the design with an off-broadside main beam, given the same number of array sensors and attached FIR coefficients, it is not as good as in the broadside main beam case.

Here we give an example with the same configuration as in Figure 5.12. The only difference is that now the desired main beam is pointing to the off-broadside direction $\theta = -30^\circ$, as specified in the following equation:

$$F_{1D}(\sin\theta) = \frac{1}{5}\sum_{m=-2}^{2} \mathrm{e}^{-jm\pi(\sin\theta - \sin\frac{\pi}{6})} \tag{5.27}$$

The design is obtained by sampling the pattern $P(\Omega_1, \Omega_2)$ by 32×32 points and then truncating them to the required dimensions of 16×20. The result is shown in Figure 5.19. Compared to Figure 5.12, the variation of the response over different frequencies is clearly visible. This problem can be explained by the considerable discontinuity of the periodic function $P(\Omega_1, \Omega_2)$ at $\Omega_2 = \cdots, -3\pi, -\pi, \pi, 3\pi, \cdots$, when the main beam is not pointing to the broadside, as shown in Figures 5.20 and 5.21. Because of the

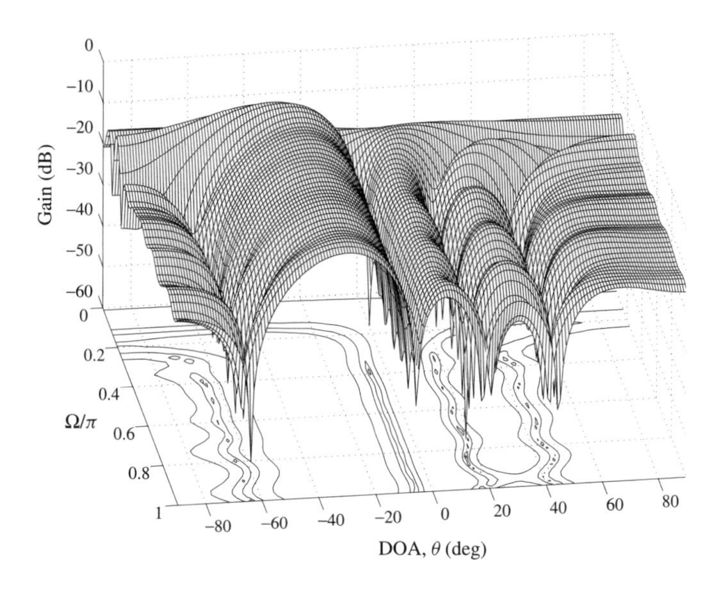

Figure 5.19 The resultant beam pattern for a linear array with its main beam at $\theta = -30°$

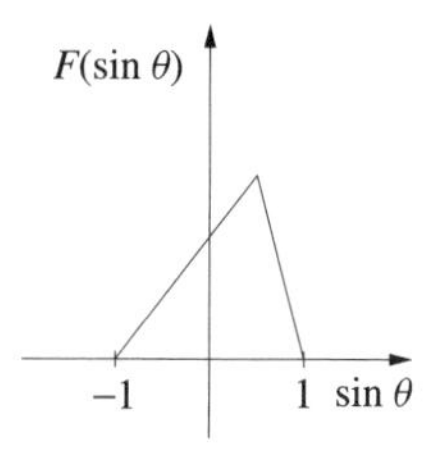

Figure 5.20 A desired frequency invariant beam pattern with an off-broadside main beam

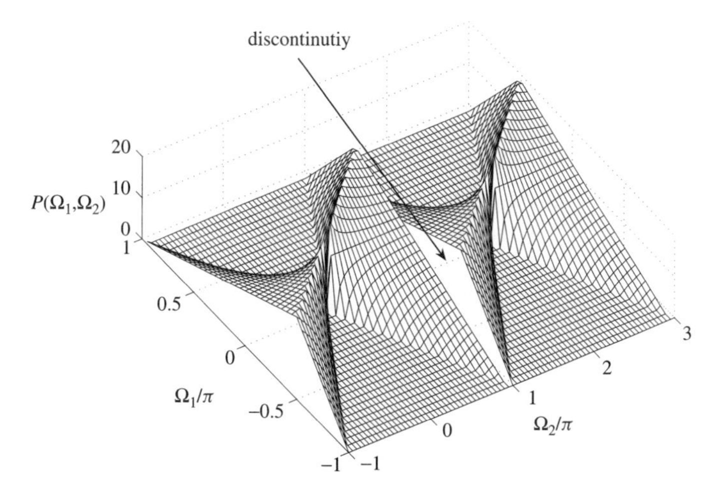

Figure 5.21 The discontinuity of $P(\Omega_1, \Omega_2) = F(\Omega_1/\Omega_2)$ when the desired main beam is off-broadside as shown in Figure 5.20

discontinuity incurred when sampling $P(\Omega_1, \Omega_2)$ and subsequently applying the inverse DFT, the response of $P(\Omega_1, \Omega_2)$ around this area cannot be controlled well. This leads to a poor performance of the proposed method, especially for frequencies close to π. This problem also occurs, although less pronounced, even if the main beam is at broadside but the beam pattern is non-symmetric with respect to it.

Since the problem with the off-broadside main beam design is due to the considerable discontinuity of the periodic function $P(\Omega_1, \Omega_2)$ at $\Omega_2 = \cdots, -3\pi, -\pi, \pi, 3\pi, \cdots$, we can sample the function $P(\Omega_1, \Omega_2)$ in the Ω_2 direction more densely and then permit more coefficients for the corresponding temporal dimension of the beamformer after truncation. For simplicity, we can sample $P(\Omega_1, \Omega_2)$ by a large number in both directions of Ω_1 and Ω_2. After the inverse DFT, we can truncate the results with a rectangular window and leave much more coefficients in the temporal dimension (TDL length) than in the spatial dimension (sensor number).

In the example shown in Figure 5.22, we sampled $P(\Omega_1, \Omega_2)$ by 256×256 points and then truncated the IDFT results to 16×128. Comparing this result with Figure 5.19, we can clearly see the significantly improved frequency invariance property. However, due to the discontinuity, no matter how many coefficients we keep for the TDLs, its performance at around $\Omega = \pi$ will never be as good as at the other frequencies.

Although a better frequency invariance property can be achieved by the proposed approach, the resultant computational complexity for implementing this FIB also increases significantly. For example, the number of coefficients for the off-broadside example in Figure 5.19 is only $16 \times 20 = 320$, while for the example in Figure 5.22, it is

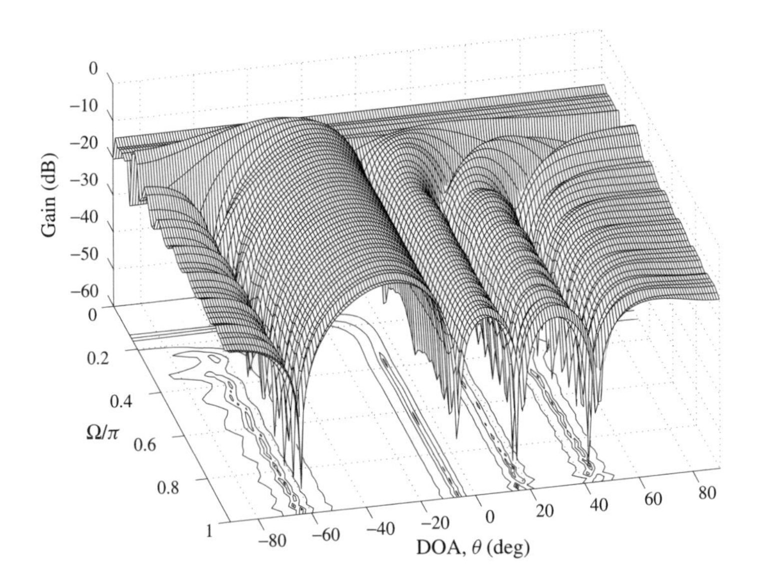

Figure 5.22 The design example with an off-broadside main beam in the direction $-30°$ obtained by increasing the temporal dimension of the beamformer

$16 \times 128 = 2048$. In general, the number of real multiplications required for implementing the fullband beamformer is given by:

$$C_{full}^{real} = MJ \tag{5.28}$$

for a real-valued input signal and:

$$C_{full}^{complex} = 2MJ \tag{5.29}$$

for a complex-valued input signal.

Another solution to the problem is to design a broadside main beam first and then steer the array to the desired direction by means of appropriate delays implemented by either some analogue devices or FIR/IIR filters. In this way, the broadside main beam is shifted to the desired direction. However, as already discussed in Section 1.4.2, for the sidelobe region, it is not a simple shift relationship and the response could be very different from the original one (Liu and Weiss, 2008c, 2009a). Moreover, the total number of coefficients is still very large taking into account both the steering filters and the original TDL coefficients of the broadside beamformer.

In Section 5.3, we will implement the design in subbands with the aim of reducing its computational complexity and also extending its frequency invariance property over a wider bandwidth.

5.2.4 Further Generalization to the FIB Design

We have now discussed the design of frequency invariant beamformers for the case with a continuous sensor and signals in Section 5.2.1 and the one with discrete sensors and signals in Section 5.2.2. The design result is in the form of temporal impulse response $d(x,t)$ or $d[m,n]$ (for linear arrays).

However, it may be advantageous to separate the design into two steps (Liu and Weiss, 2008b): we obtain the desired frequency response $D(x,\omega)$ first and then decide how we implement this frequency response, such as using an analogue filter, an FIR filter or an IIR filter, with the second step being an standard filter design problem. In this section we will briefly discuss the process of obtaining the desired frequency responses for one-dimensional, two-dimensional and three-dimensional arrays.

5.2.4.1 One-Dimensional Array

The beam response $P(\omega,\theta)$ of the continuous linear array shown in Figure 5.1 is given by Equation (5.1). By substitution of $\omega_1 = (\omega \sin\theta)/c$, we have:

$$P(\omega_1,\omega) = \int_{-\infty}^{+\infty} \mathrm{e}^{-j\omega_1 x} D(x,\omega)\mathrm{d}x \tag{5.30}$$

Then $P(\omega,\theta)$ can be obtained by first applying a 1-D Fourier transform to $D(x,\omega)$ and then re-substituting $\omega_1 = (\omega \sin\theta)/c$. To achieve a frequency invariant response $F(\sin\theta)$,

$P(\omega_1, \omega)$ must, after re-substituting $\omega_1 = (\omega \sin \theta)/c$, be identical to $F(\sin \theta)$. As discussed in the previous cases, we use the substitution $\sin \theta = c(\omega_1)/\omega$.

Now, given the desired frequency invariant response $F(\sin \theta)$, we set $P(\omega_1, \omega) = F((\omega_1 c)/\omega)$. Applying an inverse Fourier transform to $P(\omega_1, \omega)$ with respect to the variable ω_1, we obtain the desired frequency response $D(x, \omega)$ for any position x. $D(x, \omega)$ can then be realized by either an analogue filter or digital filter, which can be obtained by an appropriate filter design method.

For the uniformly spaced linear array shown in Figure 5.7, the integration in Equation (5.1) changes to summation:

$$P(\omega, \theta) = \sum_{m=-\infty}^{+\infty} D(md_x, \omega) e^{-j \frac{\omega \sin \theta m d_x}{c}} \tag{5.31}$$

With $\omega_1 = (\omega \sin \theta d_x)/c$, we have:

$$P(\omega_1, \omega) = \sum_{m=-\infty}^{+\infty} D(md_x, \omega) e^{-jm\omega_1} \tag{5.32}$$

In order to avoid aliasing, $d_x < \lambda_{\min}/2$. Based on the discussion in the continuous sensor case, the design of this discrete sensor case can be performed as follows.

Step 1. Given a desired response $F(\sin \theta)$, and the substitution $\sin \theta = (\omega_1 c)/(\omega d_x)$:

$$P(\omega_1, \omega) = \begin{cases} F(c\omega_1/(\omega d_x)) & \text{for } c|\omega_1|/|\omega d_x| \leq 1 \wedge \omega \leq \omega_{\max} \\ A(\omega_1) & \text{for otherwise} \end{cases} \tag{5.33}$$

is obtained for $\omega_1 \in [-\pi \ \pi)$, where $A(\omega_1)$ is an arbitrary function with finite values. As the Fourier transform of $D(md_x, \omega)$, $P(\omega_1, \omega)$ should be a periodic function with a period of 2π. Similar to previous cases, $A(\omega_1)$ can be modified arbitrarily without affecting the resultant beam pattern in theory and its choice should aim to generate a function $P(\omega_1, \omega)$ as smooth as possible and in the design examples, we simply choose it to be zero for the whole defined area.

Step 2. Applying the 1-D inverse Fourier transform to $P(\omega_1, \omega)$ with respect to the variable ω_1, we can obtain the desired frequency responses $D(md_x, \omega)$ for the sensor at positions md_x, $m = \ldots, -1, 0, 1, \ldots$. An inverse DFT can be used as an approximation.

Note that in the FIB design described so far, we have only specified the maximum frequency of interest $\omega_{\max}$, but not the minimum frequency of interest $\omega_{\min}$ and the design simply extends to frequencies as low as possible. However, it is straightforward to add this lower boundary in the design. For example, with the minimum frequency of interest $\omega_{\min}$, Equation (5.33) changes to:

$$P(\omega_1, \omega) = \begin{cases} F(c\omega_1/\omega d_x) & \text{for } c|\omega_1|/|\omega d_x| \leq 1 \wedge \omega \in [\omega_{\min} \ \ \omega_{\max}] \\ A(\omega_1) & \text{for otherwise} \end{cases} \tag{5.34}$$

5.2.4.2 Two-Dimensional Array

For the 2-D continuous sensor array shown in Figure 5.6, its beam response is given by Equation (5.9). With substitutions $\omega_1 = (\omega \sin\theta \cos\phi)/c$ and $\omega_2 = (\omega \sin\theta \sin\phi)/c$, we have:

$$P(\omega_1, \omega_2, \omega) = \iint_{-\infty}^{+\infty} D(x, y, \omega)\, \mathrm{e}^{-j\omega_1 x} \mathrm{e}^{-j\omega_2 y} \mathrm{d}x\mathrm{d}y \tag{5.35}$$

Suppose $F(\sin\theta\cos\phi, \sin\theta\sin\phi)$ is the desired frequency invariant response. We set $P(\omega_1, \omega_2, \omega) = F((\omega_1 c)/\omega, (\omega_2 c)/\omega)$. Applying a 2-D inverse Fourier transform to $P(\omega_1, \omega_2, \omega)$ with respect to the variables ω_1 and ω_2, we then obtain the desired frequency response $D(x, y, \omega)$ for any position (x, y).

Now considering the discrete sensor array in Figure 5.10, we have:

$$P(\omega_1, \omega_2, \omega) = \sum_{l,m=-\infty}^{+\infty} D(ld_x, md_y, \omega)\, \mathrm{e}^{-jl\omega_1} \mathrm{e}^{-jm\omega_2} \tag{5.36}$$

with $\omega_1 = (\omega \sin\theta \cos\phi d_x)/c$ and $\omega_2 = (\omega \sin\theta \sin\phi d_y)/c$. Then the 2-D frequency invariant beamformer can be designed as follows.

Step 1. With the substitutions $\sin\theta\cos\phi = (c\omega_1)/(\omega d_x)$ and $\sin\theta\sin\phi = (c\omega_2)/(\omega d_y)$, we obtain $P(\omega_1, \omega_2, \omega)$ defined over one period $\omega_1, \omega_2 \in [-\pi\ \pi)$ as:

$$P(\omega_1, \omega_2, \omega) = \begin{cases} F(c\omega_1/(\omega d_x), c\omega_2/(\omega d_y)) & \text{for } (c\omega_1/(\omega d_x))^2 + (c\omega_1/(\omega d_y))^2 \le 1 \wedge \omega \le \omega_{\max} \\ A(\omega_1, \omega_2) & \text{for otherwise} \end{cases} \tag{5.37}$$

where $A(\omega_1, \omega_2)$ is an arbitrary function with finite values and its choice should also aim to generate a smoother function $P(\omega_1, \omega_2, \omega)$.

Step 2. Applying a 2-D inverse Fourier transform to $P(\omega_1, \omega_2, \omega)$ with respect to the two variables ω_1 and ω_2 returns the desired frequency response $D(ld_x, md_y, \omega)$ for the corresponding sensors. As an approximation, the 2-D IDFT can be employed.

5.2.4.3 Three-Dimensional Array

Based on the previous analyses we can develop a similar design method for the 3-D case and it is omitted here.

5.2.4.4 Design Examples

We provide two examples with potential applications to microphone arrays. The frequency range of interest is set to be $0 < f \le 12$ kHz and the propagation speed is 340 m/s. The first example is for a linear array with 21 sensors and the second one for a planar array with

21×21 sensors, both with an adjacent sensor spacing of $34\,000/(2 \times 12\,000) = 1.42$ cm. The desired beam pattern for the linear array is given by:

$$F_{1D}(\sin\theta) = \sum_{m=-3}^{3} h_m \mathrm{e}^{-jm\pi \sin\theta} \tag{5.38}$$

where the coefficients $\{h_{-3}, h_{-2}, \ldots, h_2, h_3\}$ are obtained by the filter design program *remez* provided by MATLAB©. Their values are given by:

$$\{h_m\} = [0.0307\ \ 0.2028\ \ 0.1663\ \ 0.2004\ \ 0.1663\ \ 0.2028\ \ 0.0307] \tag{5.39}$$

They form a desired equi-ripple frequency invariant response.

We employ a 64-point 1-D IDFT on the resultant periodic function $P(\omega_1, \omega)$ and set $P(\omega_1, \omega) = 0$ for the area where there is no signal existing, as indicated in step 1 of the 1-D array design. The resultant 64 desired frequency responses $D(md_x, \omega)$, with each corresponding to an analogue or digital filter following each of the sensors, are then truncated to the size of 21 directly. Given the 21 desired frequency responses $D(md_x, \omega)$, $m = -10, \ldots, 1, 0, 1, \ldots, 10$, we can employ an appropriate filter design method to design 21 analogue or digital filters to realize them. The magnitude responses $|D(md_x, \omega)|$ are shown in Figure 5.23 to give a rough idea about the design results. As the filter design problem is not the focus of this approach, we assume that the 21 filters have been obtained with the desired frequency responses $D(md_x, \omega)$ (the same assumption for the planar array design) and then use $D(md_x, \omega)$ to calculate the resultant beam pattern, which is shown in Figure 5.24 with a clear frequency invariance property except for very low frequencies.

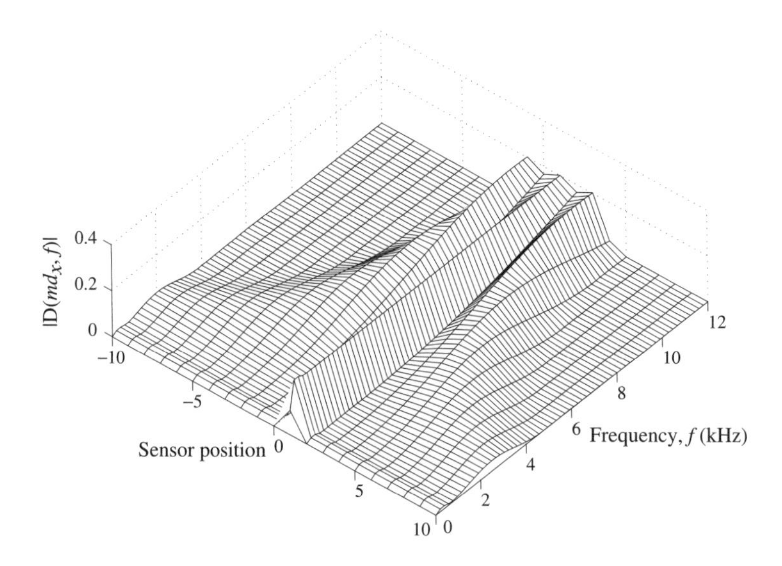

Figure 5.23 The resultant magnitude responses $|D(md_x, f)|$ of the 21 filters in the frequency domain for the linear array design result

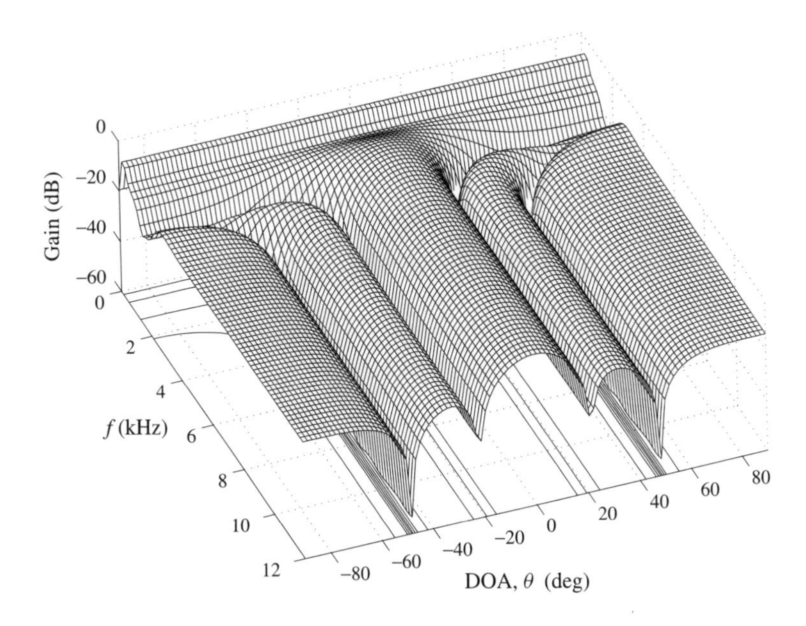

Figure 5.24 The resultant frequency invariant beam pattern for the linear array design

For the second design example, the desired beam pattern is given by:

$$F_{2D}(\sin\theta\cos\phi, \sin\theta\sin\phi) = \frac{1}{25}\sum_{l=-2}^{2}\sum_{m=-2}^{2} \mathrm{e}^{-jl\pi\sin\theta\cos\phi}\mathrm{e}^{-jm\pi\sin\theta\sin\phi} \tag{5.40}$$

According to Equation (5.37), we first obtain the response $P(\omega_1, \omega_2, \omega)$ with $A(\omega_1, \omega_2) = 0$. A 64×64 2-D IDFT is employed and the 64×64 desired frequency responses $D(ld_x, md_y, \omega)$ are then subsequently truncated to the intended array dimensions of 21×21. The resultant beam pattern is four-dimensional and here we can only provide some exemplary snapshots.

Different from the previous examples provided for 2-D arrays, we here draw the response in a cylindrical coordinates system to the frequencies $\omega = 7.2\,\text{kHz}$ and $\omega = 9.6\,\text{kHz}$, respectively, shown in Figures 5.25 and 5.26. The height axis is the magnitude response of the beam, the radial coordinate is for the elevation angle θ and the angle coordinate is for the azimuth angle ϕ. The frequency invariance property can be verified by the similarity of these two figures, and further shown by a slice of its beam pattern at $\phi = 60°$, as given in Figure 5.27.

5.3 Subband Design of Frequency Invariant Beamformers

As mentioned in Section 5.2.3, for an off-broadside main beam design, we have to increase the temporal dimension of the beamformer in combination with a denser sampling grid in the Fourier domain to achieve a better result. However, the increase of the temporal dimension leads to a much higher computational complexity in implementation.

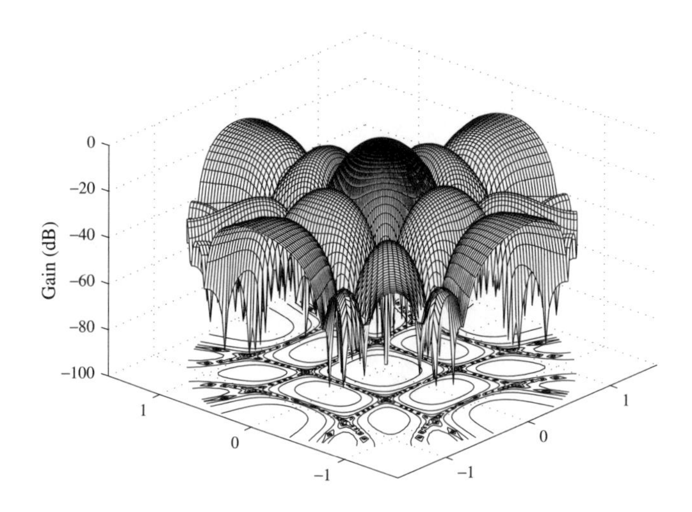

Figure 5.25 The resultant beam pattern of the planar array at $f = 7.2\,\text{kHz}$

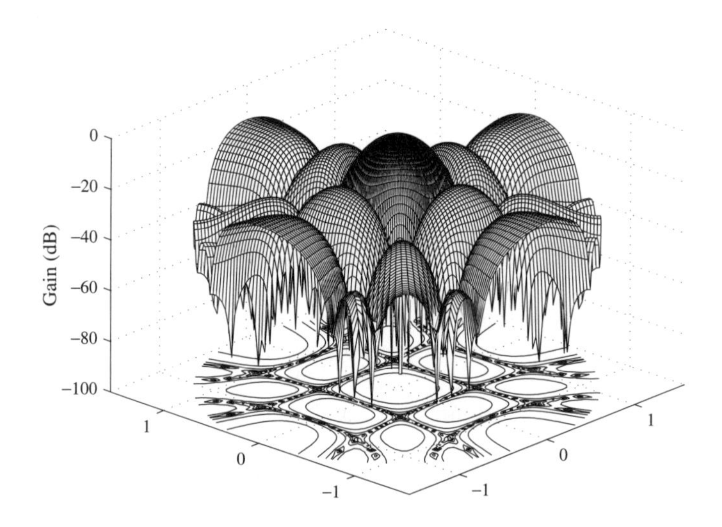

Figure 5.26 The resultant beam pattern of the planar array at $f = 9.6\,\text{kHz}$

In the past, in order to reduce the computational complexity and improve the performance of the system, subband methods have been proposed for wideband adaptive beamforming as introduced in Chapter 3. Similarly we can also implement the frequency invariant beamformer in subbands.

Two subband implementations will be introduced (Liu and Weiss, 2004a, 2009b; Zhao *et al.*, 2009b). In the first one, each received array signal is split into K decimated subbands

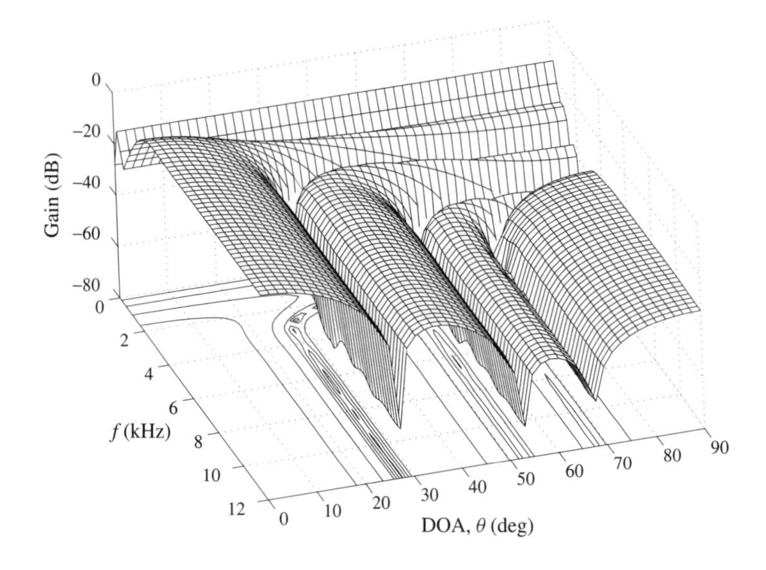

Figure 5.27 A slice of the beam pattern at $\phi = 60^\circ$

by an analysis filter bank. The corresponding subband signals form K sets of subband arrays and an FIB is operated in each of the subband arrays. The outputs of the subband FIBs are then combined together by a synthesis filter bank to form the fullband output signal. When the required spatio-temporal dimension of the fullband array becomes large enough, the subband implementation will have a much lower computational complexity. As the spatio-temporal distribution of the subband signals at each of the subband arrays is different from the original fullband signal, we cannot use the design method for fullband arrays directly. Therefore, a modified method for the subband FIB design is proposed. In a refinement, we further modify this subband FIB design for use with nested arrays.

Without loss of generality, we will only consider the design based on uniform linear arrays.

5.3.1 First Implementation

5.3.1.1 Structure

For the subband implementation of the FIB based on uniform linear arrays, each of the received array signals $x_m[n]$, $m = 0, 1, \ldots, M-1$, is split into K subbands by the analysis filter bank and the corresponding subband signals form K sets of subband arrays. An FIB is operated in each of the subband arrays and the processed subband signals are then combined together by a synthesis filter bank into the fullband output. There are in total K subband FIBs and each has M subband sensor inputs and its TDL/FIR filter length is J_{sub}. Figure 5.28 shows this implementation, where the blocks labelled 'A' are the analysis filter banks and the block labeled 'S' is the synthesis filter bank.

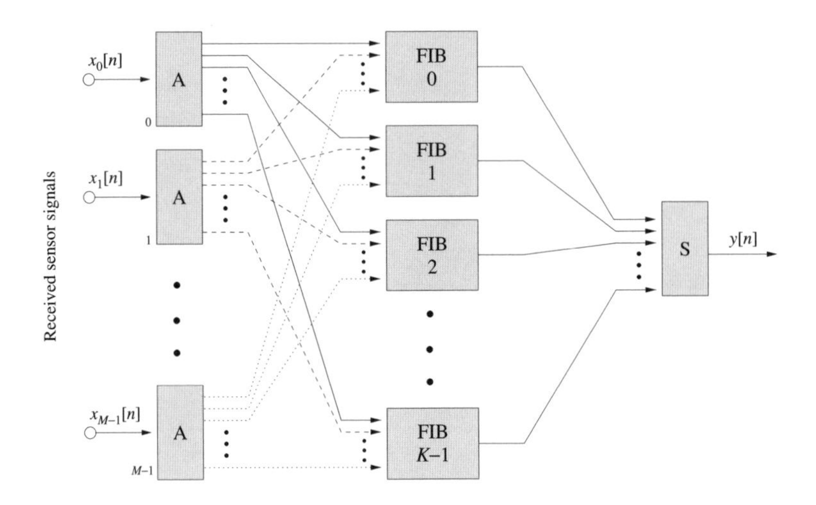

Figure 5.28 Subband implementation of the frequency invariant beamformer.

The advantage of this subband implementation is similar to the subband implementation of adaptive filters, which can be explained by the fact that N subband FIBs are trying to model the original fullband beamformer to have the same spatio-temporal filtering effect. Therefore, due to the subband decomposition and decimation, the FIR filter length J_{sub} of the subband FIBs can be shorter than the original fullband beamformer to achieve a similar performance. As indicated in Equation (3.30) of Section 3.3.1, Chapter 3, we normally choose:

$$J_{sub} = \frac{J + l_p}{N} \tag{5.41}$$

where l_p is the length of the prototype filter for the oversampled Generalized DFT (GDFT) filter banks (Weiss and Stewart, 1998), which will be used in our subband implementations.

Since the subband FIBs have a shorter length and operate at a lower rate, the computational complexity of the whole system can be much lower than the fullband implementation.

5.3.1.2 Design of Subband FIBs

For the design of the subband beamformer, because the normalized angular frequency $\tilde{\Omega}$ in decimated subbands has changed after the decimation operation in the analysis filter bank, we need to modify the method in Section 5.2.2 to fit into the new scenario.

For the design of the ith subband beamformer, at first, we need to find the relationship $\tilde{\Omega} = S_i(\Omega)$ between the decimated subband $\tilde{\Omega}$ and the fullband Ω. Suppose the spectrum of the output signal of the analysis filter bank is $X_i(e^{j\Omega})$ before decimation, then the output spectrum $Y_i(e^{j\tilde{\Omega}})$ after decimation is given by (Vaidyanathan, 1993):

$$Y_i(e^{j\tilde{\Omega}}) = \frac{1}{K}\sum_{k=0}^{K-1} X_i(e^{j(\frac{\tilde{\Omega}-2k\pi}{K})}) \tag{5.42}$$

We can find each $\tilde{\Omega} = S_i(\Omega)$ according to this equation.

For the ith subband beamformer, its response to the decimated subband input signal can be written as:

$$\tilde{P}_i(\tilde{\Omega}, \theta) = \sum_{m,n=0,0}^{M-1, J_{sub}-1} d_i[m, n] \, \mathrm{e}^{-jm\Omega \sin\theta} \, \mathrm{e}^{-jn\tilde{\Omega}} \tag{5.43}$$

where $d_i[m, n]$ are the coefficients of the ith subband beamformer. We use the same phase difference $\mathrm{e}^{-j\Omega \sin\theta}$ between adjacent subband sensor signals as that of the fullband beamformer, because it does not change after decimation. For its response to the original fullband input signal, we have:

$$\tilde{P}_i(S_i(\Omega), \theta) = \sum_{m,n=0,0}^{M-1, J_{sub}-1} d_i[m, n] \, \mathrm{e}^{-jm\Omega \sin\theta} \, \mathrm{e}^{-jnS_i(\Omega)} \tag{5.44}$$

By substituting $\Omega_1 = \Omega \sin\theta$ and $\tilde{\Omega}_2 = \tilde{\Omega}$ into Equation (5.43), we have:

$$\tilde{P}_i(\Omega_1, \tilde{\Omega}_2) = \sum_{m,n=0,0}^{M-1, J_{sub}-1} d_i[m, n] \, \mathrm{e}^{-jm\Omega_1} \, \mathrm{e}^{-jn\tilde{\Omega}_2} \tag{5.45}$$

Thus, we can obtain $d_i[m, n]$ by applying the inverse Fourier transform to the desired subband response $\tilde{P}_i(\Omega_1, \tilde{\Omega}_2)$. As $\tilde{\Omega} = S_i(\Omega)$ and $\Omega_2 = \Omega$, we also have $\tilde{\Omega}_2 = S_i(\Omega_2)$ and then:

$$\tilde{P}_i(\Omega_1, S_i(\Omega_2)) = P(\Omega_1, \Omega_2) \tag{5.46}$$

for the ith subband, where $P(\Omega_1, \Omega_2)$ is given in Equation (5.17).

From the discussion above, we obtain a modified method applicable to the subband FIB design using the inverse DFT as an approximation to the inverse Fourier transform.

We first obtain $P(\Omega_1, \Omega_2)$ from the fullband beamformer design method, as given in Equation (5.18). Then we uniformly sample Ω_1 and $\tilde{\Omega}_2$ in $(-\pi \; \pi]$ with $M_{max} \times J_{max}$ points, where $M_{max} \geq M$ and $J_{max} \geq J_{sub}$. We calculate $\tilde{P}_i(\Omega_1, \tilde{\Omega}_2)$ on these points according to Equation (5.46). Applying the inverse DFT to the result, we then obtain $d_i[m, n]$ with dimensions $M_{max} \times J_{max}$, which needs to be truncated to fit the subband FIB dimensions $M \times J_{sub}$.

However, there is one problem when calculating $\tilde{P}_i(\Omega_1, \tilde{\Omega}_2)$ according to Equation (5.46), as one value of $\tilde{\Omega}_2 = \tilde{\Omega} = S_i(\Omega_2)$ in general corresponds to K different values of $\Omega_2 = \Omega$ according to Equation (5.42). To remove this ambiguity, the bandwidth of each subband signal before decimation should be limited to no more than $(2\pi)/N$ (including both the negative and positive frequencies). Figure 5.29 shows the spectrum changes after a signal with a bandwidth $(2\pi)/3$ is decimated by a factor $N = 3$. Thus, although one value of $\tilde{\Omega}_2$ ($\tilde{\Omega}$) still corresponds to several different values of Ω_2 (Ω), which is determined by Equation (5.42) and cannot be changed, only on one value of Ω_2 (Ω), the subband signal before decimation is non-zero and on all the other values of Ω_2 (Ω), the subband signal is zero. We will then use that value to calculate the unique $\tilde{P}_i(\Omega_1, \tilde{\Omega}_2)$ according to Equation (5.46). In the example of Figure 5.29, the point $\tilde{\Omega} = (2\pi)/3$ corresponds to three different

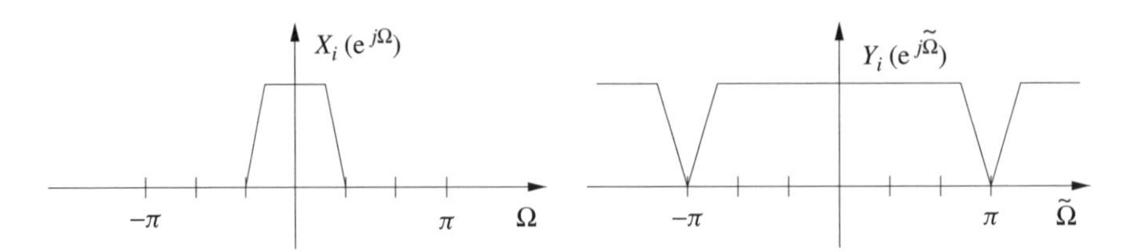

Figure 5.29 The spectrum changes after a signal with a bandwidth of $(2\pi)/3$ is decimated by a factor of $K = 3$

values of Ω, which are respectively $\Omega = -(4\pi)/9$, $(2\pi)/9$ and $(8\pi)/9$. As $X_i(e^{j\Omega}) = 0$ for $\Omega = -(4\pi)/9$ and $(8\pi)/9$, we will choose $\Omega = (2\pi)/9$ as the corresponding point of $\tilde{\Omega} = (2\pi)/3$.

5.3.1.3 Computational Complexity

With this bandwidth constraint, we cannot use the perfect-reconstruction maximally decimated ($N = K$) filter banks in our subband beamformer design, as the bandwidth of the analysis filter bank in such systems is always larger than $(2\pi)/N$. As a result, we have to employ the oversampled ($N > K$) filter banks with a very low aliasing level, such as the oversampled GDFT filter banks discussed in Section 3.1.3. The number of real multiplications required to implement the GDFT filter bank (analysis or synthesis) has been given by Equation (3.26) for a real-valued input signal and Equation (3.27) for a complex-valued input signal.

For each of the subband beamformers, the number of multiplications required is $4MJ_{sub} = 4M(J + l_p)/N$ (for the GDFT filter banks, the input to the subband FIBs is complex-valued and each of the subband FIBs is also complex-valued). There are in total M analysis filter banks and one synthesis filter bank and the signal rate for subband FIBs is $1/N$ of the fullband one. Therefore the total number of multiplications for the whole subband FIB system shown in Figure 5.28 is:

$$C_{sub}^{real} = \frac{M+1}{N}\left(l_p + 4K\log_2 K + 4K\right) + \frac{2KM(J + l_p)}{N^2} \tag{5.47}$$

for a real-valued input signal and:

$$C_{sub}^{complex} = \frac{M+1}{N}\left(2l_p + 4K\log_2 K + 8K\right) + \frac{4KM(J + l_p)}{N^2} \tag{5.48}$$

for a complex-valued input signal.

As an example, the number of multiplications $C_{sub}^{complex}$ and $C_{full}^{complex}$ (complexity for a fullband FIB with complex-valued input) as a function of the fullband FIR filter length J is shown in Figure 5.30 for the case with $M = 16$, $K = 10$, $N = 8$, and $lp = 110$. The dashed line is for the subband FIB and the dotted line is for the fullband FIB. For $J \leq 91$, the computational complexity of the subband FIB is always higher than that of the fullband one; when $J > 91$, we start to have savings with the subband FIB and

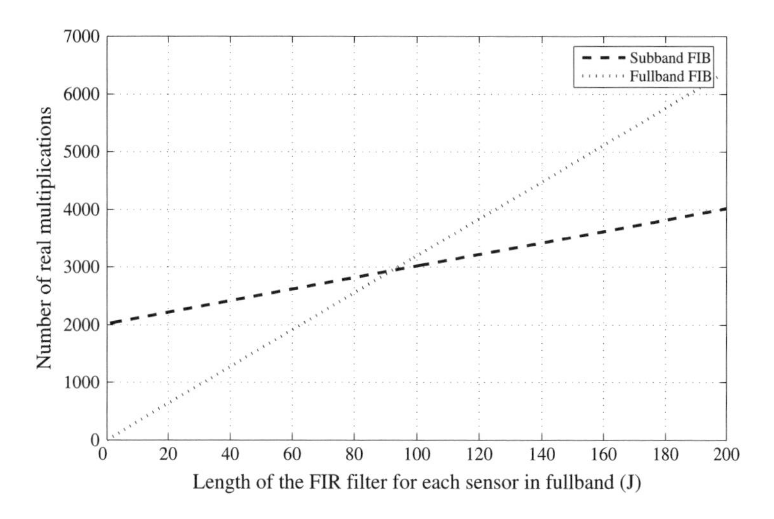

Figure 5.30 A comparison of the computational complexity between the fullband FIB and its subband implementation

the larger the fullband FIR filter length J, the more savings we have in computational complexity.

5.3.2 Second Implementation–Scaled Aperture

The spatial resolution of a beamformer is dependent on both the aperture of the sensor array and the frequency of the impinging signal and it is difficult to achieve a constant beamwidth for lower frequencies. To extend the constant beamwidth property to a frequency as low as possible, we here introduce a frequency invariant beamformer with a scaled aperture, which is shown in Figure 5.31, depicting the exemplary case for $M = 4$ sensors for each octave and beamformers operating in 4 uniformly split decimated subbands, with each having a bandwidth $\pi/4$, whereby the array signals are drawn from a total of 8 nested sensors. For the 3 octave groups of subband FIBs, FIB 0 operates on the lowest band, FIB 1 forms the second octave and the remaining two FIBs are responsible for the highest octave band covered by two subbands. As indicated in Figure 5.31, not all subbands are required for processing from each sensor.

As the array spacing is different for each octave of subband beamformers, we need to change the subband beamformer design procedure correspondingly. In Figure 5.31, beamformers 3 and 2 have a standard spacing of $d = (\lambda_{\min})/2$, so we can apply the previous design procedure directly; for beamformers 1 and 0, they have a spacing of $\lambda_{\min}$ and $2\lambda_{\min}$, respectively, and then we must consider this difference in our design. Suppose the spacing of the subband beamformer is σ times the standard spacing $(\lambda_{\min})/2$, then its response to the original fullband input signal is:

$$\tilde{P}_i(S_i(\Omega), \theta) = \sum_{m,n=0,0}^{M_i-1, J_i-1} d_i[m, n]\, \mathrm{e}^{-jm\sigma\Omega \sin\theta}\, \mathrm{e}^{-jnS_i(\Omega)} \tag{5.49}$$

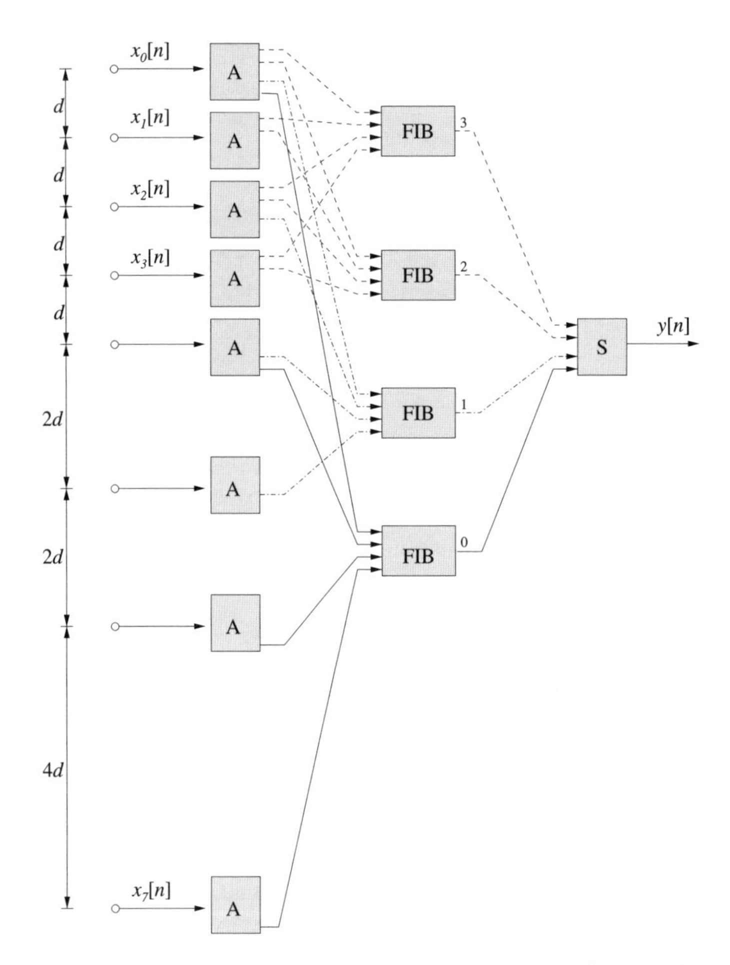

Figure 5.31 A beamforming structure with a scaled array aperture for FIBs in various octave bands

where $d_i[m, n]$ is the coefficients of the ith subband beamformer with dimension $M_i \times J_i$. By substituting $\tilde{\Omega}_1 = \sigma\Omega \sin\theta = \sigma\Omega_1$ and $\tilde{\Omega}_2 = S_i(\Omega)$ into Equation (5.49), we have:

$$\tilde{P}_i(\tilde{\Omega}_1, \tilde{\Omega}_2) = \sum_{m,n=0,0}^{M_i-1, J_i-1} d_i[m, n]\, \mathrm{e}^{-jm\tilde{\Omega}_1}\, \mathrm{e}^{-jn\tilde{\Omega}_2} \tag{5.50}$$

The relationship between the subband beamformer's response and the fullband beamformer's response is given by:

$$\tilde{P}_i(\sigma\Omega_1, S_i(\Omega_2)) = P(\Omega_1, \Omega_2) \tag{5.51}$$

Then the design can be modified as follows.

First, we uniformly sample $\tilde{\Omega}_1$ and $\tilde{\Omega}_2$ in $(-\pi\ \pi]$ with $M_{max} \times J_{max}$ points. Then we obtain the response $\tilde{P}_i(\tilde{\Omega}_1, \tilde{\Omega}_2)$ on these points according to Equation (5.51). With the inverse DFT, the temporal response $d_i[m, n]$ with dimensions $M_{max} \times J_{max}$ is obtained and then truncated to the dimensions $M_i \times J_i$.

5.3.3 Design Examples

We give one example for each of the implementations. Both are based on the same desired ideal response with a main beam pointing to the direction $\theta = -30^\circ$, as given in Equation (5.27) of Section 5.2.3.

For the first one, we employ the $K = 10$-channel oversampled GDFT filter banks with a decimation ratio $N = 8$. The length of the prototype filter is $l_p = 110$. Each of the received sensor signals is split into 10 subbands and in total there are 10 subband beamformers, each of which has dimensions 12×29 since $J_{sub} = (128 + 110)/8 \approx 29$. The resultant beam pattern is shown in Figure 5.32, which is very similar to the example given in Figure 5.22, except for the frequencies very close to $\Omega = \pi$, where it seems that the discontinuity problem has become more serious in subbands.

However, this problem can be avoided since it is only limited to frequencies extremely close to π and has no obvious effect on other frequencies. Suppose the maximum frequency of interest is $\omega_{\max}$. When we design this wideband array, we can consider its operating frequency to be $\omega_{\max} + \delta$, where $\delta > 0$ is a very small value. In this way, $\omega_{\max} + \delta$ will correspond to the normalized frequency $\Omega = \pi$ and $\omega_{\max}$ corresponds to the normalized frequency a little lower than $\Omega = \pi$. Thus, the resultant array can maintain a satisfactory frequency invariant response over the bandwidth of interest below $\omega_{\max}$.

For the computational complexity of this subband FIB system, for a real-valued input signal, it is $C_{sub}^{real} = 1761$ and for the corresponding fullband FIB, it is $C_{full}^{real} = 2048$;

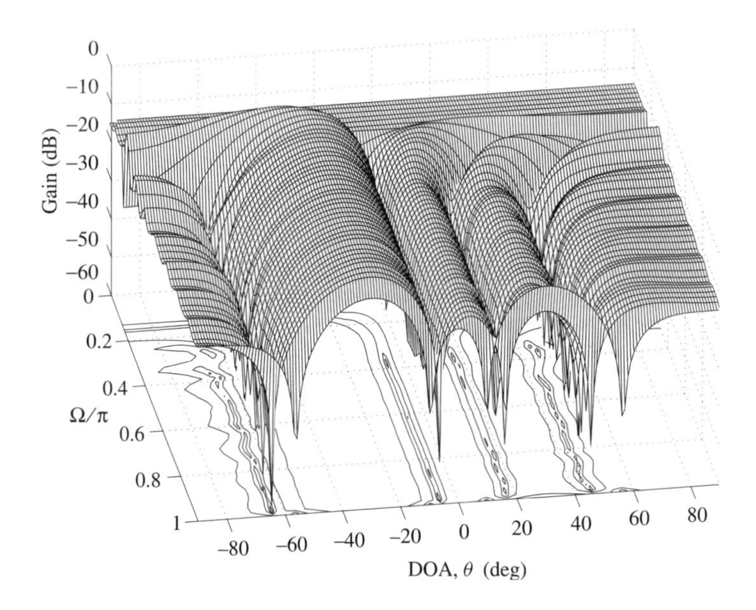

Figure 5.32 The resultant beam pattern for the uniformly spaced subband linear array

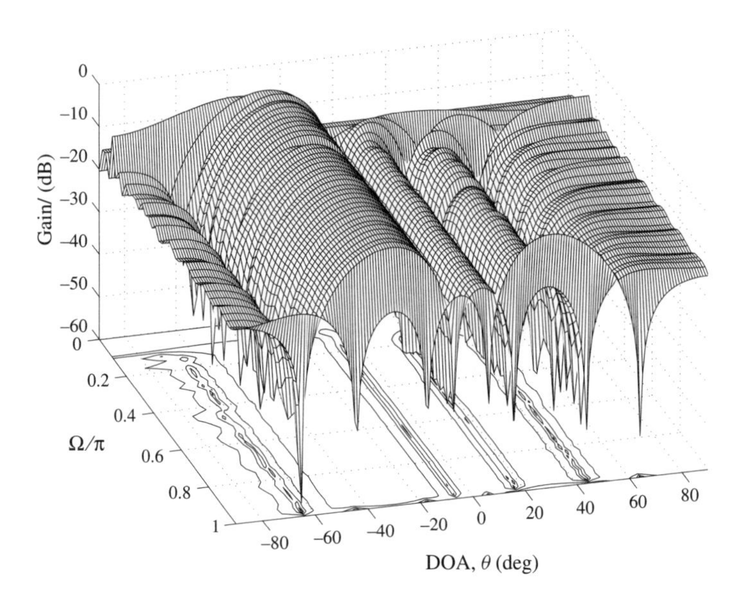

Figure 5.33 The resultant beam pattern for the scaled aperture array

for the complex-valued case, we have $C_{sub}^{complex} = 3320$ and $C_{full}^{complex} = 4096$. In both cases, a much lower computational complexity has been achieved by the subband-based beamformer.

Next, for the array with scaled apertures, we employ the same oversampled GDFT filter banks. There are two octaves. The first two channels and the last two channels belong to the first octave while the six channels in the middle belong to the second octave. For each of them, the subband beamformer has a dimension of 16×29. There are 16 sensors for each octave and in total there are $16 + 8 = 24$ sensors. As a result, we have in total 24 analysis filter banks and one synthesis filter bank, i.e. 8 more filter banks than the first example. The number of subband FIBs is the same as the first one. Then the total number of real multiplications for this implementation is $C_{scaled}^{real} = 1761 + 8 \times C_{GDFT}^{real} = 2044$ for real-valued input signals and $C_{scaled}^{complex} = 3320 + 8 \times C_{GDFT}^{complex} = 3753$ for complex-valued input signals. So, although this scaled aperture system has a higher complexity than the first subband implementation, it is still lower than the fullband one and its frequency invariance property has been extended to as low as $\Omega = 0.1\pi$, as shown in Figure 5.33.

5.4 Frequency Invariant Beamforming for Circular Arrays

In Sections 5.2 and 5.3, the focus of the design is on uniformly spaced linear arrays, rectangular arrays and cubic (the three dimensions may not be the same) arrays. In this section, we discuss an approach proposed for the design of FIBs based on uniformly spaced circular arrays (Chan and Pun, 2002). This design can be extended to concentric circular arrays (Chan and Chen, 2005, 2007; Chen *et al.*, 2007) and concentric spherical arrays (Chan and Chen, 2006; Chen and Chan, 2007).

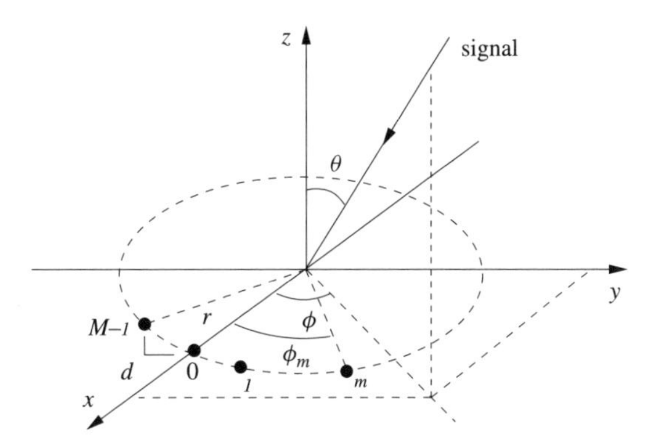

Figure 5.34 A uniformly spaced circular array with M sensors and a circumferential sensor spacing of d, where a signal impinges from the direction (θ, ϕ)

5.4.1 Phase Mode Processing

Consider the uniform circular array with M omnidirectional sensors and a circumferential sensor spacing of d, as shown in Figure 5.34. The radius r of the circular array is given by:

$$r = \frac{Md}{2\pi} \tag{5.52}$$

The position vector of the mth sensor is given by $(r \cos \phi_m, r \sin \phi_m, 0)$, $m = 0, 1, \ldots, M - 1$, where $\phi_m = m2\pi/M$ is the angle measured from the x axis to the mth sensor. The sensor spacing d is $\alpha/2$ times the wavelength $\lambda_{\min}$ of the highest frequency component of the impinging signal, i.e. $d = \alpha\lambda_{\min}/2$, where α is a scalar, then we have:

$$r = \alpha\frac{\lambda_{\min}}{2}\frac{M}{2\pi} = \alpha\beta\frac{\lambda_{\min}}{2} \tag{5.53}$$

with $\beta = M/2\pi$.

For a signal with an angular frequency ω, the phase difference between the centre of the circular array and the mth sensor is given by:

$$\begin{aligned}\Phi &= \omega\frac{r \sin\theta \cos(\phi - \phi_m)}{c}\\ &= \frac{\omega\lambda_{\min}}{2c}\alpha\beta \sin\theta \cos(\phi - \phi_m)\end{aligned} \tag{5.54}$$

where c is the signal propagation speed.

In the discrete form, if we sample the signal at the Nyquest rate, i.e. $T_s = (\lambda_{\min})/(2c)$, then we have $(\omega\lambda_{\min})/(2c) = \Omega$, where $\Omega = \omega T_s$. Then the mth element of the steering vector $\mathbf{d}(\theta, \phi, \Omega)$ of this uniform circular array is given by:

$$d_m = \mathrm{e}^{j\alpha\beta\Omega \sin\theta \cos(\phi - \phi_m)} \tag{5.55}$$

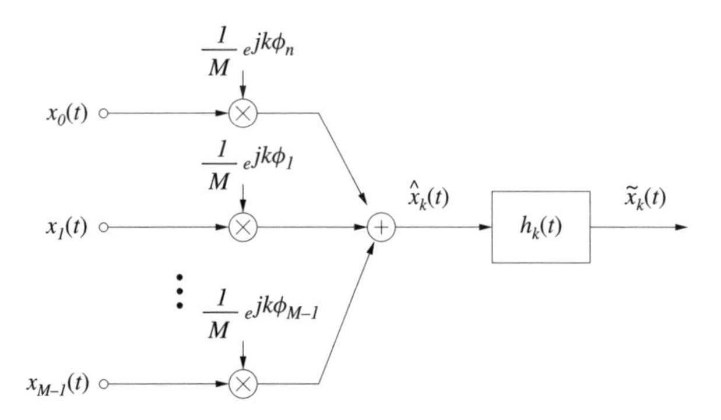

Figure 5.35 Processing of the received array signals for the kth phase mode by the weight vector $\mathbf{w}_k$ and the following filter $h_k(t)$, with $\hat{x}_k(t)$ and $\tilde{x}_k(t)$ as the respective outputs

Without loss of generality, we assume the signals arrive from the (x, y)-plane, i.e. $\theta = \pi/2$. Then d_m changes to:

$$d_m = \mathrm{e}^{j\alpha\beta\Omega\cos(\phi-\phi_m)} \tag{5.56}$$

which can be expanded to the following form (Abramowitz and Stegun, 1970):

$$d_m = \sum_{n=-\infty}^{+\infty} j^n J_n(\alpha\beta\Omega)\mathrm{e}^{jn(\phi-\phi_m)} \tag{5.57}$$

where J_n is the Bessel function of the first kind.

Applying a weight vector $\mathbf{w}_k$ to the received array signals $x_m(t)$, $m = 0, 1, \ldots, M-1$, as shown in Figure 5.35, we obtain the output $\hat{x}_k(t)$, given by:

$$\hat{x}_k(t) = \mathbf{w}_k^T \mathbf{x}(t) \tag{5.58}$$

where:

$$\begin{aligned}\mathbf{w}_k &= \frac{1}{M}[\mathrm{e}^{jk\phi_0}\ \mathrm{e}^{jk\phi_1}\ \cdots\ \mathrm{e}^{jk\phi_{M-1}}]^H \\ \mathbf{x}(t) &= [x_0(t)\ x_1(t)\ \cdots\ x_{M-1}(t)]^T\end{aligned} \tag{5.59}$$

The beam response of this beamformer with the weight vector $\mathbf{w}_k$ is given by:

$$\begin{aligned}P_k(\phi, \Omega) &= \mathbf{w}_k^T \mathbf{d}(\phi, \Omega) \\ &= \frac{1}{M}\sum_{m=0}^{M-1}\sum_{n=-\infty}^{+\infty} j^n J_n(\alpha\beta\Omega)\mathrm{e}^{jn(\phi-\phi_m)}\mathrm{e}^{jk\phi_m} \\ &= \frac{1}{M}\sum_{m=0}^{M-1}\sum_{n=-\infty}^{+\infty} j^n J_n(\alpha\beta\Omega)\mathrm{e}^{jn\phi}\mathrm{e}^{j(k-n)\phi_m}\end{aligned}$$

$$
\begin{aligned}
&= \frac{1}{M} \sum_{m=0}^{M-1} \sum_{n=-\infty}^{+\infty} j^n J_n(\alpha\beta\Omega) e^{jn\phi} e^{jm\frac{2\pi(k-n)}{M}} \\
&= \sum_{n=-\infty}^{+\infty} j^n J_n(\alpha\beta\Omega) e^{jn\phi} \frac{1}{M} \sum_{m=0}^{M-1} e^{jm\frac{2\pi(k-n)}{M}} \\
&= j^k J_k(\alpha\beta\Omega) e^{jk\phi} + \sum_{n=k+lM} j^n J_n(\alpha\beta\Omega) e^{jn\phi} \qquad (5.60)
\end{aligned}
$$

where l is a non-zero integer ($l \in \mathbf{Z}$ but $l \neq 0$) and we have used the following result:

$$
\frac{1}{M} \sum_{m=0}^{M-1} e^{jm\frac{2\pi(k-n)}{M}} = \begin{cases} 1 & \text{for } k - n = pM, \quad p \in \mathbf{Z} \\ 0 & \text{for otherwise} \end{cases} \qquad (5.61)
$$

The higher-order components in Equation (5.60) can be ignored if the absolute value of k does not exceed some threshold K. In this case, $P_k(\phi, \Omega)$ can be approximated by:

$$
P_k(\phi, \Omega) \approx j^k J_k(\alpha\beta\Omega) e^{jk\phi} \qquad (5.62)
$$

for $|k| \leq K$. According to Davies (1983) and Mathews and Zoltowski (1994), for each frequency Ω, we shall choose:

$$
K(\Omega) \approx \alpha\beta\Omega \qquad (5.63)
$$

and:

$$
K(\Omega) < \frac{M}{2} \qquad (5.64)
$$

where $K(\Omega)$ means the value of K is dependent on Ω.

To meet both conditions, the circumferential sensor spacing d is chosen to be half the wavelength of the corresponding frequency Ω (Davies, 1983; Mathews and Zoltowski, 1994). For wideband signals with a frequency range $\Omega \in [\Omega_{\min}\ \Omega_{\max}]$, we will choose $d = (\lambda_{\min})/2$, i.e. $\alpha = 1$ and $K \approx \beta\Omega_{\min}$.

Note the response function in Equation (5.62) is frequency dependent. In order to achieve a frequency independent response, we further process the output $\hat{x}_k(t)$ of the kth phase mode by a filter $h_k(t)$, which has a frequency response $H_k(\Omega) = 1/(j^k J_k(\alpha\beta\Omega))$ and its output is given by:

$$
\tilde{x}_k(t) = h_k(t) * \hat{x}_k(t) = h_k(t) * (\mathbf{w}_k^H \mathbf{x}(t)) \qquad (5.65)
$$

where $*$ denotes the convolution operation. The beam response $\hat{P}_k(\phi)$ with the weight vector $\mathbf{w}_k$ and the filter $h_k(t)$ can be expressed as:

$$
\hat{P}_k(\phi) = e^{jk\phi} \qquad (5.66)
$$

However, there is a potential problem with this arrangement. If the Bessel function $J_k(\alpha\beta\ \Omega)$ has some zero responses over the frequency range of interest $\Omega \in [\Omega_{\min}\ \Omega_{\max}]$,

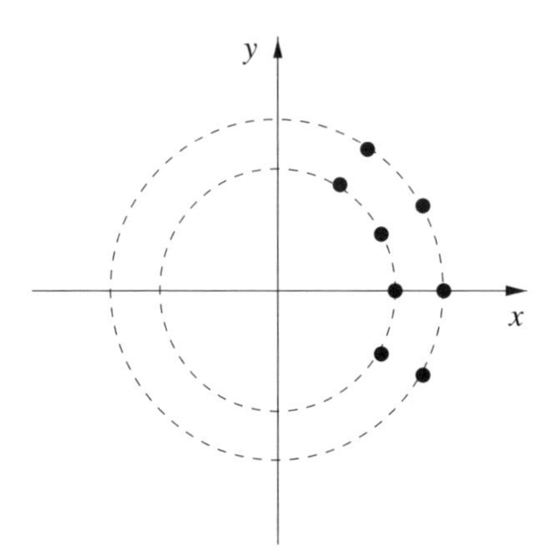

Figure 5.36 A concentric circular array with two rings as an example

the kth phase mode filter $h_k(t)$ obtained as an inverse of $J_k(\alpha\beta\Omega)$ will be unstable and have an infinite response at those zero response positions.

It is possible to solve this problem by changing the array spacing within the range $d \leq \lambda_{\text{min}}/2$ and/or the sensor number to try to move those zero points out of the frequency range of interest. Another way is to modify the response of the Bessel function $J_k(\alpha\beta\Omega)$ at those zero points and assign a small non-zero value to those points so that the resultant filter $h_k(t)$ becomes stable. But this will lead to some distortion to the output signal at those frequency points.

A third solution is to employ multiple-ring circular arrays or concentric circular arrays (Chan and Chen, 2007; Di Claudio, 2005), as shown in Figure 5.36 with two rings as an example. These two sets of circular arrays may have different number of sensors. For each ring, we can obtain a set of phase mode outputs, denoted by $\hat{x}_{0,k}(t)$ and $\hat{x}_{1,k}(t)$, respectively, where k is the phase mode order. Then the final kth phase mode output is obtained by adding them together:

$$\begin{aligned} \hat{x}_k(t) &= \hat{x}_{0,k}(t) + \hat{x}_{1,k}(t) \\ &\approx \mathrm{e}^{jk\phi} j^k (J_k(\alpha_0\beta_0\Omega) + J_k(\alpha_1\beta_1\Omega)) \end{aligned} \tag{5.67}$$

where α_0, β_0, α_1 and β_1 are the corresponding parameters for the two rings and obtained in the same way as α and β in the single-ring circular array case in Figure 5.34. Since $J_k(\alpha_0\beta_0\Omega)$ and $J_k(\alpha_1\beta_1\Omega)$ will normally have different zero response points, the sum of them will smooth out any zero response over the frequency range of interest and the frequency response of the kth phase mode filter $h_k(t)$ can be obtained as:

$$H_k(\Omega) = \frac{1}{j^k(J_k(\alpha_0\beta_0\Omega) + J_k(\alpha_1\beta_1\Omega))} \tag{5.68}$$

Although this can solve the problem, the cost is that we have to employ a much larger number of sensors.

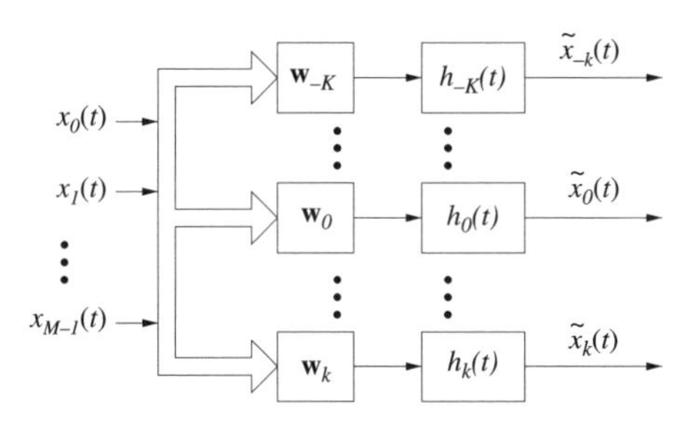

Figure 5.37 The $2K+1$ processing blocks with outputs $\tilde{x}_k(t)$, $k=-K,\ldots,-1,0,1,\ldots,K$

5.4.2 *FIB Design*

Based on the analysis to the processing block of Figure 5.35, we can build $2K+1$ such blocks with a weight vector $\mathbf{w}_k$ and the corresponding filter $h_k(t)$, $k=-K,\ldots,-1,0,1,\ldots,K$, as shown in Figure 5.37. If we consider each of the blocks as one virtual sensor, then its response to a signal arriving from the DOA angle ϕ is given by $e^{jk\phi}$, as indicated in Equation (5.66). Their outputs $\tilde{x}_k(t)$, $k=-K,\ldots,-1,0,1,\ldots,K$, can be considered as the outputs of a new virtual sensor array, with its steering vector $\mathbf{d}(\phi)$ expressed as:

$$\begin{aligned}\mathbf{d}(\phi) &= \left[e^{-jK\phi}\ e^{-j(K-1)\phi}\ \cdots\ e^{j(K-1)\phi}\ e^{jK\phi}\right]^T \\ &= e^{jK\phi}\left[e^{-j2K\phi}\ e^{-j(2K-1)\phi}\ \cdots\ e^{-j\phi}\ 1\right]^T\end{aligned} \tag{5.69}$$

If we want to design a frequency invariant beamformer $P(\phi)$, we can apply a set of coefficients w_l, $l=0,1,\ldots,2K+1$, to this virtual sensor array. The beam response is given by:

$$P(\phi) = e^{jK\phi}\sum_{l=0}^{2K} e^{-jl\phi}w_l = \mathbf{w}^H\mathbf{d}(\phi) \tag{5.70}$$

where:

$$\mathbf{w} = [w_0^*\ w_1^*\ \cdots\ w_{2K}^*]^T \tag{5.71}$$

Clearly the problem of finding the coefficients w_l to achieve a desired response $P(\phi)$ is equivalent to an FIR filter design problem and all of the existing FIR filter design techniques can be applied here directly.

5.4.3 *Design Example*

The design is based on a UCA with 10 sensors and the adjacent circumferential sensor spacing d is chosen to be half of the wavelength corresponding to the highest frequency

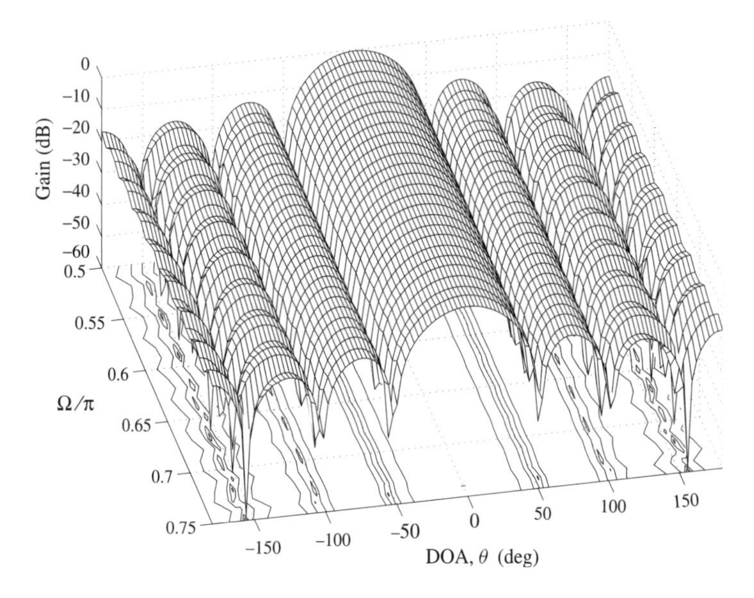

Figure 5.38 The resultant beam pattern for the frequency range $\Omega \in [0.5\pi, 0.75\pi]$

($\alpha = 1$). The frequency range of interest is $\Omega \in [0.5\pi \ 0.75\pi]$. With $\beta = M/(2\pi) = 5/\pi$, we have:

$$K \approx \beta\Omega_{\min} = \frac{5}{\pi} \times 0.5\pi = 2.5 \approx 3 \tag{5.72}$$

Therefore in total we can have $2 \times 3 + 1 = 7$ phase mode outputs. Each of the filter $h_k(t)$ is designed according to the requirement $H_k(\Omega) = 1/(j^k J_k(\alpha\beta\Omega))$ and realized by a 128-tap FIR filter. The beam pattern with a uniform weighting for $\mathbf{w}$ is shown in Figure 5.38. Its frequency invariance property can be seen clearly.

5.5 Direct Optimization for Frequency Invariant Beamforming

As a special class of fixed wideband beamformer, the frequency invariant beamformer can also be designed through the direct optimization approaches discussed in Chapter 4. The key is to introduce an additional constraint in the design to control the frequency invariance property during the optimization. In this section, we will review three of them: the convex optimization approach, the least squares approach and the eigenfilter approach (Duan *et al.*, 2008; Yan *et al.*, 2007; Zhao *et al.*, 2008, 2009a,b,c,d).

5.5.1 Convex Optimization

To design a frequency invariant beamformer, we need to incorporate a measurement of this property into the original formulation and an easy way is to add additional constraints to make sure the difference error between the response $P(\Omega, \theta)$ of the beamformer at the reference frequency Ω_0 and the other frequencies within the frequency band of interest Ω_{pb} is limited to a very small value ε.

Specifically, we can constrain the following mean squared difference error over the full frequency invariant region in the design:

$$\begin{aligned} v_{FI} &= \frac{1}{\Omega_{pb}\Theta_{FI}} \int_{\Omega_{pb}} \int_{\Theta_{FI}} |P(\Omega,\theta) - P(\Omega_0,\theta)|^2 \mathrm{d}\Omega \mathrm{d}\theta \\ &= \frac{1}{\Omega_{pb}\Theta_{FI}} \int_{\Omega_{pb}} \int_{\Theta_{FI}} |\mathbf{w}^H(\mathbf{d}(\Omega,\theta) - \mathbf{d}(\Omega_0,\theta))|^2 \mathrm{d}\Omega \mathrm{d}\theta \\ &= \mathbf{w}^H \boldsymbol{D}_{FI}\mathbf{w} \end{aligned} \tag{5.73}$$

where Θ_{FI} is the angle region where frequency invariance is required and:

$$\boldsymbol{D}_{FI} = \frac{1}{\Omega_{pb}\Theta_{FI}} \int_{\Omega_{pb}} \int_{\Theta_{FI}} (\mathbf{d}(\Omega,\theta) - \mathbf{d}(\Omega_0,\theta))(\mathbf{d}(\Omega,\theta) - \mathbf{d}(\Omega_0,\theta))^H \mathrm{d}\Omega \mathrm{d}\theta \tag{5.74}$$

With a real-valued $\mathbf{w}$, similar to the discussion in Section 4.2, we have:

$$v_{FI} = \mathbf{w}^H \boldsymbol{D}_{FIR}\mathbf{w} \tag{5.75}$$

where $\boldsymbol{D}_{FIR}$ is the real part of $\boldsymbol{D}_{FI}$; v_{FI} is also called the spatial response variation (SRV) in Duan *et al.*, (2008). The symmetric matrix $\boldsymbol{D}_{FIR}$ can be decomposed into the following form:

$$\boldsymbol{D}_{FIR} = \boldsymbol{V}_{FIR}\boldsymbol{U}_{FIR}\boldsymbol{V}_{FIR}^H \tag{5.76}$$

where $\boldsymbol{U}_{FIR}$ is a diagonal matrix with its diagonal elements being the eigenvalues of $\boldsymbol{D}_{FIR}$ and $\boldsymbol{V}_{FIR}$ is a full-rank matrix with its columns being the corresponding eigenvectors. Then we have:

$$v_{FI} = \mathbf{w}^H \boldsymbol{V}_{FIR}\boldsymbol{U}_{FIR}\boldsymbol{V}_{FIR}^H\mathbf{w} = |\boldsymbol{U}_{FIR}^{\frac{1}{2}}\boldsymbol{V}_{FIR}^H\mathbf{w}|^2 \tag{5.77}$$

Then the frequency invariant constraint can be expressed as:

$$|\boldsymbol{U}_{FIR}^{\frac{1}{2}}\boldsymbol{V}_{FIR}^H\mathbf{w}| < \varepsilon \tag{5.78}$$

This constraint can be incorporated into the original convex design formulation to achieve a frequency invariant beam pattern.

As an example, consider the minimax design described by Equation (4.10) in the convex optimization part of Section 4.1, Chapter 4. Adding the above constraint directly, we have:

$$\begin{aligned} &\min_{\mathbf{w}}\{ \max_{\substack{i=0,\ldots,I_\Omega-1 \\ j=1,\ldots,J_\theta-1}} |\boldsymbol{C}(\Omega_i,\theta_j)^T\mathbf{w}|\} \\ &\text{subject to } \boldsymbol{C}(\Omega_i,\theta_0)^T\mathbf{w} = \mathbf{f}(\Omega_i),\ i=0,\ldots,I_\Omega-1 \\ &|\boldsymbol{C}(\Omega_i,\theta_k)^T\mathbf{w}| < \delta_k,\ i=0,\ldots,I_\Omega-1 \\ &|\boldsymbol{U}_{FIR}^{\frac{1}{2}}\boldsymbol{V}_{FIR}^H\mathbf{w}| < \varepsilon \end{aligned} \tag{5.79}$$

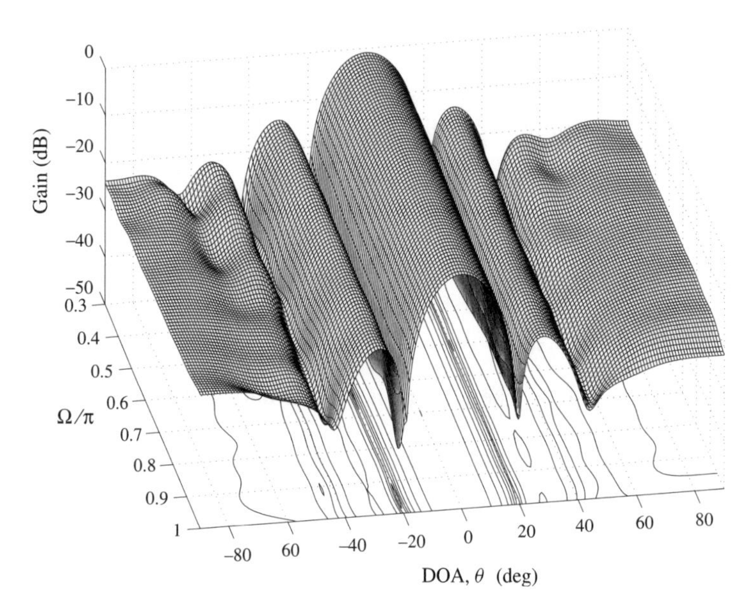

Figure 5.39 The resultant beam pattern for the frequency range $\Omega \in [0.3\pi\ \pi]$ using the convex optimization method

Suppose we want to apply the frequency invariance property over the full angle region between -90° and 90°. Then when we minimize the sidelobe response in Equation (5.79), we only need to minimize that at the reference frequency Ω_0, since with the frequency invariant constraint, the sidelobe response at the other frequencies will be minimized automatically. For the remaining constraints listed there, we can also drop those at frequencies other than Ω_0. Then the above formulation can be further simplified as:

$$\begin{aligned}
&\min_{\mathbf{w}}\{\max_{j=1,\ldots,J_\theta-1} |\boldsymbol{C}(\Omega_0,\theta_j)^T\mathbf{w}|\} \\
&\text{subject to } \boldsymbol{C}(\Omega_0,\theta_0)^T\mathbf{w} = \mathbf{f}(\Omega_0) \\
&|\boldsymbol{C}(\Omega_0,\theta_k)^T\mathbf{w}| < \delta_k \\
&|\boldsymbol{U}_{FIR}^{\frac{1}{2}}\boldsymbol{V}_{FIR}^H\mathbf{w}| < \varepsilon
\end{aligned} \tag{5.80}$$

A design example is shown in Figure 5.39 for a wideband beamformer with $M = 10$ sensors and $J = 20$ coefficients following each sensor. The look direction is the broadside, i.e. $\theta_0 = 0^\circ$, and the value of T_0/T_s in $\mathbf{f}(\Omega_0)$ is 10. The sidelobe area is from -90° to -20° and 20° to 90° and discretized into 100 points. The frequency range of interest is $\Omega_{pb} = [0.3\pi\ \pi]$ and discretized into $I_\Omega = 20$ points. When we calculate the frequency invariant constraint, we have discretized the region from -90° to 90° into 60 points. The reference frequency is $\Omega_0 = 0.65\pi$ and $\varepsilon = 0.04$. There is no constrained direction θ_k for the design. Compared to Figure 4.2, there is a clear frequency invariance property with the resultant beam pattern.

Note we can also choose to constrain the sidelobe response of the beamformer to a small value (an upper bound) and consider υ_{FI} as the cost function to be minimized. However, the problem is that we may not be able to achieve a good frequency invariance property (a small enough SRV) when the upper bound of the sidelobe response is too small.

5.5.2 Least Squares

The mean squared difference error or SRV υ_{FI} can be considered as a cost function and combined with the original least squares cost function to form a least squares approach to the FIB design.

Take the special case in Equation (4.36) in the constrained least squares design part of Section 4.2 as an example. In that case, the least squares cost function $J_{ls}(\mathbf{w})$ is given by:

$$J_{ls} = \mathbf{w}^H \boldsymbol{G}_{ls} \mathbf{w} \tag{5.81}$$

with:

$$\boldsymbol{G}_{ls} = \int_{\Omega_{pb}} \int_{\Theta_{sl}} \boldsymbol{D}_R(\Omega, \theta) \mathrm{d}\Omega \mathrm{d}\theta \tag{5.82}$$

where we have assumed $\alpha = 0$ in the original equation. Now combined with υ_{FI}, we can form a new cost function $J_{ls,FI}$ as:

$$\begin{aligned} J_{ls,FI} &= \gamma \mathbf{w}^H \boldsymbol{G}_{ls} \mathbf{w} + (1-\gamma)\upsilon_{FI} \\ &= \mathbf{w}^H \boldsymbol{G}_{ls,FI} \mathbf{w} \end{aligned} \tag{5.83}$$

where:

$$\boldsymbol{G}_{ls,FI} = \gamma \boldsymbol{G}_{ls} + (1-\gamma)\boldsymbol{D}_{FIR} \tag{5.84}$$

where γ is a trade-off factor between sidelobe attenuation and the frequency invariance property. Then the FIB design problem can be formulated as the following constrained minimization problem:

$$\min_{\mathbf{w}} J_{ls,FI}(\mathbf{w}) \qquad \text{subject to } \boldsymbol{C}^H \mathbf{w} = \mathbf{f} \tag{5.85}$$

Its solution is given by:

$$\mathbf{w}_{opt} = \boldsymbol{G}_{ls,FI}^{-1} \boldsymbol{C} (\boldsymbol{C}^H \boldsymbol{G}_{ls,FI}^{-1} \boldsymbol{C})^{-1} \mathbf{f} \tag{5.86}$$

Note if the frequency invariance property is required for the full angle region, we can calculate the value of $\boldsymbol{G}_{ls}$ for the reference frequency only in a similar consideration as in the convex optimization part.

A design example is shown in Figure 5.40. All of the parameters are the same as in the example of Figure 4.8 in Section 4.2 and here the frequency invariance property is applied only to the main beam region from 0° to 40°, which is discretized into 30 points for calculating $\boldsymbol{D}_{FIR}$. The reference frequency is $\Omega_0 = 0.65\pi$ and the trade-off factor γ is 0.2. Compared to Figure 4.8, a frequency invariance property is clearly visible at the main beam region in the new result.

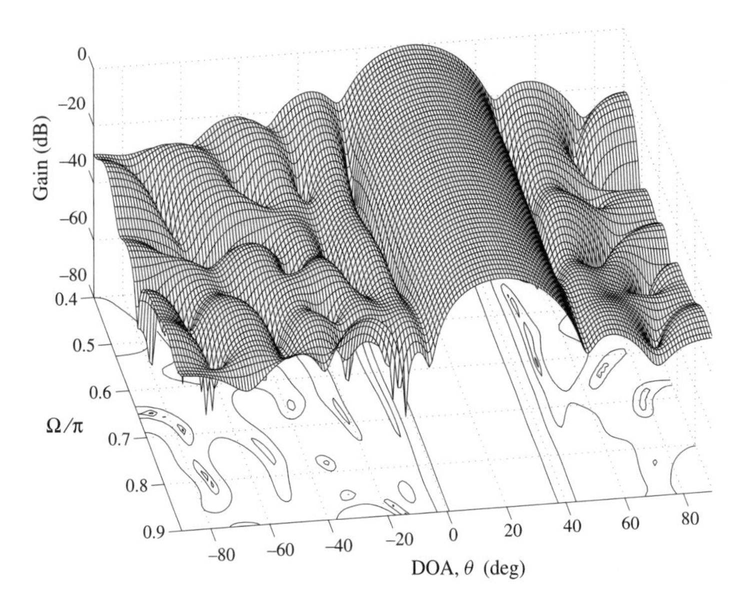

Figure 5.40 The resultant beam pattern for the frequency range $\Omega \in [0.4\pi\ 0.9\pi]$ using the least squares method

5.5.3 Eigenfilter

For the eigenfilter method, the cost function v_{FI} can be incorporated into the design in a similar way as in the least squares case.

Consider the constrained maximum energy design given in Equation (4.66) in Section 4.3 as an example. Using the result in Section 5.5.2, we can replace $\boldsymbol{G}_{sl}$ in the denominator by the newly formed $\boldsymbol{G}_{sl,FI}$, which is given by:

$$\boldsymbol{G}_{sl,FI} = \gamma \boldsymbol{G}_{sl} + (1-\gamma)\boldsymbol{D}_{FIR} \tag{5.87}$$

Then we have the following formulation for the FIB design:

$$\max_{\mathbf{w}} \frac{\mathbf{w}^H \boldsymbol{G}_{ml} \mathbf{w}}{\mathbf{w}^H \boldsymbol{G}_{sl,FI} \mathbf{w}} \quad \text{subject to} \quad \boldsymbol{C}^H \mathbf{w} = \mathbf{f} \tag{5.88}$$

Suppose we want to apply the frequency invariance property over the full angle range; then when we calculate $\boldsymbol{G}_{ml}$ and $\boldsymbol{G}_{sl}$ we only need to calculate their values at the reference frequency Ω_0, instead of the entire frequency range of interest Ω_{pb}. The solution can be obtained in the same way as described in Section 4.3.

As an example, the frequency invariant constraint is imposed on the entire beam pattern corresponding to the design shown in Figures 4.13 and 4.14 in Section 4.3. Here the entire angle region from -90° to 90° is discretized into 100 points for calculating $\boldsymbol{D}_{FIR}$. The reference frequency is $\Omega_0 = 0.65\pi$ and the trade-off factor γ is 0.01. The result is shown in Figures 5.41 and 5.42. Compared to the results without the frequency invariant constraint, although the sidelobe attenuation is not as good (about 3 dB lower) due to the trade-off between sidelobe attenuation and the frequency invariance property, we have

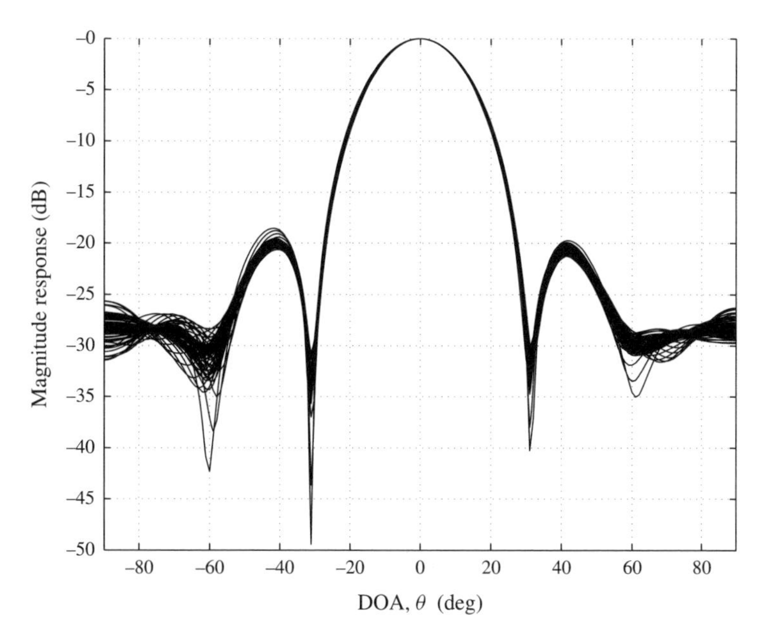

Figure 5.41 A design example using the constrained maximum energy method for the FIB design as formulated in Equation (5.88)

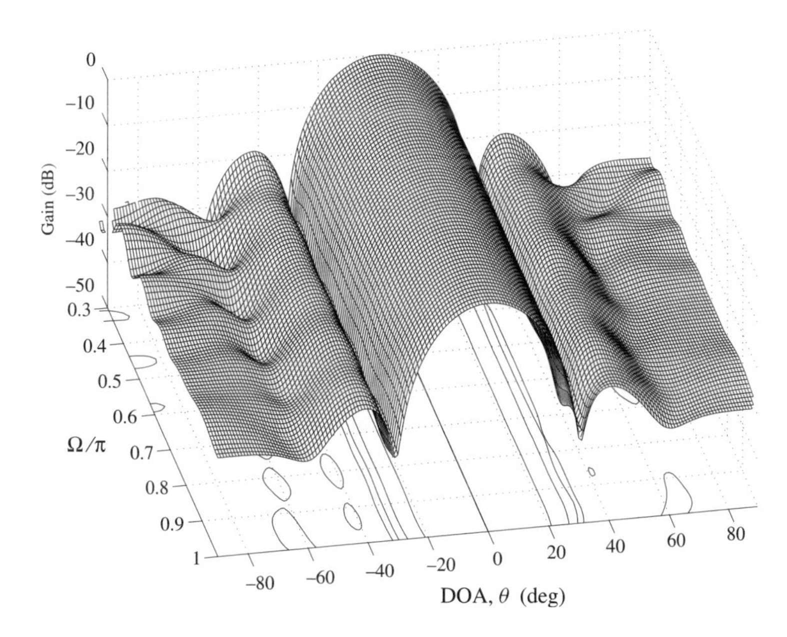

Figure 5.42 The three-dimensional beam pattern corresponding to the two-dimensional result shown in Figure 5.41

achieved a constant beamwidth with a satisfactory frequency invariant response over the entire region.

5.6 Beamspace Adaptive Wideband Beamforming

The frequency invariant beamforming technique can be used for data independent beamforming. In addition, it can also be applied to other closely related areas for an improved performance, such as DOA estimation, adaptive beamforming and blind source separation (Chan *et al.*, 2002; Liu, 2009b,d; Liu and Mandic, 2005; Liu and Weiss, 2008a; Liu *et al.*, 2005, 2007d; Sekiguchi and Karasawa, 2000; Ward *et al.*, 1998). In this section, we will discuss one of them – its application to wideband adaptive beamforming, which leads to a technique called beamspace adaptive wideband beamforming.

5.6.1 Structure

Beamspace adaptive beamforming has been studied widely in the narrowband area (Freese *et al.*, 2003; Kazanci and Krolik, 2005; Lee and Tsai, 2001; Lee and Freese, 2005; Takao and Uchida, 1989). This idea can be extended to the wideband case by employing the frequency invariant beamforming technique.

However, there are different problems arising from the wideband beamspace adaptive arrays compared to the narrowband case. In the narrowband beamspace adaptive method, the beams formed need only meet one condition in theory, i.e. all of the beams pointing to different directions should be linearly independent or otherwise it will be a waste of resources and also increase the complexity of the system unnecessarily. In the wideband case, there is an additional condition imposed, i.e., all of the beams should have a satisfactory frequency invariance property or otherwise we cannot simply use one weight for each of the beam outputs. Therefore, there will be a trade-off between the two imposed conditions to improve the beamformer's performance.

The key to the beamspace adaptive wideband beamformer is the design of FIBs and we will use the design based on multi-dimensional Fourier transforms introduced in Section 5.2.2 in our following study. In the FIB design described there, for a linear array, the starting point is the desired response $P(\sin\theta)$ which can be obtained by a narrowband beamformer design method.

Due to the close relationship between a narrowband beamformer and an FIR filter, as mentioned in Section 1.2, $P(\sin\theta)$ can also come from a 1-D prototype FIR filter $H(z)$ by the substitution $z = e^{j\pi\sin\theta}$. If $H(z)$ is a lowpass filter, then signals from the directions around $\theta = 0$ will correspond to its passband, and a beam is formed pointing to the direction $\theta = 0$. For beam directions other than the broadside, we can choose an appropriate bandpass filter.

For a planar array, we can form the desired response $P(\sin\theta\cos\phi, \sin\theta\sin\phi)$ from a 2-D prototype filter $H(z_1, z_2)$ by the substitutions $z_1 = e^{j\pi\sin\theta\cos\phi}$ and $z_2 = e^{j\pi\sin\theta\sin\phi}$. If $H(z_1, z_2)$ is a lowpass filter, then for $\theta = 0, \phi = 0$, we have $z_1 = z_2 = 1$, which corresponds to its maximum magnitude response.

In a narrowband beamspace adaptive array (Takao and Uchida, 1989), a total of N beams are formed by a beamforming network, where one is the main beam pointing to the direction of the signal of interest and the remaining $N - 1$ beams are auxiliary beams

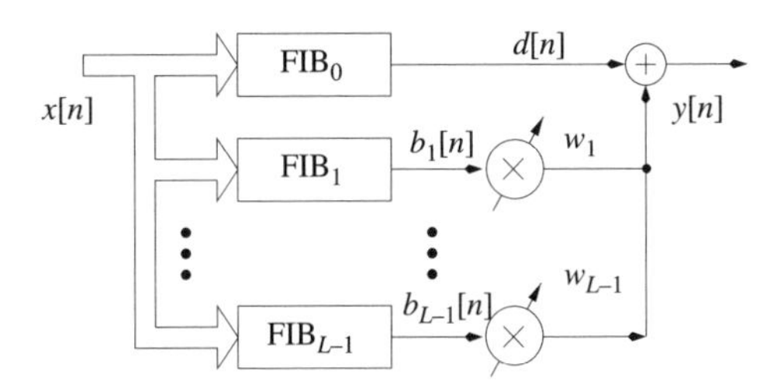

Figure 5.43 A general structure for wideband beamspace adaptive arrays

pointing to the remaining directions. The output power levels of the auxiliary beams are compared to a threshold and those higher than the threshold will be chosen in the following adaptation. In this way the resultant partially adaptive array can maintain an acceptable performance with a lower computational complexity. Extending this idea to the wideband case, we can also design N wideband beams pointing to different directions to form a wideband beamspace adaptive array. To combine the outputs of the beams with one adaptive weight for each of them, their response should be frequency invariant.

Applying the beamspace approach to wideband arrays, L frequency invariant beams will be designed pointing to the directions θ_l (θ_l, ϕ_l for planar arrays), $l = 0, 1, \ldots, L-1$. One of the beams is the main beam pointing to the direction of the signal of interest and the remaining $L-1$ beams are auxiliary beams pointing to the other directions which cover all of the possible directions of the interfering signals.

Figure 5.43 shows such a structure, where $\mathbf{x}[n]$ is the received array signal vector at time n, FIB_0 is the main beam with an output $d[n]$ and $y[n]$ is the final array output by combining the other FIB outputs $b_1[n], b_2[n], \ldots, b_{L-1}[n]$ with one adaptive coefficient w_l, $l = 1, \ldots, L-1$ for each of them, given by:

$$y[n] = d[n] - \mathbf{w}^H \mathbf{b}[n] \tag{5.89}$$

where:

$$\begin{aligned} \mathbf{b}[n] &= \left[b_1[n]\ b_2[n]\ \cdots\ b_{L-1}[n]\right]^{\mathrm{T}} \\ \mathbf{w} &= \left[w_1^*\ w_2^*\ \cdots\ w_{L-1}^*\right]^{\mathrm{T}} \end{aligned} \tag{5.90}$$

Without loss of generality, we assume the signal of interest comes from the broadside, which is known to the system. For the auxiliary beams, they have a zero response to the signal of interest and their outputs only contain noise and interfering signals. Only one adaptive weight is needed for each of the auxiliary beams and these weights can be adjusted by minimizing the mean square error $E\{|y[n]|^2\}$ between the main beam and the auxiliary beams, leading to LMS-type adaptive algorithms as described in Section 2.1.

In the discussed scenario, the classical beamforming method is the LCMV beamformer or the GSC. In Table 5.1, we list the computational complexities of the beamspace method and the GSC for the adaptive part, employing a normalized LMS algorithm as an example. The computational complexity is measured in term of the number of real multiplications and we also assumed a real-valued input signal $\mathbf{x}[n]$. Note, in Table 5.1, J is the length

Table 5.1 Comparison of the computational complexity of the adaptive part for both the beamspace adaptive beamformer and the GSC

Method	Real multiplications
Beamspace	$2L + 1$
GSC	$2(MN - 1)J + MN + 1$

of the TDL or FIR filter attached to each sensor in the GSC; the dimension of the array is $M \times N$ (planar array) with $N = 1$ for the linear array case.

5.6.2 Analysis of the Beamspace Adaptive Method

For the beamspace array to work, the frequency invariant beamforming network needs to meet two conditions.

Firstly, the beams should have a satisfactory frequency invariance property over the frequency band of interest, which is dependent on the shape of the desired beam response $P(\sin\theta\cos\phi, \sin\theta\sin\phi)$ ($P(\sin\theta)$ for linear arrays) and the temporal/spatial dimension of the corresponding beamformer. The more complicated the shape, the more coefficients we need for each of the frequency invariant beams, i.e. larger M, N and J (M and J for linear arrays).

Secondly, the beams should be linearly independent. Otherwise, some of the beam outputs will be a linear combination of the others, which causes a waste of resources and also reduces the number of effective beams. As a result, we will not be able to null out the desired number of interfering signals.

These two conditions are not independent and there is a close relationship between them. In the following, we will show that the number of independent beams formed L_{ind} cannot exceed $P = P_1 \times P_2$, where P_1 and P_2 are the dimensions of the 2-D prototype filter $H(z_1, z_2)$ for the case of planar arrays. For linear arrays, P is the length of the prototype 1-D filter $H(z)$. Note the larger the values of P_1 and P_2, the more complicated the desired frequency invariant response can be. As a result, for fixed M, N and J, the frequency invariance property of the FIBs obtained based on the more complicated prototype filter will be worse.

Next we give our proof by contradiction for planar arrays, which can be extended to the case of linear arrays in a straightforward way.

Suppose we can have $L > P$ independent beams formed by some prototype filters with dimensions $P_1 \times P_2$. These beams have a response of $H_l(e^{j\pi\sin\theta\cos\phi}, e^{j\pi\sin\theta\sin\phi})$, $l = 0, 1, \ldots, L-1$, which is derived from the corresponding prototype filter $H_l(z_1, z_2)$ with a 2-D impulse response given by

$$\boldsymbol{H}_l = \begin{bmatrix} h_{l,0,0} & \cdots & h_{l,0,P_2-1} \\ h_{l,1,0} & \cdots & h_{l,1,P_2-1} \\ \vdots & \ddots & \vdots \\ h_{l,P_1-1,0} & \cdots & h_{l,P_1-1,P_2-1} \end{bmatrix} \tag{5.91}$$

Now consider the linear combination of these impulse response matrices:

$$\mathbf{0} = \gamma_0 \boldsymbol{H}_0 + \gamma_1 \boldsymbol{H}_1 + \cdots + \gamma_{L-1} \boldsymbol{H}_{L-1} \tag{5.92}$$

where $\gamma_0, \ldots, \gamma_{L-1}$ are coefficients to be found for this equation to hold. Applying the 2-D z-transform to both sides and then using the substitutions $z_1 = e^{j\pi \sin\theta \cos\phi}$ and $z_2 = e^{j\pi \sin\theta \sin\phi}$, we arrive at:

$$0 = \sum_{l=0}^{L-1} \gamma_l H_l(e^{j\pi \sin\theta \cos\phi},\ e^{j\pi \sin\theta \sin\phi}) \tag{5.93}$$

Since the L FIBs $H_l(e^{j\pi \sin\theta \cos\phi},\ e^{j\pi \sin\theta \sin\phi})$ are independent, all of the γ_l, $l = 0, \ldots, L-1$, must be zero for Equation (5.93) to hold, and then for Equation (5.92) to hold, which means that the L matrices $\boldsymbol{H}_l$, $l = 0, 1, \ldots, L-1$ are independent. Now we form a vector $\mathbf{h}_l$, $l = 0, 1, \ldots, L-1$, using all of the P elements of matrix $\boldsymbol{H}_l$.

Consider the following equation:

$$\mathbf{0} = \gamma_0 \mathbf{h}_0 + \gamma_1 \mathbf{h}_1 + \cdots + \gamma_{L-1} \mathbf{h}_{L-1} \tag{5.94}$$

Since for Equation (5.92) to hold, all of the coefficients $\gamma_0, \ldots, \gamma_{L-1}$ must be zero and for Equation (5.94) to hold, all of the coefficients $\gamma_0, \ldots, \gamma_{L-1}$ must also be zero. However, as L is larger than $P = P_1 \times P_2$, for Equation (5.94) to hold, we can always find a set of coefficients $\gamma_0, \ldots, \gamma_{L-1}$ with at least one non-zero element. Thus, we reach a contradiction.

As the dimension of $\mathbf{h}_l$ is P, we can design at most P independent FIBs. Clearly, although we can design as many frequency invariant beams as we want, only P of them are independent and at most we can only null out $P-1$ interfering signals. Note we only apply one adaptive weight to each of the beam outputs. The reason for this is that the frequency invariance property of the beams transforms the wideband beamforming problem into a narrowband one. The better the frequency invariance property, the using of one single weight for each beam output is more justified and the performance would be better.

Therefore, the array's interference cancellation ability is dependent on both the number of independent beams and the frequency invariance property. However, for fixed M, N and J, these two factors are not independent with each other: a larger value of P means a larger number of independent FIBs; however it also means a more complicated desired beam response and therefore a worse frequency invariance property. A larger number of independent FIBs increases the interference cancellation capability, while a worse frequency invariance property reduces it. So there is a trade-off between these two factors and also an optimum point for achieving the best performance, although it is very difficult to find it analytically.

For example, in the linear array case, according to (Sekiguchi and Karasawa, 2000), the dimensions M and J of the FIB should be at least 3 times that of the prototype filter, i.e. $P \leq \min\{M/3, J/3\}$. We may choose a prototype filter with $P = \min\{M/3, J/3\}$ for a good frequency invariance property, but when the number of interferences increases and becomes larger than $(\min\{M/3, J/3\} - 1)$, the array will not be able to null out the additional interferences. Therefore we may need to sacrifice the frequency invariance property a little to increase P (but still not larger than the sensor number M) and design more independent beams. The loss in frequency invariance property can be compensated by the gain in the increasing number of independent beams. As a result, the interference cancellation ability of the array is improved. We will give some results to show this trade-off in the simulations.

The next question is, provided the length of the prototype filter P, how to design P independent frequency invariant beams. We will introduce a DFT matrix based method in the next section with the beam directions uniformly distributed in the spatial space and their independence guaranteed inherently.

5.6.3 Design of Independent FIBs

To make sure that the $L \leq P$ frequency invariant responses (or the frequency responses of the prototype filters) are independent to each other, we give a sufficient condition for it for planar arrays and then extend it to linear arrays.

As long as for each of the L prototype responses $H_l(z_1, z_2)$, there exists a point $(z_{1,l}, z_{2,l})$ ($z_{1,l} = \mathrm{e}^{j\Omega_{1,l}}$ and $z_{2,l} = \mathrm{e}^{j\Omega_{2,l}}$), for which we have $H_l(z_{1,l}, z_{2,l}) \neq 0$ and all the remaining $L-1$ responses $H_m(z_{1,l}, z_{2,l}) = 0$ for $m \neq l$, then the set of frequency responses H_l, $l = 0, 1, \ldots, L-1$, and hence the set of frequency invariant beams formed by them will be linearly independent.

Consider Equation (5.92) again. Applying the 2-D z-transform to both sides we have:

$$0 = \sum_{l=0}^{L-1} \gamma_l H_l(z_1, z_2) \tag{5.95}$$

First, we put $z_1 = z_{1,0}$ and $z_2 = z_{2,0}$ into the above equation, then:

$$0 = \sum_{l=0}^{L-1} \gamma_l H_l(z_{1,0}, z_{2,0}) \tag{5.96}$$

As only $H_0(z_{1,0}, z_{2,0}) \neq 0$ and all the other $H_{m\neq 0}(z_{1,0}, z_{2,0})$ are zero, we have $\gamma_0 = 0$. Similarly, we have $\gamma_l = 0$, $l = 1, \ldots, L-1$. As a result, for Equation (5.95) to hold, all the L coefficients must be zero, i.e. the L prototype responses are linearly independent.

For linear arrays, this condition can be stated as: **as long as for each of the L prototype responses $H_l(z)$, there exists a point $z = z_l = \mathrm{e}^{j\Omega_l}$, for which we have $H_l(z_l) \neq 0$ and all the remaining $L-1$ responses $H_m(z_l) = 0$ for $m \neq l$, then the set of frequency responses H_l, $l = 0, 1, \ldots, L-1$, and hence the set of frequency invariant beams formed by them will be linearly independent.**

The easiest way to exploit this property is to use the 2-D DFT matrix as given below:

$$\boldsymbol{D}_{l,m} = \begin{bmatrix} 1 & Z_2^m & \cdots & Z_2^{m(P_2-1)} \\ Z_1^l & Z_1^l Z_2^m & \cdots & Z_1^l Z_2^{m(P_2-1)} \\ \vdots & \vdots & \ddots & \vdots \\ Z_1^{l(P_1-1)} & Z_1^{l(P_1-1)} Z_2^m & \cdots & Z_1^{l(P_1-1)} Z_2^{m(P_2-1)} \end{bmatrix} \tag{5.97}$$

where $Z_1 = \mathrm{e}^{-j\frac{2\pi}{P_1}}$, $Z_2 = \mathrm{e}^{-j\frac{2\pi}{P_2}}$, $l = 0, 1, \ldots, P_1 - 1$ and $m = 0, 1, \ldots, P_2 - 1$. $\boldsymbol{D}_{0,0}$ is the 2-D prototype filter for the broadside main beam designed for the signal of interest and the remaining matrices are the prototype filters for the auxiliary beams. For linear arrays, $P_1 = P$ and $P_2 = 1$.

Note that the $H_l(\mathrm{e}^{j\pi \sin\theta \cos\phi}, \mathrm{e}^{j\pi \sin\theta \sin\phi})$ obtained by the 2-D DFT matrix is complex-valued for different θ and ϕ, that is:

$$H_l(\mathrm{e}^{j\pi \sin\theta \cos\phi}, \mathrm{e}^{j\pi \sin\theta \sin\phi}) = A_l(\theta, \phi)\mathrm{e}^{jB_l(\theta,\phi)} \tag{5.98}$$

where $A_l(\theta, \phi)$ and $B_l(\theta, \phi)$ are some real functions. The change of both $A_l(\theta, \phi)$ and $B_l(\theta, \phi)$ with respect to different θ and ϕ will lead to a unnecessarily complicated desired frequency invariant response. As $A_l(\theta, \phi)$ contains enough information about the shape of this response, we can ignore the phase part $B_l(\theta, \phi)$ and use only $A_l(\theta, \phi)$ as the simplified desired frequency invariant response in the design. Design results show that in this way we can significantly improve the frequency invariance property of the beams with the same array dimensions.

5.6.4 *Simulations*

The first set of simulations is based on a 15×15 uniformly spaced planar array. The beamspace adaptive beamformer employs 25 FIBs, each of which has dimensions of $15 \times 15 \times 17$ and is based on a 5×5 prototype 2-D filter obtained by the corresponding simplified DFT matrix. The signal of interest comes from the broadside and six interfering signals are from the directions $(\theta = 20^\circ, \phi = 30^\circ)$, $(\theta = 30^\circ, \phi = 70^\circ)$, $(\theta = 40^\circ, \phi = 110^\circ)$, $(\theta = 50^\circ, \phi = 150^\circ)$, $(\theta = 60^\circ, \phi = 190^\circ)$ and $(\theta = 70^\circ, \phi = 230^\circ)$, respectively. All of the signals have a bandwidth of $[0.5\pi\ 0.9\pi]$. The SIR is about -20 dB and the SNR is about 20 dB. We use a normalized LMS algorithm for adaptation. Its performance is compared to the GSC, where the TDL length is 17 for each sensor. The stepsize is 0.06 for the beamspace method and 0.1 for the GSC, which is empirically chosen to achieve roughly the same steady-state value of the mean square residual error.

The learning curves for these two beamformers are shown in Figure 5.44, where the beamspace method reaches its steady state in about 400 iterations, whereas it takes the GSC 6000 iterations. The much faster convergence speed is mainly due to the fact that the beamspace method only needs one adaptive weight for each of the FIB outputs and as a result the adaptive filter length of the beamspace method is much shorter than the GSC. In this case the adaptive filter length for the beamspace method is $25 - 1 = 24$ and for the GSC it is $(15 \times 15 - 1) * 17 = 3808$.

Since in the beamspace method the beamformer output is a linear combination of the outputs of the FIBs by one single coefficient, the resultant beam pattern of the beamspace adaptive beamformer will be frequency invariant, which can be verified by examining the beam pattern for different frequencies. As an example, the beam pattern in cylindrical coordinates for two different frequencies $\Omega = 0.5\pi$ and $\Omega = 0.9\pi$ is shown in Figures 5.45 and 5.46, respectively.

In the second set of simulations, we change the SIR value to -26 dB and the SNR to 14 dB. We still use the same set of signals as in the previous simulations. However, the last two interfering signals (from directions $(\theta = 60^\circ, \phi = 190^\circ)$ and $(\theta = 70^\circ, \phi = 230^\circ)$,

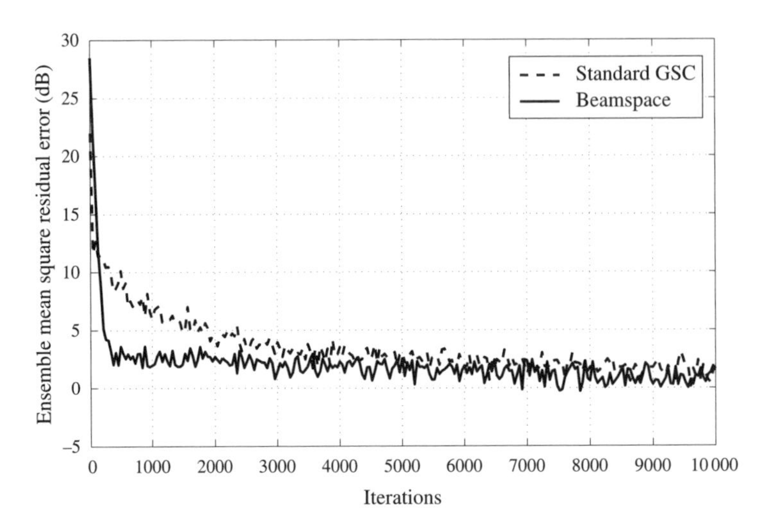

Figure 5.44 The learning curves for the beamspace adaptive beamformer and the GSC in the first set of simulations

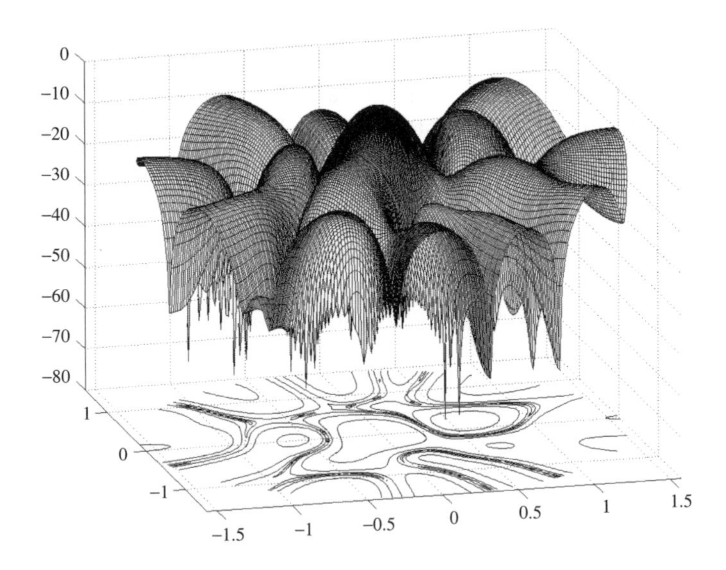

Figure 5.45 The resultant beam pattern of the beamspace adaptive beamformer for the frequency $\Omega = 0.5\pi$

respectively) do not appear until at iteration number $n = 5100$. In this way, we can easily see the tracking capability of the proposed method compared to the traditional GSC. The learning curves of the two methods are shown in Figure 5.47 with the same stepsize of 0.06. At about $n = 5100$, due to the sudden appearance of the two new interfering signals, we see a significant increase of the ensemble mean square residual error. However, the beamspace method has adjusted to the changed environment very quickly in only about

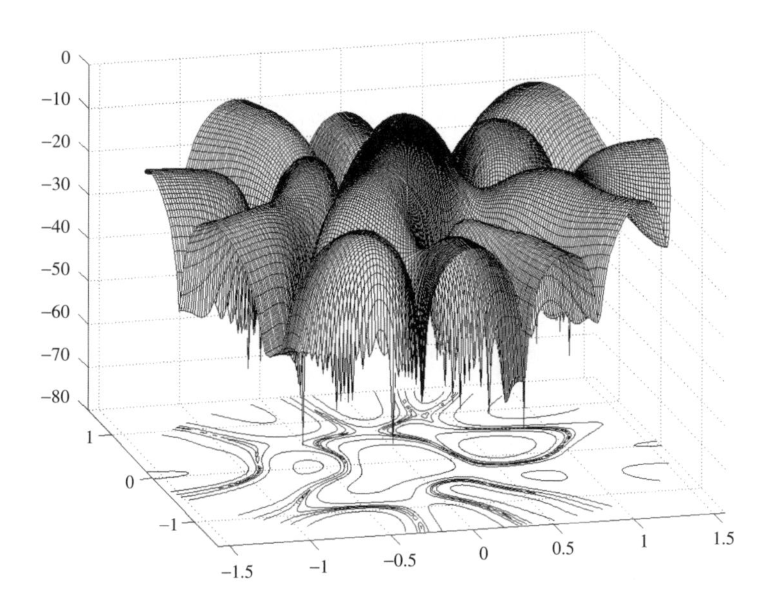

Figure 5.46 The resultant beam pattern of the beamspace adaptive beamformer for the frequency $\Omega = 0.9\pi$

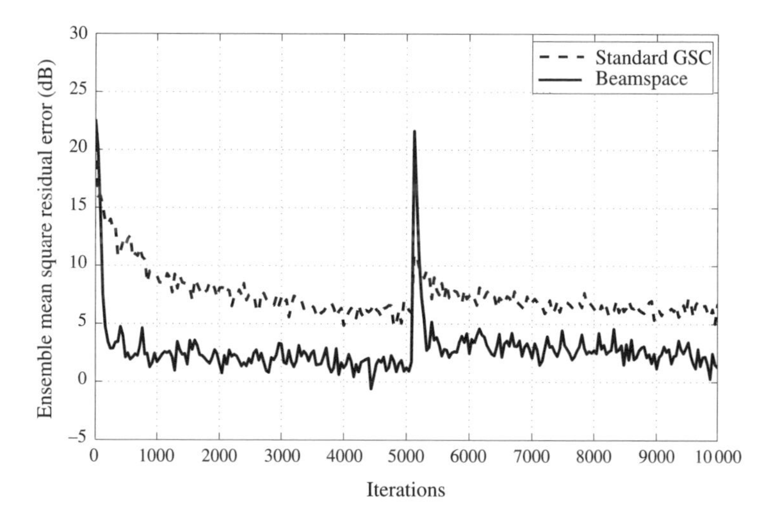

Figure 5.47 The learning curves for the beamspace adaptive beamformer and the GSC in the second set of simulations

500 iterations, whereas for the GSC it takes another 4000 iterations to reach a relatively steady state.

To show the trade-off between the frequency invariance property and the number of linear independent beams, we study the performance of a beamspace linear array, where the spatial and temporal dimensions of the FIBs are fixed as $M = 14$ and $J = 16$.

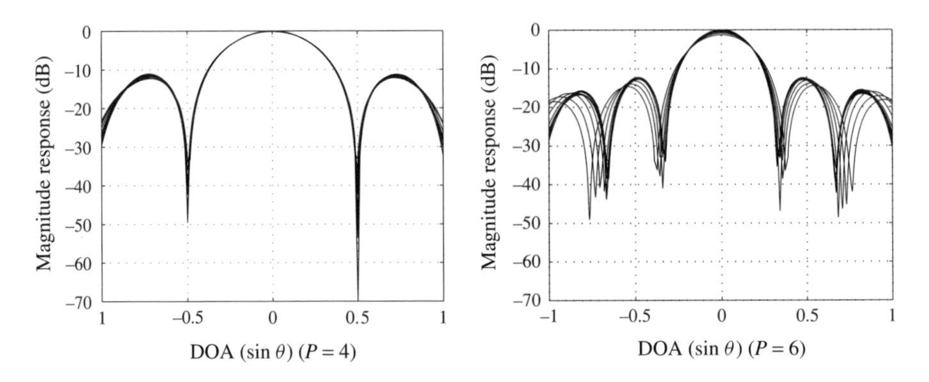

Figure 5.48 The magnitude response of the main beam over the bandwidth of $[0.4\pi\ 0.9\pi]$, based on a 4-tap and a 6-tap prototype filter, respectively

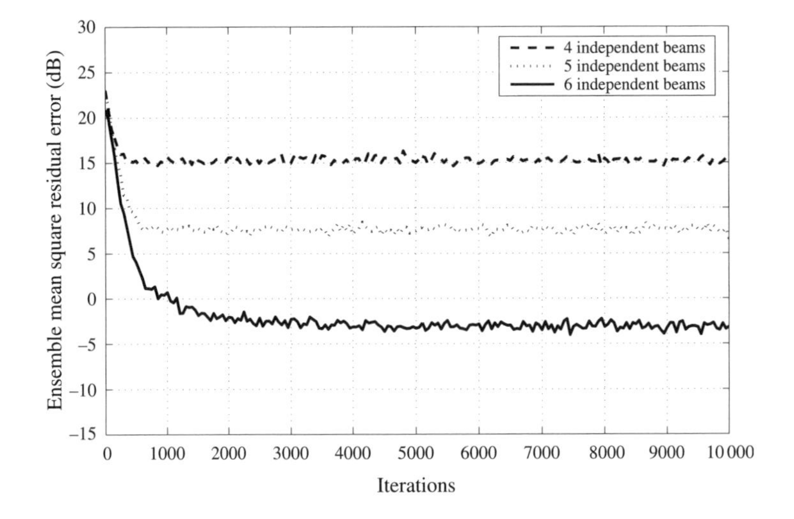

Figure 5.49 The learning curves for different numbers of independent beams

The signal of interest comes from the broadside and with an SIR of -20 dB and an SNR of 20 dB. Five interfering signals come from the angles of $20°$, $-25°$, $45°$, $-50°$, and $-80°$, respectively. Both the interfering signals and the signal of interest have a bandwidth of $[0.4\pi\ 0.9\pi]$. According to (Sekiguchi and Karasawa, 2000), ideally we should use a prototype filter of length $\lfloor 14/3 \rfloor = 4$ for the design of the 4 FIBs. Figure 5.48 shows the pattern of the main beam based on a 4-tap filter.

The learning curve with a stepsize of 0.01 is shown by the dashed line in Figure 5.49. As the number of interfering signals are 5, which is larger than $4 - 1 = 3$, the number of auxiliary beams, the 4-beam adaptive array cannot null out all of the interferences, although all of the beams have a very good frequency invariant response over the required bandwidth. As a result, the learning curve only reaches a level of 15 dB.

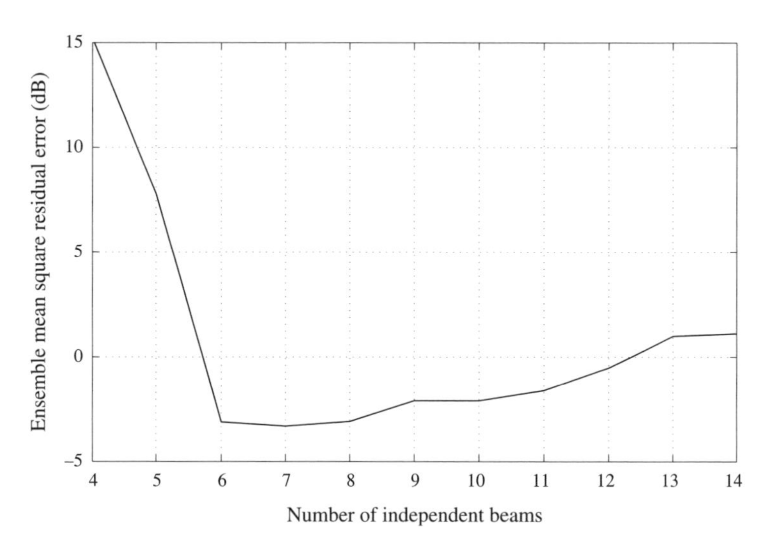

Figure 5.50 The steady-state ensemble mean square residual errors for different numbers of independent beams

In order to improve its performance, we need to sacrifice the frequency invariance property a little. So, we increased the length of the prototype filter to 5, and 5 independent beams are obtained. The learning curve of this new system with the same stepsize is shown by the dotted line in Figure 5.49. Compared to the 4-beam array, the ensemble mean square residual error has been reduced to about 8 dB.

We can further improve the performance of the system by designing 6 independent beams based on a 6-tap prototype filter. The frequency invariance property of the main beam in this case is also shown in Figure 5.48, which is clearly not as good as that of $P = 4$. However, as there are more independent beams formed in this array, a further improvement of more than 10 dB has been achieved, as shown by the solid line in Figure 5.49.

Figure 5.50 shows the ensemble mean square residual error for different number of independent beams. We see that with increased number of independent beams from the 7-beam case, the performance is getting worse and worse. The reason is, although there are more and more independent beams, their frequency invariance property has been sacrificed too much and the loss due to this cannot be compensated by the gain due to the increased number of independent beams.

5.7 Summary

In this chapter, we have discussed the design and application of a special class of fixed wideband beamformers – the frequency invariant beamformer.

We first studied the FIB design based on multi-dimensional inverse Fourier transforms, which is derived by exploiting the relationship between the array's spatial and temporal parameters and its beam pattern. This approach can be applied to 1-D, 2-D and 3-D arrays, for both continuous and discrete sensors and signals. A generalization to this approach was introduced by separating the design into two steps: the desired frequency response for each sensor is obtained first and then the problem is reduced to the design of normal FIR/IIR filters, for which the existing filter design methods can be applied.

To reduce the computational complexity of the FIB and improve the design for low frequencies, two subband implementations were discussed. In the first one, each received array signal is split into decimated subbands by an analysis filter bank and the corresponding subband signals form a series of subband arrays and an FIB is operated in each of them. Due to the change of signal frequency after decimation, a modified method for the subband FIB design was proposed accordingly. In a refinement, we further modified this subband FIB structure for use with a scaled aperture.

We also studied the FIB design problem for circular arrays based on the classical phase-mode transformation, which can be extended to concentric circular arrays and concentric spherical arrays.

As a special class of fixed wideband beamformers, the frequency invariant beamformer can also be designed through the direct optimization approaches discussed in Chapter 4. The key is to introduce an additional constraint in the design to control the frequency invariance property during the optimization. Three approaches were briefly reviewed, including the convex optimization approach, the least squares approach and the eigenfilter approach.

At the end of this chapter, an application of the FIBs to adaptive beamforming was studied, which is referred to as the beamspace adaptive wideband beamformer, with advantages of a faster convergence speed and a lower computational complexity. We gave a detailed analysis of the system for both linear arrays and planar arrays by showing the close relationship between the number of independent FIBs and the dimensions of the prototype filters for the FIB design. There is a clear trade-off between the two conditions imposed on the beamspace adaptive wideband beamformer: to improve the interference cancellation capability of the array, we may need to sacrifice the frequency invariance property of the beams to some degree for more linearly independent ones.

6

Blind Wideband Beamforming

For the adaptive wideband beamforming approaches introduced in previous chapters, we normally assume that a reference signal or the DOA angle of the signal of interest is available. However, it is possible to exploit the signal statistical properties to extract one or all of the signals of interest without knowing or estimating the signal DOA angles. The technique involved in this process is called blind source separation (BSS) and the approach to extracting the signal of interest from the received array signals employing the BSS technique is called blind beamforming. For wideband signals, it leads to the technique called blind wideband beamforming, which is the topic of this chapter.

The BSS technique will be reviewed first as an approach to recover the source signals from all kinds of their mixtures; then we will give an analysis of the signal mixing problem in the context of array signal processing and show that the beamforming problem can be solved by the BSS technique. Finally we will present an efficient blind wideband beamforming method based on the previously introduced frequency invariant beamforming technique.

6.1 Blind Source Separation

6.1.1 Introduction

Blind source separation has wide applications in the areas of biomedical engineering, sonar, radar, speech enhancement, telecommunications, etc., and has been studied extensively in recent years (Cichocki and Amari, 2003; Haykin, 2000a; Hyvarinen *et al.*, 2001). It is a technique to recover the original sources from all kinds of their mixtures, without the knowledge of the mixing process and the sources themselves. A block diagram of the BSS problem is shown in Figure 6.1. There are L source signals $s_0, s_1, \ldots, s_{L-1}$, which are passed through an unknown mixing system with added noise; by M sensors we acquire M received mixed signals $x_0, x_1, \ldots, x_{M-1}$. With appropriate separation algorithms, the original signals are then separated from their mixtures subject to the ambiguities of permutation and scaling.

The source signals can be mixed in a linear or nonlinear manner. Much of the current research and development has been focused on the linear mixing case, which is further divided into two categories: instantaneous mixing and convolutive mixing.

For instantaneous mixing, the mixtures are modeled as weighted sums of individual sources without dispersion or time delay, given by:

$$\mathbf{x}[n] = \mathbf{A}\mathbf{s}[n] + \mathbf{v}[n] \tag{6.1}$$

Wideband Beamforming Wei Liu and Stephan Weiss

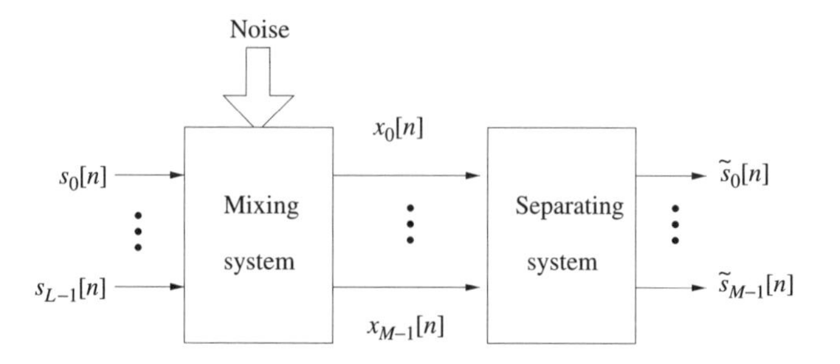

Figure 6.1 A general diagram of the blind source separation problem

with:

$$\begin{aligned}\mathbf{s}[n] &= \left[s_0[n]\ s_1[n]\ \cdots\ s_{L-1}[n]\right]^{\mathrm{T}} \\ \mathbf{x}[n] &= \left[x_0[n]\ x_1[n]\ \cdots\ x_{M-1}[n]\right]^{\mathrm{T}} \\ [\boldsymbol{A}]_{m,l} &= a_{m,l},\ m = 0, \ldots, M-1,\ l = 0, \ldots, L-1 \end{aligned} \tag{6.2}$$

where $\mathbf{v}[n]$ is the noise vector. We normally assume that the sources are zero-mean and the elements of $\mathbf{v}[n]$ are white Gaussian and independent of the source signals. To recover the sources successfully, we normally require $\mathrm{rank}[\boldsymbol{A}] = L$ and $L \leq M$.

For convolutive mixing, the received signals $\mathbf{x}[n]$ are sums of filtered versions of the sources, which can be expressed as:

$$\mathbf{x}[n] = \sum_{i=-\infty}^{+\infty} \boldsymbol{A}_i \mathbf{s}[n-i] + \mathbf{v}[n] \tag{6.3}$$

where $\boldsymbol{A}_i$ is the matrix impulse response of the linear time invariant multichannel mixing filter.

The source signals can be recovered in ways based on different assumptions on their statistical properties, such as nonstationarity (Matsuoka *et al.*, 1995; Pham and Cardoso, 2001), periodicity (Jafari *et al.*, 2006), temporal correlation (Barros and Cichocki, 2001; Belouchrani *et al.*, 1997; Cichocki and Thawonmas, 2000; Liu *et al.*, 2006, 2007a,b; Molgedey and Schuster, 1994), sparseness (Georgiev *et al.*, 2005) and constant modulus (Johnson *et al.*, 1998). We can also assume that the source signals are spatially independent and non-Gaussian; therefore, the aim of separation is identical to the estimation of the independent components from the received mixtures. Therefore, for this case, we can use independent component analysis (ICA) methods such as the fastICA and other kurtosis-related algorithms (Amari, 1998; Cichocki *et al.*, 1997; Delfosse and Loubaton, 1995; Hyvarinen and Oja, 1997; Liu and Mandic, 2006).

In general, by blind source separation we obtain all the L sources simultaneously, but we can also choose to extract a single source or a subset of sources from their mixtures and repeat this process until we extract the last source or the last one from a desired subset of sources (Barros and Cichocki, 2001; Cichocki and Thawonmas, 2000; Cichocki *et al.*, 1997; Cruces-Alvarez *et al.*, 2004; Delfosse and Loubaton, 1995; Li and Wang,

2002; Liu and Mandic, 2006; Liu *et al.*, 2006, 2007a,b, 2008a,b; Malouche and Macchi, 1998). The BSS approach operating in this way is called blind source extraction (BSE) (Cichocki and Amari, 2003).

To illustrate the concept of BSS, in the next section we will use the BSE as an example and present a simple kurtosis based algorithm.

6.1.2 A Blind Source Extraction Example

For independent source signals, we can use ICA methods to perform the extraction. As the mixtures of independent/uncorrelated Gaussian sources mixed by an orthogonal mixing matrix are still independent/uncorrelated, the ICA approach is only applicable to the case with non-Gaussian sources, or at most one source is Gaussian. Under this condition, the extraction of one original source signal is equivalent to extracting an independent component from the mixtures.

For simplicity, we assume that the mixed signals are instantaneous mixtures and noise-free. To estimate one of the independent sources, we consider a linear combination of the mixed signals given by:

$$y[n] = \mathbf{w}^H \mathbf{x}[n] = \mathbf{g}^H \mathbf{s}[n] \tag{6.4}$$

where $\mathbf{g}^H = \mathbf{w}^H A$ is the global demixing vector.

By the central limit theorem, under some mild conditions, a sum of independent random variables approaches to a Gaussian distribution. If we try to maximize the non-Gaussianity of $y[n]$, it will give the original source signal with the maximum non-Gaussianity. A classical measure of non-Gaussianity is the kurtosis, which for a zero-mean random variable y is defined as (Hyvarinen *et al.*, 2001):

$$kt(y) = E\{y^4\} - 3(E\{y^2\})^2 \tag{6.5}$$

where $E\{\cdot\}$ denotes the statistical expectation operator.

The kurtosis of a Gaussian random variable is zero, whereas for most non-Gaussian random variables it is non-zero. Random variables are called super-Gaussian for a positive kurtosis and sub-Gaussian for a negative kurtosis. Suppose x_1 and x_2 are two independent random variables; then we have:

$$kt(x_1 + x_2) = kt(x_1) + kt(x_2) \tag{6.6}$$

Another important property is:

$$kt(\alpha y) = \alpha^4 kt(y) \tag{6.7}$$

where α is a constant.

6.1.2.1 Algorithm

In the following kurtosis-based algorithm, we minimize the cost function (Cichocki *et al.*, 1997)

$$C(\mathbf{w}) = -\frac{\beta}{4} \kappa_4(y) \tag{6.8}$$

where $k_4(y) = kt(y)/(E\{y^2\})^2$ is the normalized kurtosis of y and $\beta = 1$ for the extraction of source signals with positive kurtosis and -1 for sources with negative kurtosis. Applying the standard gradient descent method, we obtain the following update equation:

$$\mathbf{w}[n+1] = \mathbf{w}[n] + \mu\phi(y[n])\mathbf{x}[n] \tag{6.9}$$

where μ is the stepsize and:

$$\phi(y[n]) = \beta\left[\frac{m_2(y)}{m_4(y)}y^3[n] - y[n]\right] \tag{6.10}$$

with $m_q(y)$ being estimated as follows:

$$m_q(y)[n] = (1-\lambda)m_q(y)[n-1] + \lambda|y[n]|^q, \quad q = 2, 4 \tag{6.11}$$

λ is the forgetting factor.

6.1.2.2 Performance Index

A performance index PI is introduced in order to evaluate the performance of different BSS algorithms.

For an instantaneous BSE algorithm, we can define PI based on the global demixing vector $\mathbf{g}$, given by:

$$PI = 1 - \frac{\max\{g_0^2, g_1^2, \cdots, g_{L-1}^2\}}{\sum_{l=0}^{L-1} g_l^2} \tag{6.12}$$

where g_l is the lth element of the vector $\mathbf{g}$. An alternative definition according to Cichocki and Amari (2003) is:

$$PI = \frac{1}{L-1}\left(\sum_{l=0}^{L-1}\frac{g_l^2}{\max\{g_0^2, g_1^2, \cdots, g_{L-1}^2\}} - 1\right) \tag{6.13}$$

If the source signal has been successfully extracted, then $\mathbf{g}$ will be a vector with only one nonzero element and PI in both Equations (6.12) and (6.13) will be zero. In the following simulation, we will use the PI defined in Equation (6.13).

6.1.2.3 Simulation

Three source signals with binary, uniform and Gaussian distributions, respectively, are shown in Figure 6.2. Their corresponding kurtosis values are $\{-2, -1.1833, -0.0169\}$. As they have negative values, $\beta = -1$. When we minimize the kurtosis of the extracted signal, in theory, we will recover the first source signal s_0, as it has the smallest kurtosis value.

The mixing matrix was randomly generated and given by:

$$\mathbf{A} = \begin{bmatrix} 0.9575 & 0.5207 & 0.9248 \\ -0.9356 & -0.4131 & 0.6338 \\ -0.4264 & -0.9840 & 0.6787 \end{bmatrix} \tag{6.14}$$

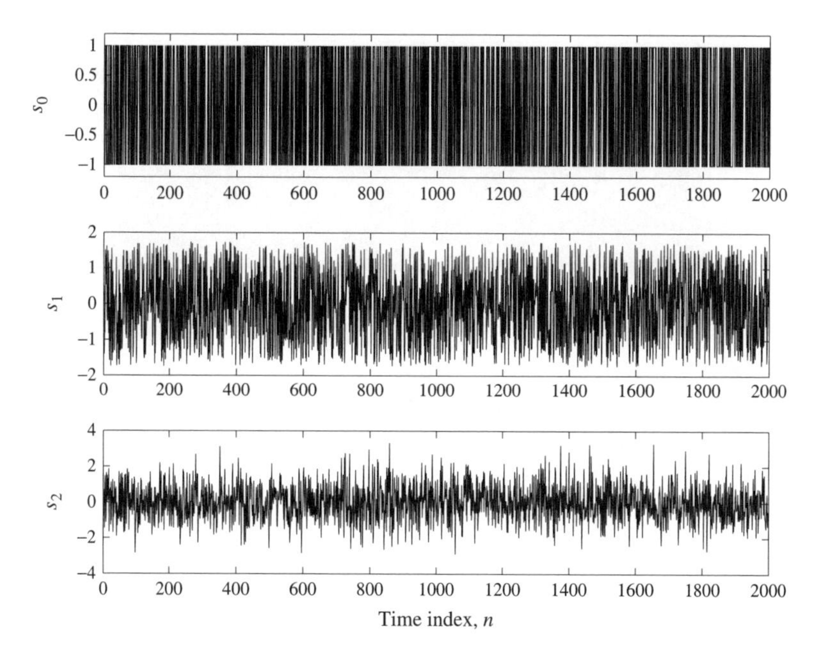

Figure 6.2 Source signals with Gaussian (s_0), binary (s_1) and uniform distributions (s_2)

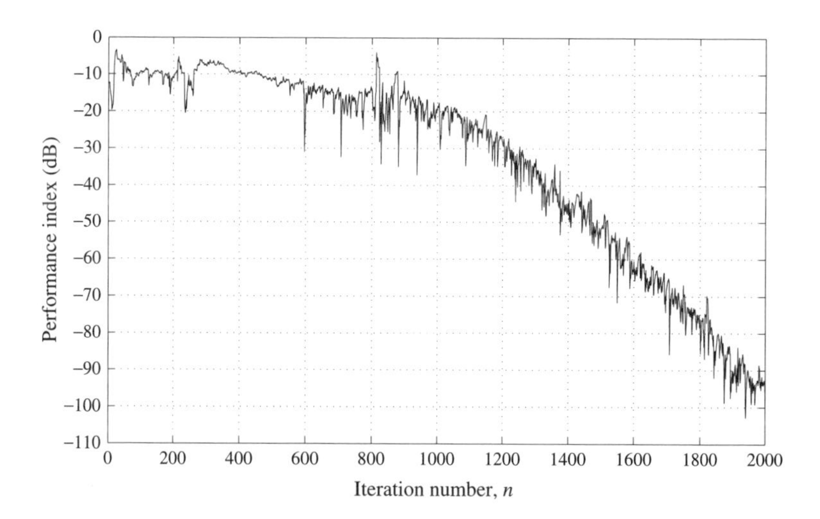

Figure 6.3 The performance index curve using the kurtosis based algorithm

In adaptation, the forgetting factor was $\lambda = 0.02$ and the stepsize $\mu = 0.08$. As shown in Figure 6.3, the performance index has reached a level of almost $-100\,\text{dB}$, indicating a successful extraction. The extracted signal is given in Figure 6.4, which matches the first source signal very well.

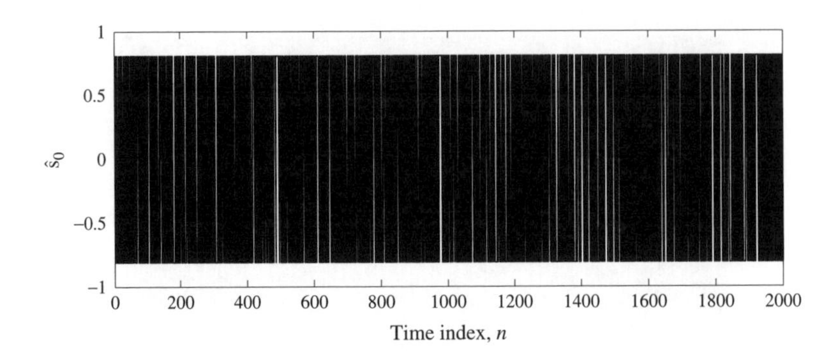

Figure 6.4 The extracted source signal using the kurtosis based algorithm

6.2 Blind Wideband Beamforming

Both the instantaneous and the convolutive mixing problems in BSS can be found in the context of array signal processing.

Consider an array of sensors with a plane-wave input as shown in Figure 6.5. Let $s_l(t)$, $l = 0, \ldots, L-1$ be the L impinging plane-wave signals that would be received at the origin of the coordinate system. Then, the signal received at the mth sensor will be:

$$x_m(t) = \sum_{l=0}^{L-1} s_l(t - \tau_{m,l}) \tag{6.15}$$

where $\tau_{m,l}$ denotes the delay from the mth sensor to the origin of the coordinate system for signal $s_l(t)$. For narrowband signals with a specific frequency ω, we have:

$$s_l(t - \tau_{m,l}) = \mathrm{e}^{-j\omega\tau_{m,l}} s_l(t) \tag{6.16}$$

Then Equation (6.15) becomes:

$$x_m(t) = \sum_{l=0}^{L-1} \mathrm{e}^{-j\omega\tau_{m,l}} s_l(t) \tag{6.17}$$

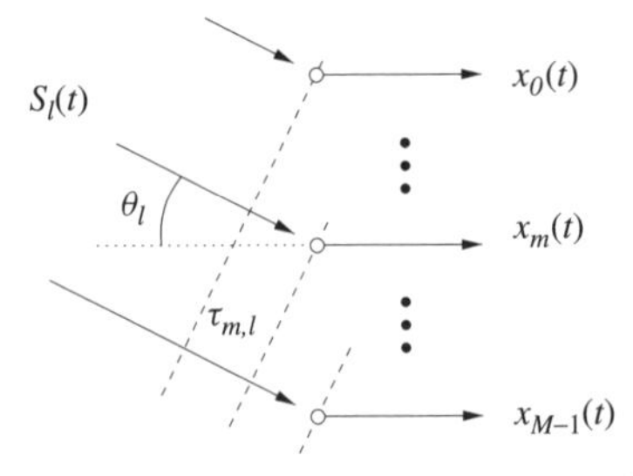

Figure 6.5 A general sensor array with a plane wave $s_l(t)$ impinging from the angle θ_l

Obviously each received sensor signal $x_m(t)$ becomes a weighted sum of the L source signals, which represents an instantaneous mixing problem.

However, for wideband signals, there are many different frequency components for each of them and we cannot have the simple form $s_l(t - \tau_{m,l}) = \alpha s_l(t)$ as shown in Equation (6.16), where α is a constant. As a result, each of the received sensor signals cannot be expressed in the form given in Equation (6.17), and we must consider a more complicated expression, namely:

$$x_m(t) = \sum_{l=0}^{L-1} \delta(t - \tau_{m,l}) * s_l(t) \tag{6.18}$$

where '$*$' denotes the convolution operator. In a compact form, we have:

$$\mathbf{x}(t) = \boldsymbol{A} * \mathbf{s}(t) \tag{6.19}$$

where:

$$\mathbf{s}(t) = \left[s_0(t)\ s_1(t)\ \cdots\ s_{L-1}(t)\right]^{\mathrm{T}}$$

$$\mathbf{x}(t) = \left[x_0(t)\ x_1(t)\ \cdots\ x_{M-1}(t)\right]^{\mathrm{T}}$$

$$[\boldsymbol{A}]_{m,l} = a_{m,l} = \delta(t - \tau_{m,l})$$

$$m = 0, \ldots, M-1, \quad l = 0, \ldots, L-1 \tag{6.20}$$

For convenience, we consider the discrete-time form of Equation (6.19), given by:

$$\mathbf{s}[n] = \left[s_0[n]\ s_1[n]\ \cdots\ s_{L-1}[n]\right]^{\mathrm{T}}$$

$$\mathbf{x}[n] = \left[x_0[n]\ x_1[n]\ \cdots\ x_{M-1}[n]\right]^{\mathrm{T}} \tag{6.21}$$

Now the elements $a_{m,l}$ of $\boldsymbol{A}$ become discrete-time filters and the problem can be formulated as:

$$\mathbf{x}[n] = \sum_{i=-\infty}^{+\infty} \mathbf{A}_i \mathbf{s}[n-i] \tag{6.22}$$

where the $M \times L$ matrix $\boldsymbol{A}_i$ is the matrix impulse response of the linear time invariant multichannel mixing filter with its (m, l)th element $a_{i,m,l}$ given by:

$$a_{i,m,l} = \frac{\sin\left((i - \tau_{m,l}/T)\pi\right)}{(i - \tau_{m,l}/T)\pi} \tag{6.23}$$

where T is the sampling period.

To show this convolutive nature more clearly, we can rewrite this formulation in the frequency domain. Suppose we have L signals $s_l(\omega, t), l = 0, \ldots, L-1$ with a specific angular frequency ω (i.e. a narrowband signal) arriving at the array from different directions, then the received array signals can be expressed as:

$$\mathbf{x}(\omega, t) = \boldsymbol{A}(\omega)\mathbf{s}(\omega, t) \tag{6.24}$$

$$\mathbf{s}(\omega, t) = \left[s_0(\omega, t)\ s_1(\omega, t)\ \cdots\ s_{L-1}(\omega, t)\right]^{\mathrm{T}}$$

$$\mathbf{x}(\omega, t) = \left[x_0(\omega, t)\ x_1(\omega, t)\ \cdots\ x_{M-1}(\omega, t)\right]^{\mathrm{T}} \tag{6.25}$$

and $\boldsymbol{A}(\omega)$ is a $M \times L$ matrix given by

$$\boldsymbol{A}(\omega) = \begin{bmatrix} \mathrm{e}^{-j\omega\tau_{0,0}} & \mathrm{e}^{-j\omega\tau_{0,1}} & \ldots & \mathrm{e}^{-j\omega\tau_{1,L-1}} \\ \mathrm{e}^{-j\omega\tau_{1,0}} & \mathrm{e}^{-j\omega\tau_{1,1}} & \ldots & \mathrm{e}^{-j\omega\tau_{1,L-1}} \\ \vdots & \vdots & \ddots & \vdots \\ \mathrm{e}^{-j\omega\tau_{M-1,0}} & \mathrm{e}^{-j\omega\tau_{M-1,1}} & \ldots & \mathrm{e}^{-j\omega\tau_{M-1,L-1}} \end{bmatrix} \tag{6.26}$$

We can see for a signal with a specific frequency ω, the received signal is an instantaneous mixture of the original signals. However, if the impinging signals change their frequency, $\boldsymbol{A}(\omega)$ will change as it is a function of ω. Then, for wideband signals, the mixing matrix for different frequency components will be different. In the temporal domain, it will be a convolutive mixing problem, as given in Equation (6.22). To recover the original source signals, we can apply the existing BSS algorithms to the the received mixed signals $x_m[n], m = 0, 1, \ldots, M-1$ directly, which leads to the concept of blind beamforming (Cardoso and Souloumiac, 1993; Gönen and Mendel 1997; Huang *et al.*, 2007; Liu, 2009b; Liu and Mandic, 2005; Sheinvald, 1998; Yang *et al.*, 2004).

For wideband signals, an instantaneous BSS algorithm will not work and we have to employ a convolutive BSS algorithm (Douglas and Sun, 2003; Matsuoka and Nakashima, 2001). However, the problem with the convolutive BSS algorithms is their extremely high computational complexity and one solution is to first transform the received sensor signals from the time domain into the frequency domain. Since convolutive mixing in the time domain corresponds to instantaneous mixing in the frequency domain (Parra and Spence, 2000; Smaragdis, 1998), existing separation algorithms for instantaneous mixtures can be applied directly to the frequency-domain signals. Notice, however, that frequency-domain BSS introduces the well-known permutation problem among different frequency bins, and an online implementation of these algorithms is extremely difficult for a large number of mixtures (Sawada *et al.*, 2004).

6.3 Blind Beamforming Based on Frequency Invariant Transformation

In this section we introduce a simple method for blind wideband beamforming based on a frequency invariant transformation (Liu, 2009b; Liu and Mandic, 2005). The received array signals are first passed through a network of frequency invariant beamformers. We will show that after this FIB network, the convolutive mixing problem will be transformed into an instantaneous one by means of its frequency invariance property. An appropriate BSS algorithm for instantaneous mixtures can then be applied to obtain the original sources, according to source signal statistics. In this way, the convolutive or the frequency-domain BSS algorithms are avoided.

6.3.1 Structure

An FIB network for instantaneous BSS is shown in Figure 6.6, where we have N sets of array coefficients $\boldsymbol{W}_i$, $i = 0, \ldots, N-1$. For a linear array, it is given by:

$$\boldsymbol{W}_i = \begin{bmatrix} w_{i,0,0} & w_{i,0,1} & \cdots & w_{i,0,J-1} \\ w_{i,1,0} & w_{i,1,1} & \cdots & w_{i,1,J-1} \\ \vdots & \vdots & \ddots & \vdots \\ w_{i,M-1,0} & w_{i,M-1,1} & \cdots & w_{i,M-1,J-1} \end{bmatrix} \quad (6.27)$$

Each set $\boldsymbol{W}_i$ forms a frequency invariant beamformer with a response $P_i(\theta)$, for which its output $y_i[n]$ is given by:

$$y_i[n] = \sum_{m=0}^{M-1} \sum_{k=0}^{J-1} w_{i,m,k} \times x_m[n-k] \quad (6.28)$$

The time-domain quantity corresponding to $P_i(\theta)$ is $P_i(\theta)\delta(n)$, where n is the time index. There is no delay in these responses which means that the whole beamforming network is noncausal. To make it causal, we can simply shift the system along the time index by some delay Δn to obtain the response $P_i(\theta)\delta(n - \Delta n)$. Since the same delay can be used for all the N FIBs, this delay can be ignored. Let the DOA angles for the L sources be respectively $\theta_0, \theta_1, \ldots, \theta_{L-1}$, which are unknown. Then the outputs of this network of N FIBs can be expressed as:

$$\mathbf{y}[n] = \boldsymbol{A}\mathbf{s}[n] \quad (6.29)$$

with:

$$\mathbf{y}[n] = \left[y_0[n] \; y_1[n] \; \cdots \; y_{N-1}[n]\right]^{\mathrm{T}}$$

$$\boldsymbol{A} = \begin{bmatrix} P_0(\theta_0) & P_0(\theta_1) & \cdots & P_0(\theta_{L-1}) \\ P_1(\theta_0) & P_1(\theta_1) & \cdots & P_1(\theta_{L-1}) \\ \vdots & \vdots & \ddots & \vdots \\ P_{N-1}(\theta_0) & P_{N-1}(\theta_1) & \cdots & P_{N-1}(\theta_{L-1}) \end{bmatrix} \quad (6.30)$$

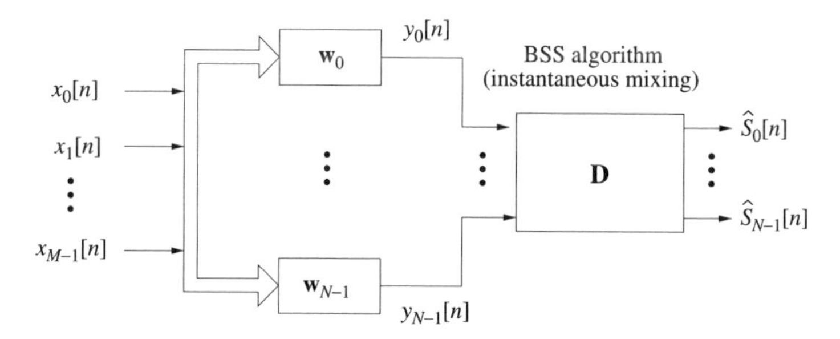

Figure 6.6 A general instantaneous BSS structure based on the frequency invariant beamforming network for wideband array signals

Notice that each row of the matrix $\boldsymbol{A}$ represents a sampling of the frequency invariant response of the corresponding frequency invariant beam at those directions $\theta_0, \theta_1, \ldots, \theta_{L-1}$ of the source signals. As desired, the problem has now been transformed into an instantaneous mixing problem, where $y_i[n], n = 0, 1, \ldots, N-1$ are the corresponding instantaneous mixtures and $\boldsymbol{A}$ is the new mixing matrix. Note the DOA angles of the impinging signals are unknown to the system, i.e. $\boldsymbol{A}$ is unknown. To find the inverse of $\boldsymbol{A}$ and recover the original source signals, we can now employ the standard instantaneous BSS algorithms.

6.3.2 The Algorithm

In the next section we will run two sets of simulations and in the first set, we will use a density matching BSS algorithm to separate some speech signals based on the natural gradient adaptation (Amari, 1998; Douglas, 2002; Jafari *et al.*, 2004), for which the update is given by

$$\boldsymbol{D}[n+1] = \boldsymbol{D}[n] + \mu\left[\boldsymbol{I} - \mathbf{f}(\hat{\mathbf{s}}[n])\hat{\mathbf{s}}^T[n]\right]\boldsymbol{D}[n] \tag{6.31}$$

with:

$$\hat{\mathbf{s}}[n] = [\hat{s}_0[n], \ldots, \hat{s}_{N-1}[n]]^T = \boldsymbol{D}[n]\mathbf{y}[n] \tag{6.32}$$

and:

$$\mathbf{f}(\hat{\mathbf{s}}[n]) = \left[\mathrm{sign}(\hat{s}_0[n]), \ldots, \mathrm{sign}(\hat{s}_{N-1}[n])\right]^T \tag{6.33}$$

where symbol $\boldsymbol{D}$ denotes the separation matrix and $\hat{\mathbf{s}}[n]$ are the recovered sources, as shown in Figure 6.6.

In the second set of simulations, we will use the normalized kurtosis based BSE algorithm to extract one of the source signals, as given by Equation (6.9) in Section 6.1.2.

It is well known that, in general, we can only successfully separate at most M sources from M sensors, that is, we can only design $N \leq M$ independent FIBs. The N FIBs considered will have their main beam directions equally distributed over the DOA range $[-90° \; 90°)$. After this transformation, we have N instantaneous mixtures in total. In general, to achieve a successful separation, the number of beams N should not be less than L.

As the FIB network can serve not only as a transformation, but also as an initial separation, we would like the N beamformers to have a sidelobe level as low as possible. However, the shape of these beamformers only affects the value of the new instantaneous mixing matrix $\boldsymbol{A}$. As long as the N beam responses are independent (not necessarily having a low sidelobe level), hence $\boldsymbol{A}$ is of full rank, the instantaneous separation algorithm will be able to recover the sources successfully. This can be seen by the method proposed in Liu (2009b), where the frequency invariant transformations are based on phase mode processing and there are no beams formed at all.

6.3.3 Simulations

The first set of simulations is based on a uniformly spaced linear array with $M = 19$ sensors, which receives five plane-wave sources coming from the DOA angles of

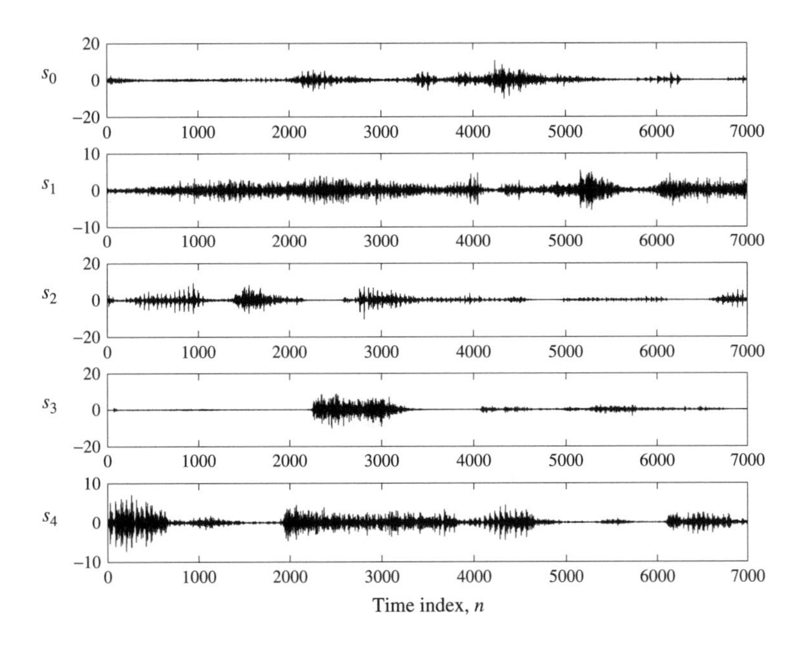

Figure 6.7 The five source signals for the first set of simulations

$-60°, -20°, 0°, 20°$ and $60°$, respectively, as shown in Figure 6.7. Their time-delayed versions are mixed together at the sensors according to Equation (6.19). As we have only five sources, there is no need to design 19 FIBs to transform them into instantaneous mixtures. For simplicity, we design only five FIBs with their desired beam responses shown in Figure 6.8. Each of the five sets of coefficients $\boldsymbol{W}_i$ has dimensions of 19×25.

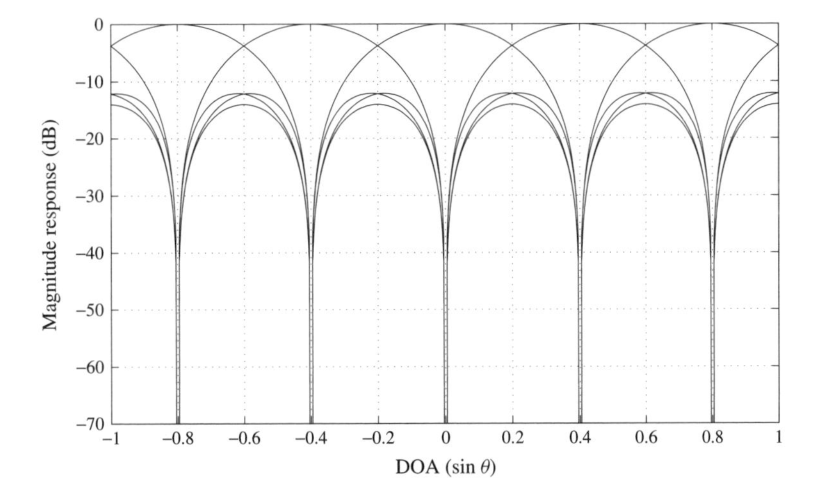

Figure 6.8 The beam shapes of the five frequency invariant beamformers

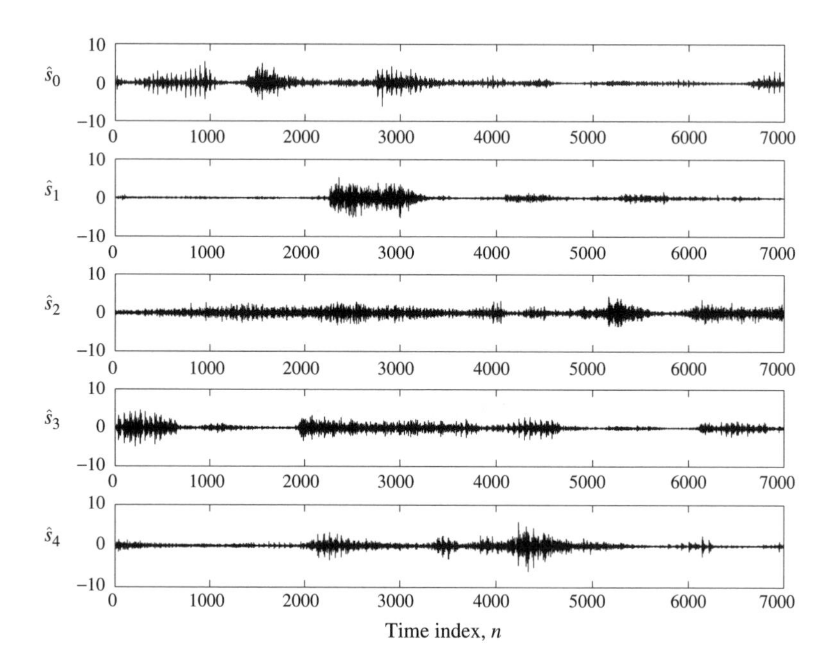

Figure 6.9 The five separated source signals for the first set of simulations

The outputs $y_0[n]$, $y_1[n]$, $y_2[n]$, $y_3[n]$, and $y_4[n]$ of this transformation network are now the instantaneously mixed signals. As the frequency invariance property of the five beamformers is not good enough in both the lower frequency band and the high frequency band around π, we have filtered the source signals, which limits their bandwidth to $[0.4\pi \ 0.9\pi]$.

The separation is performed by the algorithm in Equation (6.31) with a stepsize of 0.000 02. After the algorithm has reached its steady state, the resulting separation matrix $\boldsymbol{D}$ was used to recover the source signals. The five separated sources are shown in Figure 6.9. There is a clear match between the original sources and our separated sources, indicating a successful separation.

To show the robustness of this approach, we have run a second set of simulations. There are three band-limited source signals with binary, uniform and Gaussian distributions, respectively, similar to the source signals shown in Figure 6.2. We use the algorithm in Equation (6.9) to minimize the kurtosis of the extracted signal ($\beta = -1$), where $\mu = 0.006$ and $\lambda = 0.025$. These three source signals arrive from different directions and 100 simulations have been run. Each time the three source signals were randomly generated and their DOAs were also randomly distributed in the range $[-90^\circ \ 90^\circ]$.

As the three signals could come from any direction, we designed seven FIBs with their desired beam responses shown in Figure 6.10. The learning curve of the output SIR for the extracted signal is shown in Figure 6.11, which is the averaged result of 100 independent simulations. Note when the three signals come from directions very close to one another, we will have an ill-conditioned mixing problem and it will be very difficult to separate them. Bearing this in mind, we can say the learning curve shows a very robust separation capability of the introduced approach.

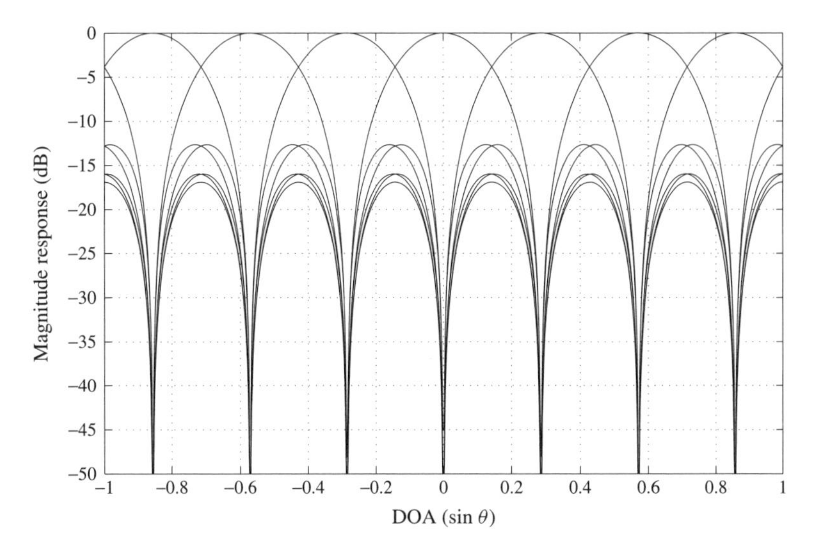

Figure 6.10 The desired beam responses of the seven frequency invariant beamformers in the network

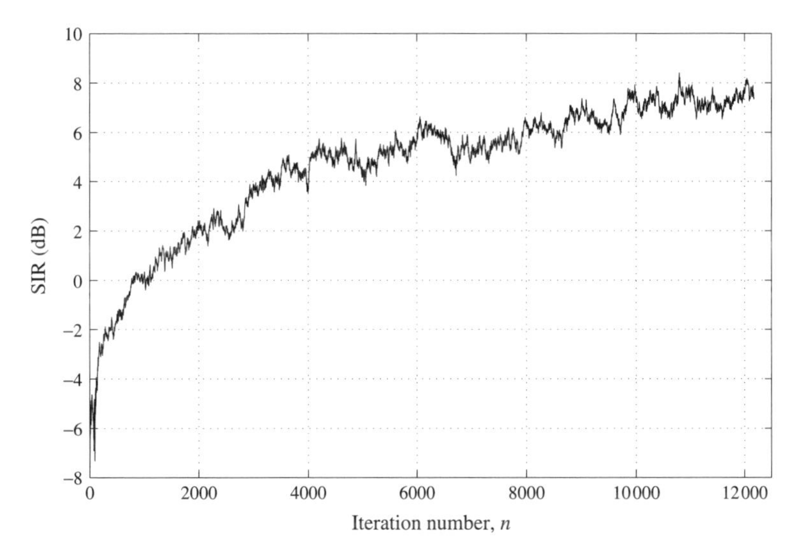

Figure 6.11 The SIR learning curve for the extracted signal in the second set of simulations, averaged over 100 simulations

6.4 Summary

In this chapter we have focused on a different class of adaptive beamformers, which is based on the concept of blind source separation. For this class of beamformers, neither a reference signal nor the DOA information of the desired signal is needed and only some assumptions on the statistical properties of the source signals are required.

We first gave a review of the blind source separation technique, with a kurtosis-based BSE algorithm as an example. BSS is a technique to recover the original sources from all kinds of their mixtures, without the knowledge of the mixing process and the sources themselves. For the linear mixing problem, it can be further divided into two categories: instantaneous mixing and convolutive mixing, both of which can be found in the context of array signal processing. We have shown that, in principle, for narrowband signals, we will have an instantaneous mixing problem, while for wideband signals, the problem is convolutive in nature. As a result, the blind wideband beamforming problem can be solved directly by the existing convolutive BSS methods, although the computational complexity of such a system would be extremely high even for a modest number of sensors.

An efficient way to solve the blind wideband beamforming problem is to employ the frequency invariant beamforming technique introduced in Chapter 5. In this approach, the received array signals are first passed through a network of frequency invariant beamformers. It has been shown that after this FIB network, the convolutive mixing problem is transformed into an instantaneous one and an appropriate BSS algorithm for instantaneous mixtures can then be applied to obtain the original sources.

7

Wideband Beamforming with Sensor Delay-Lines

In this chapter we will introduce a recently proposed wideband beamforming structure (Liu, 2007, 2009a). A special property of this new structure is that there is no any form of temporal processing of the received array signals required, such as tapped delay-lines or FIR/IIR filters. The key to this structure is to use an additional spatial processing dimension to replace the traditionally required temporal processing one, which leads to a sensor delay-line/spatial filtering based system. This is a new approach to wideband beamforming and actually most of the work done based on the traditional structure can be applied to this new structure without much difficulty.

In the following, we will first give a general review of this structure in Section 7.1 and then study the fixed beamformer design problem in Section 7.2, where it is mainly focused on the design of frequency invariant beamformers. The general adaptive case is dealt with in Section 7.3 with the beamspace adaptive method described in Section 7.4.

7.1 Sensor Delay-Line Based Structures

7.1.1 *Introduction*

A common feature of the wideband beamforming structures studied in the previous chapters is the use of tapped delay-lines (TDLs) or FIR/IIR filters in its discrete form. Figure 7.1 shows such a traditional structure for wideband beamforming based on a linear array system with M sensors, where a signal $s_l(t)$ arrives from the direction ϕ_l. TDLs and FIR/IIR filters are equivalent structures, since they both perform a temporal filtering process to form a frequency dependent response for each of the received wideband sensor signals to compensate the phase difference for different frequency components.

In the temporal filtering based system, the length J of the TDLs or IIR/FIR filters is dependent on the bandwidth of the impinging signals (Compton, 1988c; Vook and Compton, 1992). In general, the larger the bandwidth, the longer the TDLs or the IIR/FIR filters (Yu *et al.*, 2007). The delays between samples/taps become shorter and shorter with increasing signal bandwidth and as a result, very high speed circuits may need to be employed. As an example, consider a general ultra-wideband system with a frequency range from 0 Hz to 6 GHz. If we perform beamforming in a digital form, then the sampling

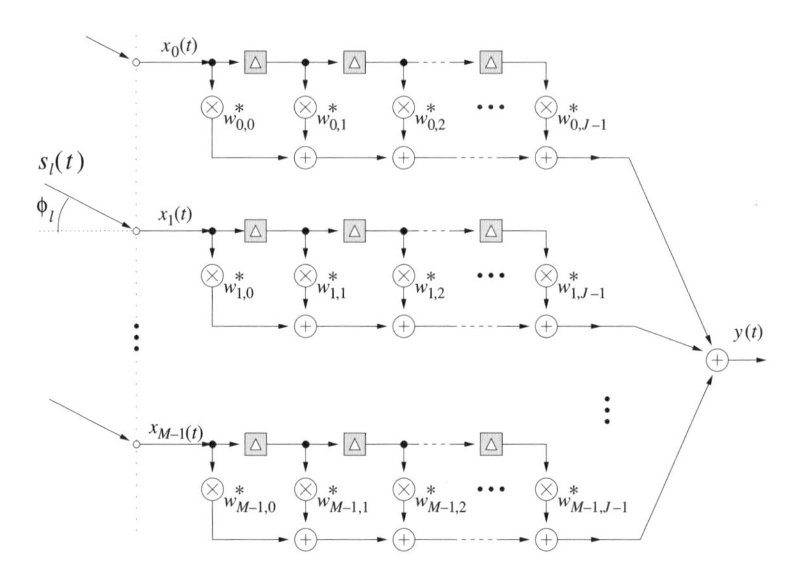

Figure 7.1 A traditional structure for wideband beamforming based on a linear array system with M sensors, where a signal $s_l(t)$ arrives from the direction ϕ_l

rate should be at least 12 GHz. With 16 bits per sample, it is an extremely difficult task and also very expensive to build such a high-speed circuit with current technology. Another example is the passive millimetre wave security imaging system. The millimetre wave band extends from approximately 30 GHz to 300 GHz and although digital beamforming can provide many obvious advantages for such an imaging system, such as simplicity, flexibility, personnel throughput enhancement and significantly reduced volume, it is still a challenging open problem due to the extremely high sampling frequency required (even after downconversion to an intermediate frequency) (Salmon *et al.*, 2007).

To avoid the difficulty associated with the traditional temporal filtering based system and also provide an alternative to it, we can replace the wired delays by spatial propagation delays by positioning more sensors behind the original array system. We consider this new system as a wideband beamforming structure with sensor delay-lines (SDLs) or simply spatial filters (Liu, 2007, 2009a).

This approach can be applied to any array systems. For the original linear array with TDLs or FIR/IIR filters, we will have a rectangular array without temporal filtering, as shown in Figure 7.2, to replace the structure in Figure 7.1; for an original rectangular array with TDLs or FIR/IIR filters, we will have a cubic array without temporal filtering, as shown in Figure 7.3; for a semicircular array, we will have a new system shown in Figure 7.4, or a cylindrical form for a circular array, etc. Moreover, these SDL sensors could be positioned not necessarily in a regular spacing, but arbitrary.

This idea is actually not limited to wideband array signal processing. It can be generalized to any wideband signal processing scenarios when we are processing some spatially propagated signals, i.e. instead of employing traditional FIR/IIR filters or TDLs to process the signal, we can construct a spatial filter using multiple sensors placed at different spatial positions to perform the same signal processing task.

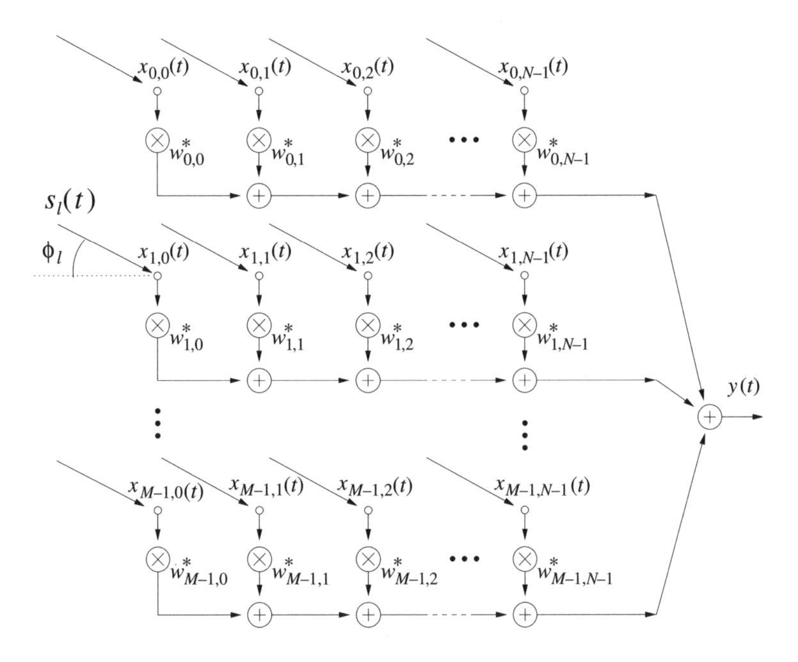

Figure 7.2 The rectangular array without temporal filtering to replace the traditional linear array with temporal filtering

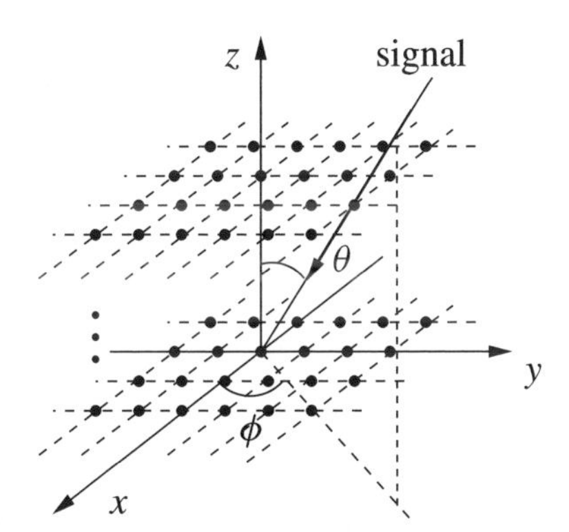

Figure 7.3 The cubic array without temporal filtering to replace the traditional rectangular array with temporal filtering

The advantage of the proposed SDL structure is twofold:

1. Since only one coefficient (constant) is required for each received array signal, the function of the original temporal filtering process is now achieved by spatial filtering using multiple sensors so that a complicated wideband beamforming system is able to be implemented by simple analogue circuits. This is useful especially when the signal frequency (and bandwidth) are very high and a digital implementation becomes extremely difficult or even not viable at all, as mentioned in the two examples.

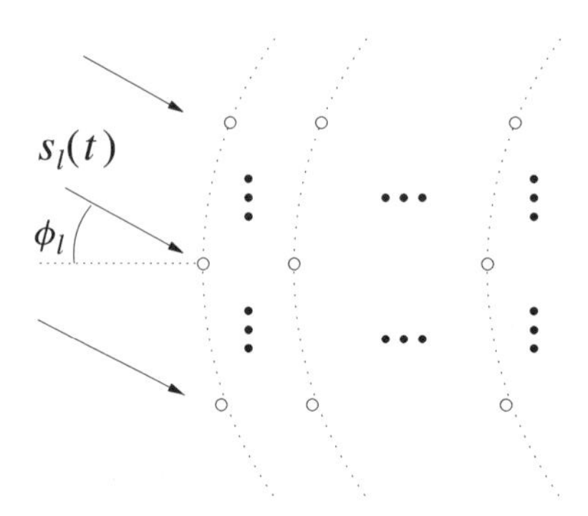

Figure 7.4 The concentric semicircular array without temporal filtering to replace the traditional semicircular array with temporal filtering

2. The second advantage is its potential to avoid the beam widening effect at high off-boresight angels. For example, for the traditional TDL structure in Figure 7.1, the beamwidth will increase significantly when a broadside main beam is steered to an angle ϕ closer to 90°. However, for the corresponding SDL structure in Figure 7.2, we can rotate the set of coefficients by 90° to form a main beam pointing to the direction $\phi = 90^\circ$ and the new beam will have the same beamwidth, because its effective aperture will not go to zero as in the linear array case and it has a full coverage over the 360 degrees azimuth range.

Note if we want to implement this beamformer in the digital domain, we will still require the same sampling rate as the traditional one. However, we may need the proposed structure for those applications where we want to avoid the beam widening effect at high off-boresight angles and using more sensors may well be justified by meeting the requirements of the corresponding applications.

This idea is derived from a recently proposed frequency invariant beamforming design method for rectangular arrays (Ghavami, 2002). A special characteristic of this fixed beamformer is that there are no TDLs involved and only one single weight is attached to each sensor. By examining its beam pattern, we can find that this rectangular array is approximating the response of a frequency invariant linear array with TDLs (Liu and Weiss, 2004b, 2008b, 2009b; Liu *et al.*, 2007c; Sekiguchi and Karasawa, 2000) by employing spatial propagation delays instead of wired delays. The general concept of replacing the temporal filtering process by spatial filtering implemented by sensor delay-lines is then derived and extended to the adaptive case and also to arbitrary beam patterns and sensor position patterns (Liu, 2007, 2009a).

In the next we will focus on the widely used linear array, which corresponds to the rectangular array in the new approach, and give a simple discussion about its wideband beamforming capability.

7.1.2 Wideband Response of the SDL-Based Structure

As discussed in the previous section, the idea is to use sensor delays to replace the wired delays in the traditional adaptive wideband beamforming structure. For the linear array shown in Figure 7.1, the corresponding new structure will be a planar array and it has been shown in Figure 7.2, where each of the TDLs with length J is replaced by a SDL with N sensors, and the delay between adjacent taps is thus replaced by the spatial propagation delay between adjacent sensors.

In this new structure, there are in total $M \times N$ sensors, with the received sensor signals denoted by $x_{m,n}(t)$, $m = 0, 1, \ldots, M-1, n = 0, 1, \ldots, N-1$. As there is only one coefficient for each received sensor signal, there are no TDLs involved and fixed/adaptive beamforming can be performed by very simple analogue circuits. Note that there is no temporal information used in the beamforming process. Therefore it can be considered as a wideband beamforming structure with spatial-only information.

In Figure 7.2, the output $y(t)$ can be expressed as:

$$y(t) = \mathbf{w}^H \mathbf{x}(t) \tag{7.1}$$

where:

$$\begin{aligned}\mathbf{w} &= \left[w_{0,0} \ldots w_{M-1,0}\ w_{0,1} \ldots w_{M-1,1} \ldots w_{M-1,N-1}\right]^T \\ \mathbf{x}(t) &= \left[x_{0,0}(t)\ \ldots\ x_{M-1,0}(t)\ x_{0,1}\ \ldots\ x_{M-1,1}(t)\ \ldots\ x_{M-1,N-1}\right]^T\end{aligned} \tag{7.2}$$

Suppose all of the signals are on the same plane as the rectangular array, i.e. the elevation angle $\theta = 90^\circ$. Then the response of the array with respect to temporal frequency ω rad/s and angle of arrival ($\theta = 90^\circ, \phi$) of the impinging signal is given by:

$$P(\omega, \phi) = \sum_{m,n=0,0}^{M-1,N-1} w_{m,n}^* e^{-j\frac{m\omega \sin \phi d_y}{c}} e^{-j\frac{n\omega \cos \phi d_x}{c}} \tag{7.3}$$

where c is the wave propagation speed, d_x is the spacing between adjacent SDL sensors and d_y is the spacing between sensors (m, n) and $(m+1, n)$ with $m = 0, 1, \ldots, M-1$ and $n = 0, 1, \ldots, N-1$.

With the following substitutions:

$$\begin{aligned}\omega_1 &= \frac{\omega \cos \phi d_x}{c} \\ \omega_2 &= \frac{\omega \sin \phi d_y}{c}\end{aligned} \tag{7.4}$$

we have:

$$P(\omega_1, \omega_2) = \sum_{m,n=0,0}^{M-1,N-1} w_{m,n}^* e^{-jm\omega_2} e^{-jn\omega_1} \tag{7.5}$$

Then, given a set of coefficients $w^*_{m,n}$, the array's response can be obtained by first applying the two-dimensional Fourier transform to them as in Equation (7.5) and then the two substitutions as in Equation (7.4).

From Equation (7.4), we have:

$$\frac{\omega_2 d_x}{\omega_1 d_y} = \tan\phi \tag{7.6}$$

Thus, ϕ is given by:

$$\phi = \arctan\frac{\omega_2 d_x}{\omega_1 d_y} \tag{7.7}$$

From Equation (7.4), we also have:

$$\omega = \frac{c\omega_2}{\sin\phi d_y} \tag{7.8}$$

Then, given any desired wideband response $P(\omega,\phi)$, we can use the substitutions in Equations (7.7) and (7.8) to express it in the form of $P(\omega_1,\omega_2)$; then the desired coefficients $w^*_{m,n}$ can be obtained by applying the inverse Fourier transform to $P(\omega_1,\omega_2)$. This provides a theoretical explanation about why this structure can form a wideband beam response. Note a desired frequency invariant response $P(\phi)$ can be considered as a special case and the corresponding coefficients $w^*_{m,n}$ can be obtained in the same way.

For simplicity, we consider a special case with $d_x = d_y = d$. From Equation (7.4), we have:

$$\omega_1^2 + \omega_2^2 = \frac{\omega^2 d^2}{c^2} \tag{7.9}$$

and:

$$\omega_2 = \omega_1 \tan\phi \tag{7.10}$$

On the (ω_1,ω_2) plane, the response of the array to the signal with frequency ω and direction of arrival angle ϕ is located on the point shown in Figure 7.5.

Corresponding to the traditional temporal filtering based systems, the coefficients for this new structure can be determined in similar ways, depending on the specific requirement. In the next sections, both fixed and adaptive beamformers will be discussed.

7.2 Frequency Invariant Beamforming

For wideband beamforming with a fixed beam response based on the new structure, most of the design methods described in Chapters 4 and 5, including those for the frequency invariant beamformers, can be applied to this new structure directly without any difficulty and the only change is the steering vector. An exception is the approaches proposed in Sections 5.2, 5.3 and 5.4, since those approaches are based on the specific structure of the array system.

As an example, Figure 7.6 shows a frequency invariant beamformer design result based on a 15×15 rectangular array for the frequency range $\Omega \in [0.5\pi\ \pi]$ using the

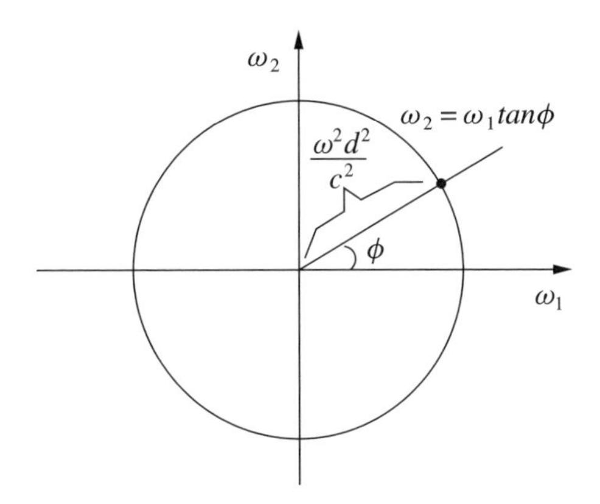

Figure 7.5 The location of the response of the array to the signal with frequency ω and direction of arrival angle ϕ

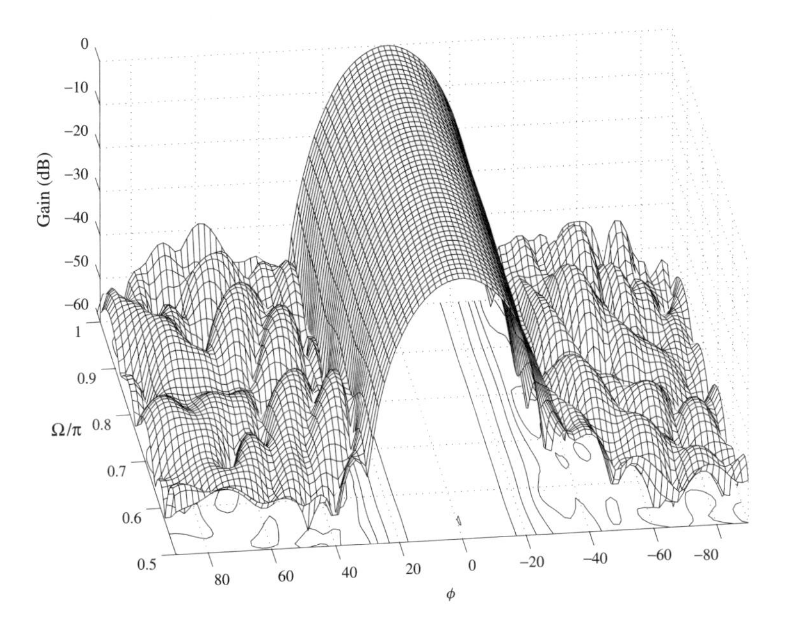

Figure 7.6 A design example with a main beam range [−30° 30°] using the convex optimization method based on a 15 × 15 rectangular array for the frequency range $\Omega \in [0.5\pi\ \pi]$

convex optimization method (Zhao *et al.*, 2008). The main beam range is [−30° 30°] and frequency invariance is applied over the range $\phi \in [-90^\circ\ 90^\circ]$ based on the minimax criterion in the sidelobe area.

In this section we will not repeat those techniques already described in Chapters 4 and 5 and only provide a frequency invariant beamformer design approach based on the specific SDL structure (Liu, 2009c; Liu and Weiss, 2008a; Liu *et al.*, 2009). The idea is very similar to the one described in Section 5.2 in Chapter 5 and can be applied to

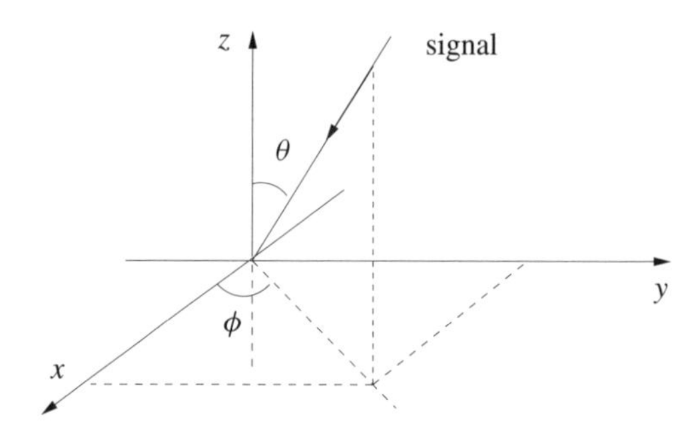

Figure 7.7 A continuous 2-D sensor array, where the signal impinges from the direction (θ, ϕ)

two-dimensional and three-dimensional uniformly spaced arrays shown in Figures 7.2 and 7.3. In the following, we will discuss these two cases separately.

7.2.1 2-D Arrays

Figure 7.7 shows an ideal continuous planar array in the (x, y) plane with a signal coming from a direction (θ, ϕ). The array's beam response with respect to temporal frequency ω rad/s and angle of arrival (θ, ϕ) is given by:

$$P(\omega, \theta, \phi) = \iint_{-\infty}^{+\infty} D(x, y) \mathrm{e}^{-j\frac{\omega \sin\theta \cos\phi}{c}x} \mathrm{e}^{-j\frac{\omega \sin\theta \sin\phi}{c}y} \mathrm{d}x\mathrm{d}y \tag{7.11}$$

where $D(x, y)$ is the response of the sensor at the position (x, y), and c is the wave propagation speed. $D(x, y)$ is a constant and independent of frequency, since there is no frequency dependent processing for each received sensor signal.

With the following substitutions:

$$\begin{aligned} \hat{\omega}_1 &= \frac{\omega \sin\theta \cos\phi}{c} \\ \hat{\omega}_2 &= \frac{\omega \sin\theta \sin\phi}{c} \end{aligned} \tag{7.12}$$

we have:

$$\hat{P}(\hat{\omega}_1, \hat{\omega}_2) = \iiint_{-\infty}^{+\infty} D(x, y) \mathrm{e}^{-j\hat{\omega}_1 x} \mathrm{e}^{-j\hat{\omega}_2 y} \mathrm{d}x\mathrm{d}y \tag{7.13}$$

Note that here $\hat{\omega}_1$ and $\hat{\omega}_2$ are actually the wavenumber vector elements k_x and k_y in Equation (1.8), respectively.

Obviously, the beam pattern of such a 2-D array can be obtained by first applying a 2-D Fourier transform to the array sensors' coefficients $D(x, y)$ according to Equation (7.13) and then using the above substitutions in Equation (7.12). Since all the two substitutions are functions of ω, the resultant beam pattern in general will also be frequency dependent. Alternatively, from the desired beam pattern $P(\omega, \theta, \phi)$, we can express it in the form

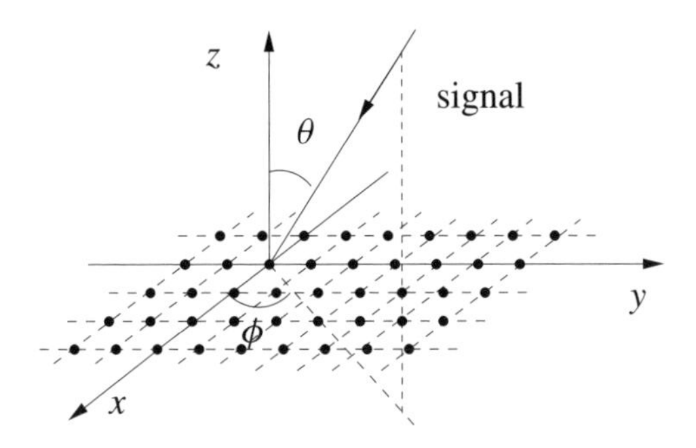

Figure 7.8 A uniformly spaced rectangular array, where the signal impinges from the direction (θ, ϕ)

of $\hat{\omega}_1$ and $\hat{\omega}_2$ and then apply a 2-D inverse Fourier transform to obtain the coefficients $D(x, y)$.

From Equation (7.12), we have:

$$\frac{\hat{\omega}_2}{\hat{\omega}_1} = \tan \phi \tag{7.14}$$

Then, for a frequency invariant beam pattern $P(\theta)$, which is also independent of the elevation angle θ, we can express it as a function of $\hat{\omega}_1$ and $\hat{\omega}_2$ according to Equation (7.14), and inverse transform to obtain $D(x, y)$.

For an equally spaced rectangular array shown in Figure 7.8, its beam response is given by:

$$P(\omega, \theta, \phi) = \sum_{k,l=-\infty}^{+\infty} D(kd_x, ld_y) e^{-j\frac{k\omega \sin\theta \cos\phi d_x}{c}} e^{-j\frac{l\omega \sin\theta \sin\phi d_y}{c}} \tag{7.15}$$

where d_x and d_y are the element spacing in the x- and y-directions, respectively, $D(kd_x, ld_y)$ is the response of the sensor at the position $(kd_x, ld_y), k, l = \ldots, -1, 0, 1, \ldots$. $D(kd_x, ld_y)$ is a constant and independent of frequency.

With the following substitutions:

$$\begin{aligned} \omega_1 &= \hat{\omega}_1 d_x \\ \omega_2 &= \hat{\omega}_2 d_y \end{aligned} \tag{7.16}$$

we have:

$$P(\omega_1, \omega_2) = \sum_{k,l=-\infty}^{+\infty} D(kd_x, ld_y) e^{-jk\omega_1} e^{-jl\omega_2} \tag{7.17}$$

Similar to the continuous case, the beam pattern of such a rectangular array can be obtained by first applying a 2-D Fourier transform to the array's coefficients $D(kd_x, ld_y)$ according to Equation (7.17) and then using the above substitutions in Equation (7.16).

From Equation (7.16), we have:

$$\frac{\omega_2 d_x}{\omega_1 d_y} = \tan \phi \tag{7.18}$$

Thus, the substitution ϕ is given by:

$$\phi = \arctan \frac{\omega_2 d_x}{\omega_1 d_y} \tag{7.19}$$

To achieve a frequency invariant beam pattern $P(\phi)$, we can express $P(\phi)$ as a function of ω_1 and ω_2 by the substitution in Equation (7.19) and then apply the 2-D inverse Fourier transform to obtain $D(kd_x, ld_y)$ (Ghavami, 2002; Liu and Weiss, 2008a). Since:

$$\phi = \arctan \frac{\omega_2 d_x}{\omega_1 d_y} = \arctan \frac{-\omega_2 d_x}{-\omega_1 d_y} \tag{7.20}$$

the resultant response $P(\omega_1, \omega_2)$ has a symmetric response on the (ω_1, ω_2) plane, which leads to real-valued coefficients $D(kd_x, ld_y)$ after applying the inverse transform. However, the problem is, the function $\tan \phi$ is periodic with a period of π. Therefore, the resultant beam pattern of the proposed design is also a periodic function of the azimuth angle ϕ with a period of 180°.

A simple remedy to this problem is to modify the substitution in Equation (7.19) (Liu, 2009c). Assume that the signal frequency $\omega \geq 0$. According to Equation (7.16), we have:

$$\begin{aligned} &\omega_1 \geq 0 \text{ for } \phi \in [-\frac{\pi}{2}\ \frac{\pi}{2}] \\ &\omega_1 \leq 0 \text{ for } \phi \in [-\pi\ \ -\frac{\pi}{2}] \cup [\frac{\pi}{2}\ \pi] \\ &\omega_2 \geq 0 \text{ for } \phi \in [0\ \pi] \\ &\omega_2 \leq 0 \text{ for } \phi \in [-\pi\ 0] \end{aligned} \tag{7.21}$$

Then the new set of substitutions is given by:

$$\phi = \begin{cases} \arctan \omega_2 d_x/\omega_1 d_y & \text{for } \omega_1 > 0 \\ \frac{\pi}{2} & \text{for } \omega_1 = 0 \text{ and } \omega_2 > 0 \\ -\frac{\pi}{2} & \text{for } \omega_1 = 0 \text{ and } \omega_2 < 0 \\ \arctan \omega_2 d_x/\omega_1 d_y - \pi & \text{for } \omega_1 < 0 \text{ and } \omega_2 < 0 \\ \arctan \omega_2 d_x/\omega_1 d_y + \pi & \text{for } \omega_1 < 0 \text{ and } \omega_2 > 0 \\ a & \text{for } \omega_1 = 0 \text{ and } \omega_2 = 0 \end{cases} \tag{7.22}$$

where a is a scalar with an arbitrary value since $\omega_1 = 0$ & $\omega_2 = 0$ corresponds to the case $\omega = 0$ and the array is not supposed to form any beam to the DC (direct current) signals.

Now given the desired response $P(\phi)$, the design of a uniformly spaced rectangular array can be simply described as follows:

1. Using the substitutions in Equation (7.22) in $P(\phi)$, we obtain $P(\omega_1, \omega_2)$, defined over one period $\{\omega_1, \omega_2\} \in [-\pi\ \pi)$.

2. Applying a 2-D inverse Fourier transform to $P(\omega_1, \omega_2)$ returns the desired coefficients $D(kd_x, ld_y)$ for the corresponding sensors. As an approximation, we can employ the 2-D inverse discrete Fourier transform by sampling $P(\omega_1, \omega_2)$ on the (ω_1, ω_2) plane over the range $[-\pi\ \pi)$.

One key issue in this design is to find a realisable desired frequency invariant beam pattern $P(\phi)$. We cannot obtain it through a narrowband linear array design method since the resultant beam pattern will be a periodic function of ϕ with a period π. One solution is to use an FIR filter design method to obtain a lowpass filter with a response $H_0(\Omega)$, which has a maximum response at the normalized frequency $\Omega = 0$. Then the desired response $P(\phi)$ is obtained by directly replacing Ω in $H_0(\Omega)$ by $\phi - \phi_0$, namely:

$$P(\phi) = H_0(\phi - \phi_0) \tag{7.23}$$

where ϕ_0 is the main beam direction. Certainly we can also design a bandpass FIR filter $H_1(\Omega)$ with a maximum response at $\Omega = \Omega_0$ and then use the substitution $P(\phi) = H_1(\phi)$ to obtain the desired response with a main beam pointing to $\phi = \Omega_0$. Note in this case the response of the FIR filter $H_1(\Omega)$ should not be symmetric about the point $\Omega = 0$; otherwise, the resultant $P(\phi)$ will have two main beams: one in the direction $\phi = \Omega_0$ and one in $\phi = -\Omega_0$. This requires $H_1(\Omega)$ to be the response of a complex-valued FIR filter.

7.2.1.1 Design Examples

Now we give two design examples based on a 19×19 uniformly spaced rectangular array. The frequency range of interest is between 400 Hz and 1600 Hz with a signal propagation speed $c = 340$ m/s and an array spacing $d_x = d_y = \lambda_{min}/2 = 34\,000/(2 \times 1600)$ cm $\approx$ 10 cm. As a simple example, we consider the following desired response $P(\phi)$:

$$P(\phi) = \frac{1}{7} \sum_{n=-3}^{3} e^{jn\phi} \tag{7.24}$$

which is the response of a lowpass 5-tap FIR filter with a uniform weighting. The desired response in this case has a main beam pointing to $\phi = 0°$.

With the substitution in Equation (7.22), a 64×64-point 2-D IDFT is applied to the resultant response $P(\omega_1, \omega_2)$ to obtain a set of coefficients with dimensions of 64×64; then the result is truncated to the array size 19×19. The beam pattern for the elevation angle $\theta = 90°$ over the frequency range of interest is shown in Figure 7.9, which exhibits a satisfactory frequency invariance property.

Next, we show an example with a main beam direction $\phi_0 = \pi/2$ (90°). The desired response is given by:

$$P(\phi) = \frac{1}{7} \sum_{n=-3}^{3} e^{jn(\phi-\phi_0)} \tag{7.25}$$

The resultant beam pattern is shown in Figure 7.10, with again a clear frequency invariance property.

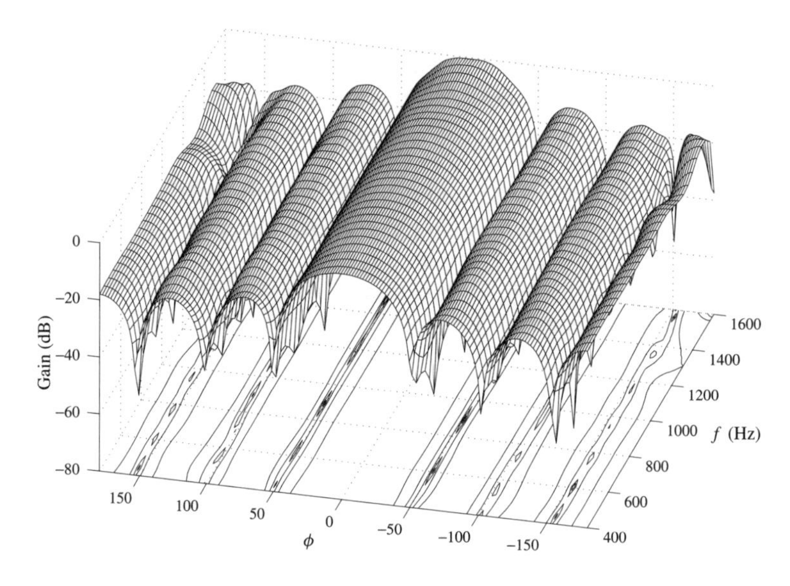

Figure 7.9 A design example with a broadside main beam ($\theta = 90$)

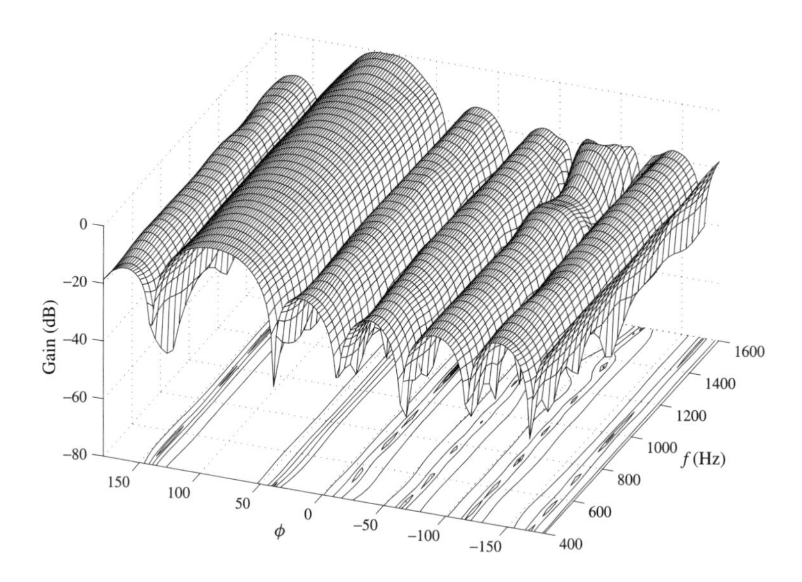

Figure 7.10 A design example with an off-broadside main beam ($\theta = 90$)

7.2.2 3-D Arrays

To derive the design method for a uniformly spaced 3-D array, we also start by studying the response of the ideal continuous 3-D array shown in Figure 7.7, where the array extends in all the three dimensions of the (x, y, z) space. The array's beam response is given by:

$$P(\omega,\theta,\phi)=\iiint_{-\infty}^{+\infty}D(x,y,z)\mathrm{e}^{-j\frac{\omega\sin\theta\cos\phi}{c}x}\mathrm{e}^{-j\frac{\omega\sin\theta\sin\phi}{c}y}\mathrm{e}^{-j\frac{\omega\cos\theta}{c}z}\mathrm{d}x\mathrm{d}y\mathrm{d}z \tag{7.26}$$

where $D(x, y, z)$ is a constant representing the response of the sensor at the position (x, y, z).

With the following substitutions:

$$\begin{aligned}\hat{\omega}_1&=\frac{\omega\sin\theta\cos\phi}{c}\\ \hat{\omega}_2&=\frac{\omega\sin\theta\sin\phi}{c}\\ \hat{\omega}_3&=\frac{\omega\cos\theta}{c}\end{aligned} \tag{7.27}$$

we have:

$$\hat{P}(\hat{\omega}_1,\hat{\omega}_2,\hat{\omega}_3)=\iiint_{-\infty}^{+\infty}D(x,y,z)\ \mathrm{e}^{-j\hat{\omega}_1x}\mathrm{e}^{-j\hat{\omega}_2y}\mathrm{e}^{-j\hat{\omega}_3z}\mathrm{d}x\mathrm{d}y\mathrm{d}z \tag{7.28}$$

Therefore, the beam pattern of such a 3-D array can be obtained by first applying a 3-D Fourier transform to the array sensors' coefficients $D(x, y, z)$ according to Equation (7.28) and then using the above substitutions in Equation (7.27). Since all the three substitutions are functions of ω, the resultant beam pattern in general will also be frequency dependent. Furthermore, we can express a desired beam pattern $P(\omega, \theta, \phi)$ in term of $\hat{\omega}_1$, $\hat{\omega}_2$ and $\hat{\omega}_3$ and then apply a 3-D inverse Fourier transform to find the coefficients $D(x, y, z)$.

Likewise, to achieve a frequency invariant beam pattern $P(\theta, \phi)$, we again express $P(\theta, \phi)$ as a function of $\hat{\omega}_1$, $\hat{\omega}_2$ and $\hat{\omega}_3$ and then inverse transform to find $D(x, y, z)$. The key is to find a way to eliminate ω in the expressions for θ and ϕ; otherwise, the resultant $D(x, y, z)$ will still be a function of ω and not a constant as required.

From Equation (7.27), we can have:

$$\begin{aligned}\frac{\hat{\omega}_1}{\hat{\omega}_2}&=\cot\phi\\ \frac{\hat{\omega}_3^2}{\hat{\omega}_1^2+\hat{\omega}_2^2}&=\frac{\cos^2\theta}{\sin^2\theta}=\cot^2\theta\end{aligned} \tag{7.29}$$

Thus, we obtain the following pair of substitutions for θ and ϕ:

$$\begin{aligned}\phi&=\mathrm{arcctg}\frac{\hat{\omega}_1}{\hat{\omega}_2}\qquad(\phi\in[0\ \ \pi))\\ \theta&=\mathrm{arcctg}\frac{\hat{\omega}_3}{\sqrt{\hat{\omega}_1^2+\hat{\omega}_2^2}}\qquad(\theta\in[0\ \ \pi])\end{aligned} \tag{7.30}$$

For $\phi \in [\pi \;\; 2\pi)$, we will have $\phi = \mathrm{arcctg}(\hat{\omega}_1/\hat{\omega}_2) + \pi$. The exact value of ϕ should also be decided by the signs of both $\hat{\omega}_1$ and $\hat{\omega}_2$, as in Equation (7.22).

Then, given the desired frequency invariant response $P(\theta, \phi)$, we can use the above two substitutions to generate the intermediate function $\hat{P}(\hat{\omega}_1, \hat{\omega}_2, \hat{\omega}_3)$ and $D(x, y, z)$ can be obtained by applying an inverse Fourier transform to it. Since in the two substitutions of Equation (7.30), there is no ω term, $\hat{P}(\hat{\omega}_1, \hat{\omega}_2, \hat{\omega}_3)$ obtained in this way will not be a function of ω. As a result, $D(x, y, z)$ will be independent of ω too.

For a uniformly spaced 3-D array, as shown in Figure 7.3, where the spacings of the array elements in the x-, y- and z-directions are d_x, d_y and d_z, respectively, the integration in Equation (7.26) changes to a summation:

$$P(\omega, \theta, \phi) = \sum_{k,l,m=-\infty}^{+\infty} D(kd_x, ld_y, md_z) \mathrm{e}^{-j\frac{k\omega \sin\theta \cos\phi d_x}{c}} \mathrm{e}^{-j\frac{l\omega \sin\theta \sin\phi d_y}{c}} \mathrm{e}^{-j\frac{m\omega \cos\theta d_z}{c}} \quad (7.31)$$

where $D(kd_x, ld_y, md_z)$ are the responses of the sensor at the position (kd_x, ld_y, md_z), k, $l, m = \ldots, -1, 0, 1, \ldots$.

With the following substitutions:

$$\begin{aligned} \omega_1 &= \hat{\omega}_1 d_x \\ \omega_2 &= \hat{\omega}_2 d_y \\ \omega_3 &= \hat{\omega}_3 d_z \end{aligned} \quad (7.32)$$

we have:

$$\hat{P}(\omega_1, \omega_2, \omega_3) = \sum_{k,l,m=-\infty}^{+\infty} D(kd_x, ld_y, md_z) \mathrm{e}^{-jk\omega_1} \mathrm{e}^{-jl\omega_2} \mathrm{e}^{-jm\omega_3} \quad (7.33)$$

Based on the discussion in the continuous sensor case, and similar to Equation (7.29), we have:

$$\begin{aligned} \frac{\omega_1 d_y}{\omega_2 d_x} &= \cot\phi \\ \frac{\omega_3^2/d_z^2}{\omega_1^2/d_x^2 + \omega_2^2/d_y^2} &= \frac{\cos^2\theta}{\sin^2\theta} = \cot^2\theta \end{aligned} \quad (7.34)$$

Then we arrive at the following pair of substitutions for θ and ϕ in the discrete sensor case:

$$\begin{aligned} \phi &= \mathrm{arcctg}\frac{\omega_1 d_y}{\omega_2 d_x} \qquad (\phi \in [0 \;\; \pi)) \\ \theta &= \mathrm{arcctg}\frac{\omega_3/d_z}{\sqrt{\omega_1^2/d_x^2 + \omega_2^2/d_y^2}} \qquad (\theta \in [0 \;\; \pi]) \end{aligned} \quad (7.35)$$

Again for $\phi \in [\pi \;\; 2\pi)$, we should have $\phi = \mathrm{arcctg}(\omega_1 d_y)/(\omega_2 d_x) + \pi$ and the exact value of ϕ should also be decided by the signs of both ω_1 and ω_2.

Now given the desired frequency invariant response $P(\theta, \phi)$, which is supposed to be a continuous function of θ and ϕ, the design of the 3-D uniformly spaced array can be described as follows:

1. Using the substitutions of Equation (7.35) in $P(\theta, \phi)$, we obtain $\hat{P}(\omega_1, \omega_2, \omega_3)$, defined over one period $\omega_1, \omega_2, \omega_3 \in [-\pi \; \pi)$.
2. Applying a 3-D inverse Fourier transform to $\hat{P}(\omega_1, \omega_2, \omega_3)$ returns the desired coefficients $D(kd_x, ld_y, md_z)$ for the corresponding sensors and again we can employ the 3-D inverse discrete Fourier transform as an approximation.

There is an effective region for the array's response in the $(\omega_1, \omega_2, \omega_3)$ domain for a given frequency range of interest. From Equation (7.32), we can have:

$$\frac{c^2\omega_1^2}{d_x^2} + \frac{c^2\omega_2^2}{d_y^2} + \frac{c^2\omega_3^2}{d_z^2} = \omega^2 \tag{7.36}$$

Suppose the range of the frequency of interest is $\omega \in [\omega_{\min} \; \omega_{\max}]$ and then we have:

$$\frac{c^2\omega_1^2}{d_x^2} + \frac{c^2\omega_2^2}{d_y^2} + \frac{c^2\omega_3^2}{d_z^2} \in [\omega_{\min}^2 \; \omega_{\max}^2] \tag{7.37}$$

Therefore, $\hat{P}(\omega_1, \omega_2, \omega_3)$ can take any value without affecting the array's response to the frequency range of interest for the following two regions:

$$0 < \frac{c^2\omega_1^2}{d_x^2} + \frac{c^2\omega_2^2}{d_y^2} + \frac{c^2\omega_3^2}{d_z^2} < \omega_{\min}^2 \tag{7.38}$$

and:

$$\frac{c^2\omega_1^2}{d_x^2} + \frac{c^2\omega_2^2}{d_y^2} + \frac{c^2\omega_3^2}{d_z^2} > \omega_{\max}^2 \tag{7.39}$$

One easy choice is to assign a constant value to it for those two regions. Intuitively, with this approach, the modified $\hat{P}(\omega_1, \omega_2, \omega_3)$ will become smoother than the original one and so the 3-D IDFT can lead to an improved design result. Note this analysis is also valid to the 2-D case.

In the above design, we first need to have the desired frequency invariant response which could be obtained by a design method for narrowband rectangular arrays if appropriate. Normally, the $P(\theta, \phi)$ thus obtained will be a function of the form $F(\sin\theta\cos\phi, \sin\theta\sin\phi)$. To avoid aliasing, $d_x, d_y, d_z \le \lambda_{min}/2$, where λ_{min} is the wavelength of the maximum frequency of interest $\omega_{\max}$, and we set $d_x = d_y = d_z = \lambda_{min}/2$. Then for $\omega > 0$, from Equations (7.32) and (7.36), we can easily derive the following substitutions:

$$\begin{aligned} \sin\theta\cos\phi &= \frac{\omega_1}{\sqrt{\omega_1^2 + \omega_2^2 + \omega_3^2}} \\ \sin\theta\sin\phi &= \frac{\omega_2}{\sqrt{\omega_1^2 + \omega_2^2 + \omega_3^2}} \end{aligned} \tag{7.40}$$

So $\hat{P}(\omega_1, \omega_2, \omega_3)$ can be obtained as:

$$\hat{P}(\omega_1, \omega_2, \omega_3) = F(\frac{\omega_1}{\sqrt{\omega_1^2 + \omega_2^2 + \omega_3^2}}, \frac{\omega_2}{\sqrt{\omega_1^2 + \omega_2^2 + \omega_3^2}}) \tag{7.41}$$

7.2.2.1 Design Example

Here we provide a design example for an acoustic array. The frequency range of interest is between 500 Hz and 1500 Hz and the propagation speed is 340 m/s. The dimensions of the 3-D equally spaced array are $19 \times 19 \times 19$ and the adjacent sensor spacing is set to be $34000/(2 \times 1500) = 11.33$ cm. The desired beam pattern is given by:

$$F(\sin\theta\cos\phi, \sin\theta\sin\phi) = \frac{1}{9}\sum_{l,m=-1,-1}^{1,1} \mathrm{e}^{-jl\pi\sin\theta\cos\phi}\,\mathrm{e}^{-jm\pi\sin\theta\sin\phi} \tag{7.42}$$

which forms a main beam towards the broadside ($\theta = \phi = 0$). We can see this desired frequency invariant response is obtained by a narrowband rectangular array with uniform weighting.

Note that after the substitutions in Equations (7.40) and (7.41), the resultant $\hat{P}(\omega_1, \omega_2, \omega_3)$ is symmetric with $\hat{P}(\omega_1, \omega_2, \omega_3) = \hat{P}(-\omega_1, -\omega_2, -\omega_3)$. Hence the coefficients $D(kd_x, ld_y, md_z)$ obtained after the inverse Fourier transform will be real-valued and the resultant frequency invariant beamformer is not only without any temporal filtering process, but also without phase-shifters, which significantly simplify the implementation. However, for a general desired beam pattern, the resultant beamformer coefficients will be complex-valued.

We employed a $32 \times 32 \times 32$-point 3-D IDFT on the resultant function $\hat{P}(\omega_1, \omega_2, \omega_3)$ and set $\hat{P}(\omega_1, \omega_2, \omega_3) = 0$ for the area outside the frequency range [300 Hz 1700 Hz] according to Equations (7.36) and (7.40). We left the region between 300 Hz and 500 Hz and the region between 1500 Hz and 1700 Hz as the transition bands. Since the beam pattern is four-dimensional, we can only provide some exemplary snapshots.

Figures 7.11, 7.12 and 7.13 give the array's response in cylindrical coordinates to the frequencies $f = 500$ Hz, 1 kHz and 1.5 kHz, respectively. The height axis is the magnitude response of the beam in dB, the radial coordinate is for the elevation angle θ (in radians) and the angle coordinate is for the azimuth angle ϕ (in radians). The frequency invariance property can be verified by the clear similarity of these three figures.

In addition, the responses with respect to θ and f for three different values of ϕ are given in Figures 7.14, 7.15 and 7.16, with $\phi = 60°, \phi = 120°$ and 240°, respectively, where θ is in degrees from 0° to 90°. The frequency invariance property is again clearly visible.

7.3 Adaptive Beamforming

For adaptive wideband beamforming based on the new structure, its coefficients can be determined in a similar way as in the traditional temporal filtering based systems and here we give two examples.

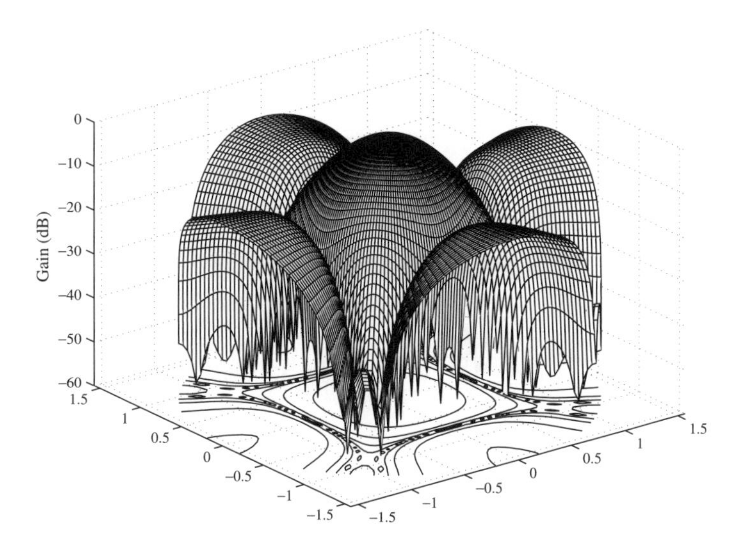

Figure 7.11 The resultant beam pattern of the 3-D array at $f = 500\,\text{Hz}$, with respect to azimuth (angle coordinate) and elevation (radial coordinate)

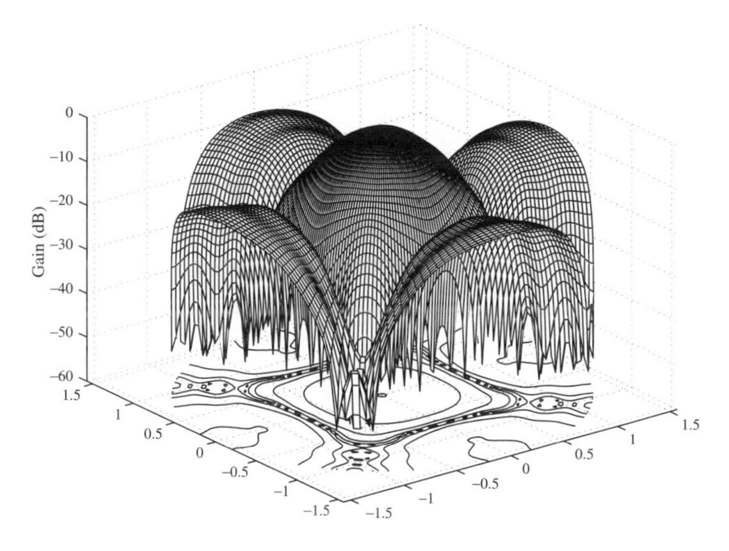

Figure 7.12 The resultant beam pattern of the 3-D array at $f = 1\,\text{kHz}$, with respect to azimuth (angle coordinate) and elevation (radial coordinate)

7.3.1 Reference Signal Based Beamformer

The first one is the case for which a reference signal $r(t)$ is available and the weights are adjusted to minimize the mean square error between the beamformer output $y(t)$ and the reference signal $r(t)$. This is the simplest case and its structure is shown in Figure 7.17. It is a classical adaptive filtering problem and can be solved by any existing adaptive algorithms such as the LMS or RLS algorithms, or their subband implementations.

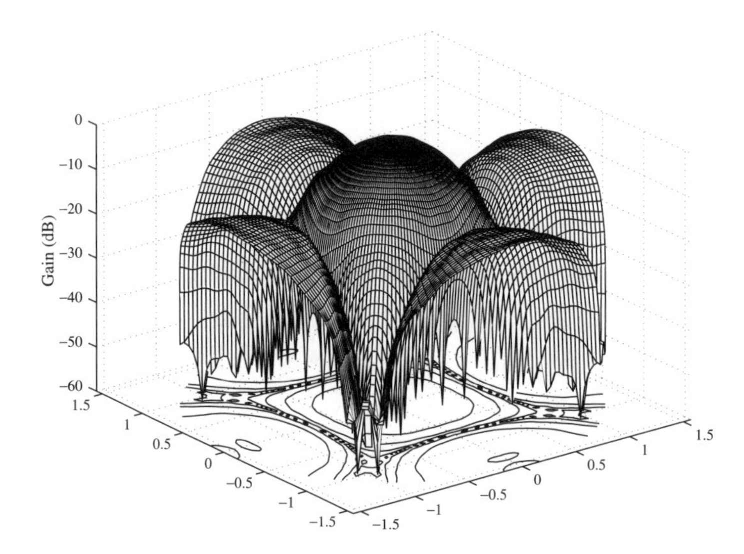

Figure 7.13 The resultant beam pattern of the 3-D array at $f = 1.5\,\text{kHz}$, with respect to azimuth (angle coordinate) and elevation (radial coordinate)

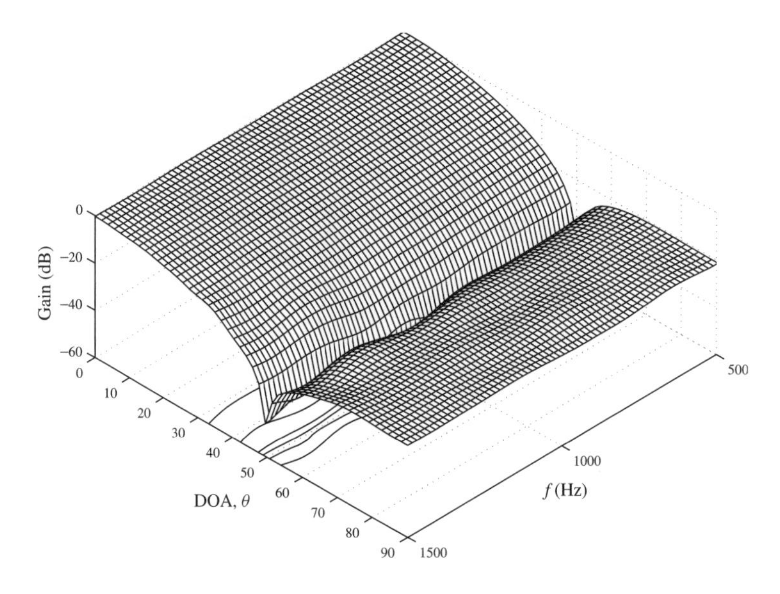

Figure 7.14 A slice of the beam pattern (θ, f) at azimuth angle $\phi = 60°$

7.3.2 Linearly Constrained Minimum Variance Beamformer

If a desired reference signal is not available, but we know the DOA angle of the signal of interest and their bandwidth range, then we can impose some constraints on the array coefficients and then adaptively minimize the variance $E\{y(t)^* y(t)\}$ of the beamformer output subject to the imposed constraints. This leads to the well-known LCMV

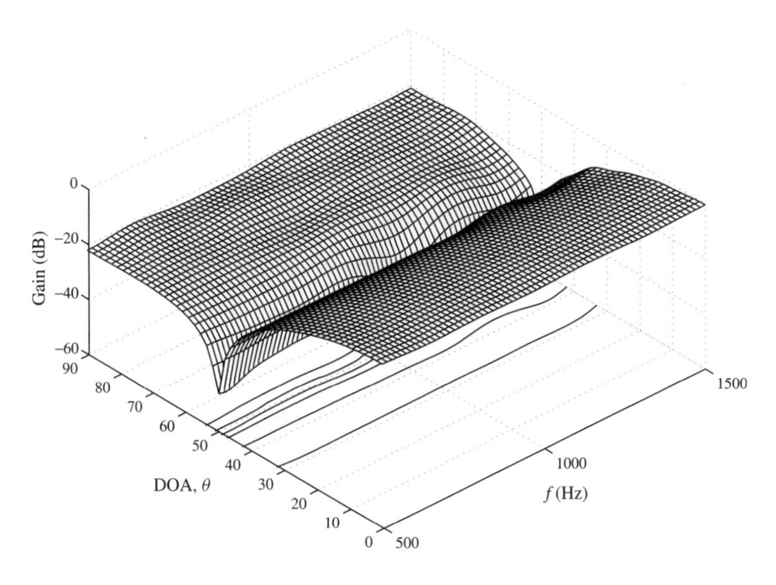

Figure 7.15 A slice of the beam pattern (θ, f) at the azimuth angle $\phi = 120^{\circ}$

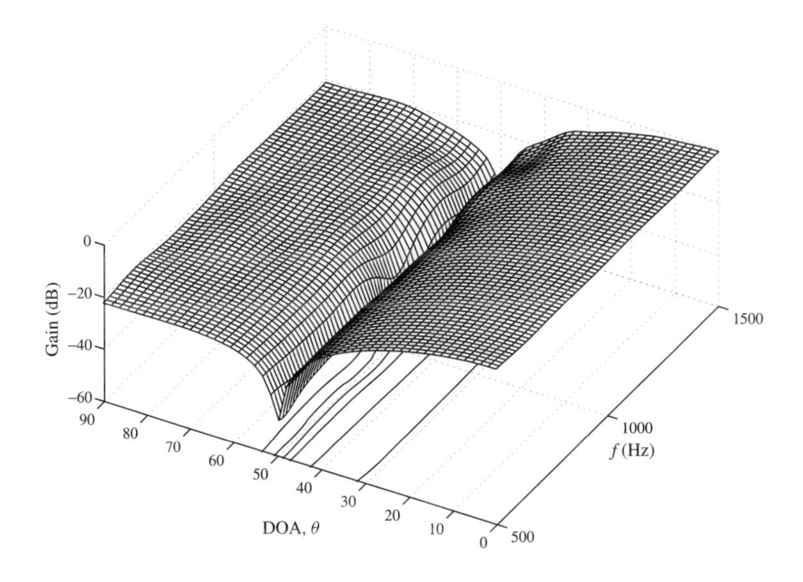

Figure 7.16 A slice of the beam pattern (θ, f) at the azimuth angle $\phi = 240^{\circ}$

beamformer (Frost, 1972). The LCMV problem can be formulated as:

$$\min_{\mathbf{w}} \mathbf{w}^H \boldsymbol{R}_{xx} \mathbf{w} \qquad \text{subject to} \qquad \boldsymbol{C}^H \mathbf{w} = \mathbf{f} \tag{7.43}$$

where $\boldsymbol{R}_{xx}$ is the covariance matrix of observed array data in $\mathbf{x}$, $\boldsymbol{C}$ is the constraint matrix and $\mathbf{f}$ is the response vector. The constraint will ensure that no matter how to adjust the

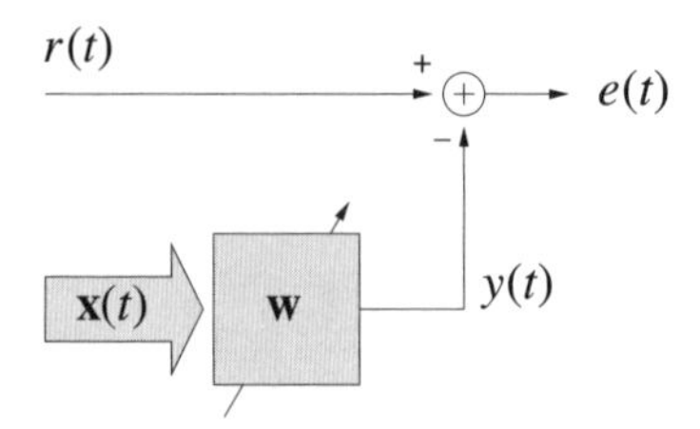

Figure 7.17 A general structure for the first implementation of the SDL-based wideband beamformer

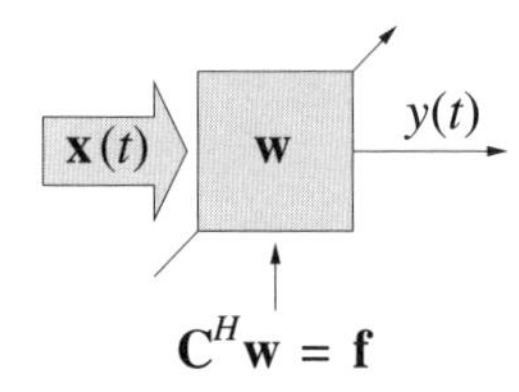

Figure 7.18 A general structure for the LCMV beamformer, where $\boldsymbol{C}^H\mathbf{w} = \mathbf{f}$ is the constraint imposed on the adaptive process

array coefficients, the resultant beamformer will have the desired response set out by the constraint equation $\boldsymbol{C}^H\mathbf{w} = \mathbf{f}$.

Figure 7.18 shows a general structure for such a LCMV beamfomer. As discussed in Chapter 2, the constrained adaptive optimization in Equation (7.43) can be conveniently solved using a generalized sidelobe canceller (GSC) (Griffiths and Jim, 1982), which performs a projection of the data onto an unconstrained subspace by means of a blocking matrix $\boldsymbol{B}$ and a quiescent vector $\mathbf{w}_q$ as shown in Figure 2.11. Thereafter, standard unconstrained adaptive algorithms such as the LMS and RLS algorithms can be invoked to minimize the variance of $y(t)$. For details, please refer to Section 2.4 of Chapter 2.

7.3.3 Discussions

As already seen, based on this SDL structure, we have formed some novel array structures, such as the concentric semicircular system in Figure 7.4, which does not make sense in the traditional paradigm of wideband array processing. However, for some other structures, such as the cubic array (corresponding to the traditional wideband rectangular array with temporal filtering) and the rectangular array (corresponding to the traditional widebband linear array with temporal filtering), we have already seen them in the traditional system. Then, in the latter case, what is the 'new wine in old bottles'?

Take the proposed rectangular array system without temporal filtering as an example. In appearance, it is indeed the same as a traditional rectangular array working for narrowband beamforming, because neither of them has TDLs or FIR/IIR filters involved and they have the same structure as shown in Figure 7.2. To explain their differences clearly, we place the rectangular array in the coordinate system shown in Figure 7.8.

In the traditional narrowband rectangular array system, we normally assume that the signal of interest comes from the broadside, i.e. $\theta = 0^\circ$ in Figure 7.8, and the interfering signals are from other directions with different θ and ϕ. If the signal of interest are not from the broadside, we can impose appropriate time delays, or phase shifts immediately after each sensor output, such that the signals incident on the array from directions of interest other than the broadside appear as identical replicas of one another at the outputs of the steering delay elements.

However, in the new approach, we shall assume the signal of interest is from the direction $\phi = 0^\circ, \theta > 0$, and the interfering signals are from other azimuth angles $\phi \neq 0$. As a result, for the case of an LCMV beamformer, the constraint matrix $\boldsymbol{C}$ will be different from that of the traditional approach. However, in the reference signal based case, we will not really see much difference in the implementation except that one is for wideband signals and one is for narrowband signals.

7.3.4 Simulations

In this part, we give some simple simulation results to show the performance of the SDL-based array system. A detailed analysis of its performance for the case $M = 2$ can be found in Lin *et al.* (2007, 2008, 2010).

Here the SDL beamforming structure with $M = N = 10$ is compared with the traditional TDL structure with $M = J = 10$. To run the simulations on a computer, we discretize the continuous signals and the sampling frequency is twice the highest frequency of the signal of interest. The spacing between adjacent sensors is half-wavelength of the signal component with the highest possible frequency (corresponding to a normalized frequency π). The signal of interest comes from the broadside and four interfering signals from the directions $\theta = 90^\circ$ and $\phi = 20^\circ, 40^\circ, -30^\circ, -60^\circ$, respectively. All of the signals have a bandwidth of $[0.4\pi \;\; \pi]$. The signal to interference ratio is about -20 dB and the signal to noise ratio is about 20 dB. We use a normalized LMS algorithm with a stepsize of 0.45 for the reference signal based case and 0.20 for the LCMV case.

In the first set of simulations, we assume a reference signal is available and the implementation in Figure 7.17 is used. The learning curves for the ensemble mean square output error are shown in Figure 7.19 and we can see that although the convergence rate of the proposed structure is lower than the traditional structure, it has reached a lower steady-state error for this specific scenario. The resultant beam pattern for the frequency range of $[0.4\pi \;\; \pi]$ for the proposed structure is shown in Figure 7.20, where the four nulls at the interference directions of $\phi = 20^\circ, 40^\circ, -30^\circ, -60^\circ$ are clearly visible, indicating a successful beamforming operation.

For the second implementation, i.e. no reference signal is available, both structures have a similar convergence speed, which is shown in Figure 7.21 with a stepsize of 0.20, implemented by a GSC. However, the SDL system again has achieved a lower steady-state error for this special case. The resultant beam pattern is shown in Figure 7.22. Similar to Figure 7.20, we can also see the nulls at the interference directions.

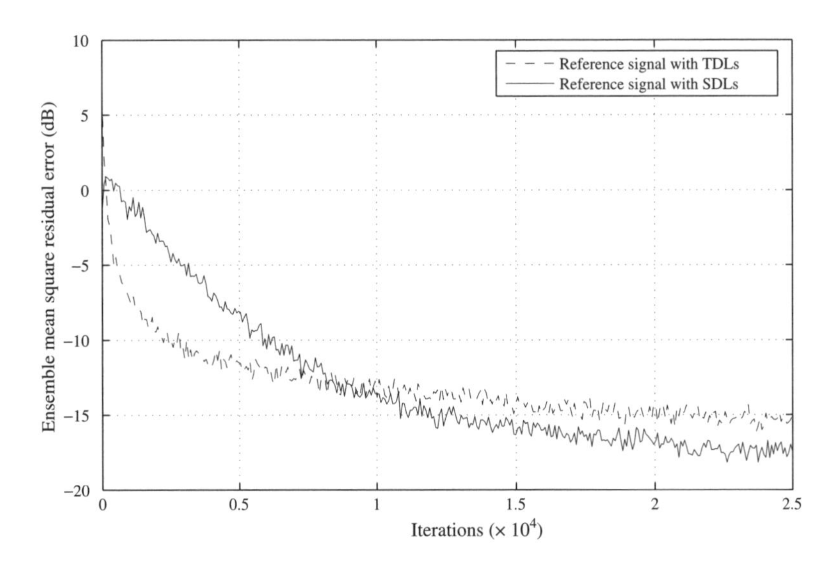

Figure 7.19 The two learning curves for the reference signal based beamformer with a stepsize of 0.45

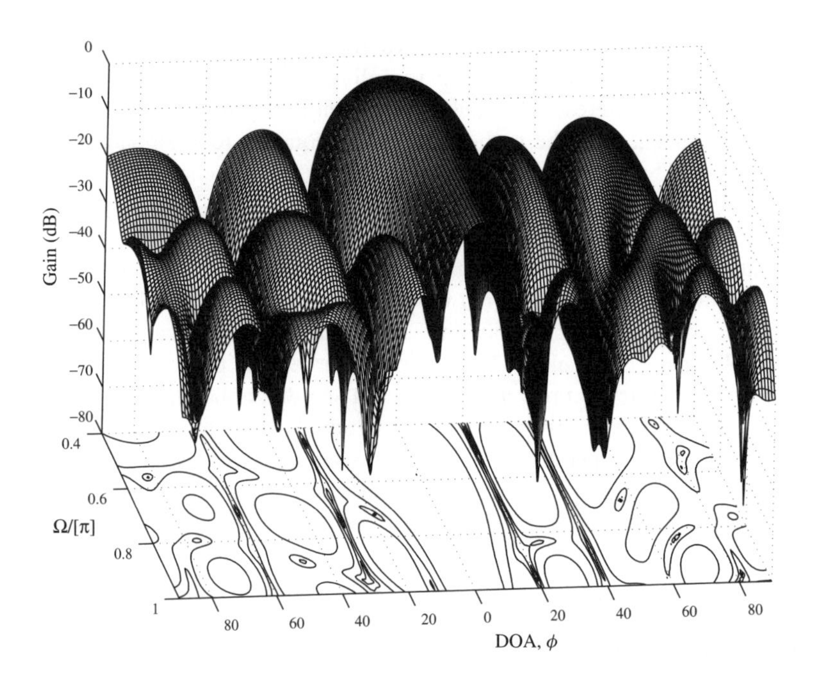

Figure 7.20 The resultant beam pattern for the bandwidth of interest $[0.4\pi\ \pi]$ for the SDL-based structure with a reference signal

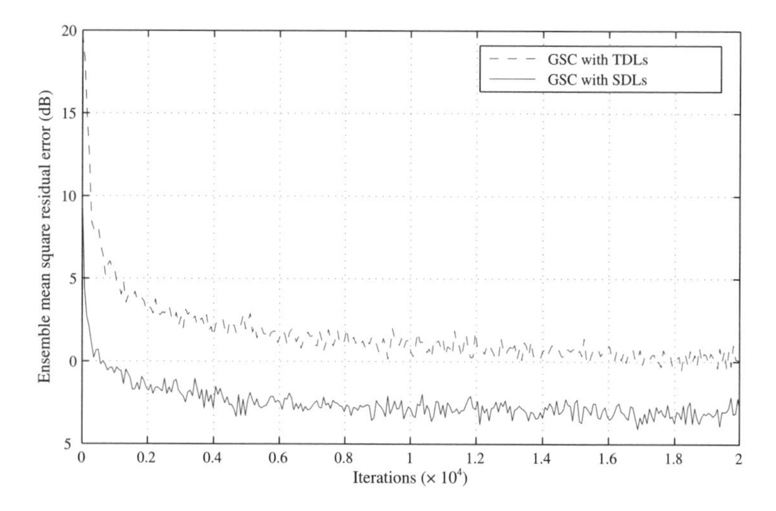

Figure 7.21 The learning curves based on a GSC with a stepsize of 0.20

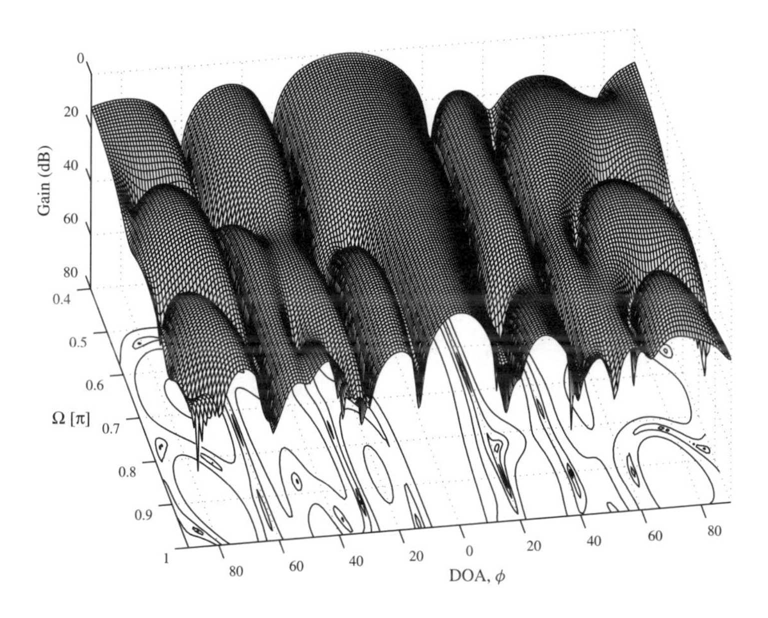

Figure 7.22 The resultant beam pattern for the SDL-based structure with a GSC

7.4 Beamspace Adaptive Beamforming

7.4.1 Structure

To increase the convergence speed and reduce the computational complexity of the SDL-based adaptive wideband beamformer, we can also adopt the beamspace adaptive approach, as introduced in Liu and Weiss (2008a).

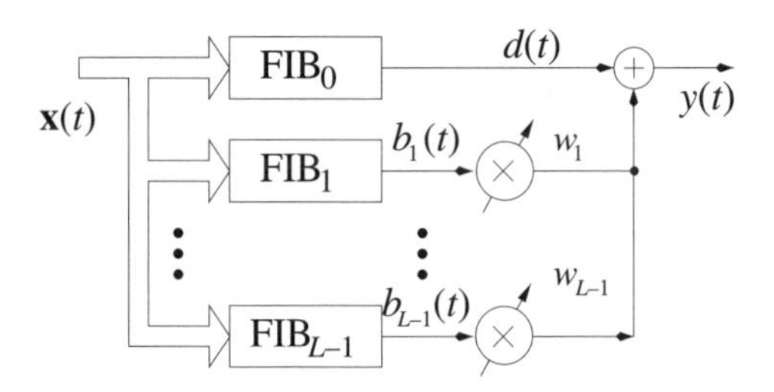

Figure 7.23 A general structure for beamspace adaptive beamforming

In this structure, L frequency invariant beams are designed pointing to the directions $\phi_l, l = 0, 1, \ldots, L-1$. One of the beams is the main beam pointing to the direction of the signal of interest and the remaining $L-1$ beams are auxiliary beams pointing to the other directions which cover all of the possible directions of the interfering signals, as described in Section 5.6 of Chapter 5. Figure 7.23 shows such a structure, where $\mathbf{x}(t)$ is the received array signal vector at time t, FIB_0 is the main beam with an output $\text{d}(t)$ and $y(t)$ is the final array output by combining the FIB outputs with one adaptive coefficient $w_l, l = 1, \ldots, L-1$ for each of them, given by:

$$y(t) = d(t) - \mathbf{w}^T \mathbf{b}(t) \tag{7.44}$$

where:

$$\begin{aligned} \mathbf{b}(t) &= \left[b_1(t)\ b_2(t)\ \cdots\ b_{L-1}(t)\right]^{\mathrm{T}} \\ \mathbf{w} &= \left[w_1\ w_2\ \cdots\ w_{L-1}\right]^{\mathrm{T}} \end{aligned} \tag{7.45}$$

Without loss of generality, we can assume the signal of interest comes from the broadside; then $\phi_0 = 0$. For the auxiliary beams, they have a zero response to the signal of interest and their outputs only contain noise and interfering signals. The coefficients $\mathbf{w}$ are adjusted by standard adaptive algorithms (Haykin, 1996).

7.4.2 Simulations

The simulations are based on a 19×19 uniformly spaced rectangular array with an inter-element spacing of 4.25 cm. The proposed beamspace adaptive beamformer employs 5 frequency invariant beams, which are derived from the corresponding desired responses given in Figure 7.24, where the responses repeat themselves for the other half of the azimuth range.

Its performance is compared with that of the traditional GSC based on the same rectangular array. The signal of interest comes from the direction $(\theta = 90°, \phi = 90°)$ and three interfering signals from the directions $(\theta = 90°, \phi = 30°)$, $(\theta = 90°, \phi = 70°)$ and $(\theta = 90°, \phi = 130°)$, respectively. All of the signals have a bandwidth [1.2 kHz 3.8 kHz]. The signal to interference ratio is about -20 dB and the signal to noise ratio is about 20 dB. We use a normalized LMS algorithm for adaptation. The stepsize is 0.06 for the beamspace approach and 0.15 for the traditional GSC, which are empirically chosen to achieve roughly the same steady-state value of the mean square residual error.

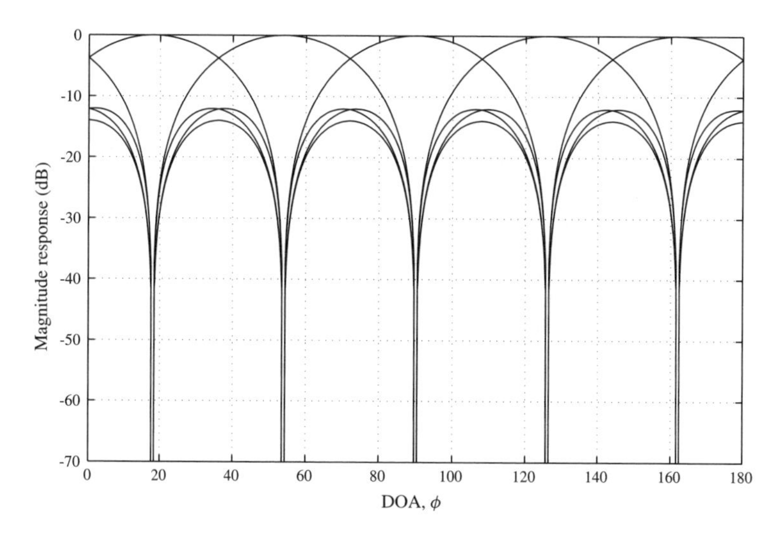

Figure 7.24 The desired responses for designing the five frequency invariant beams

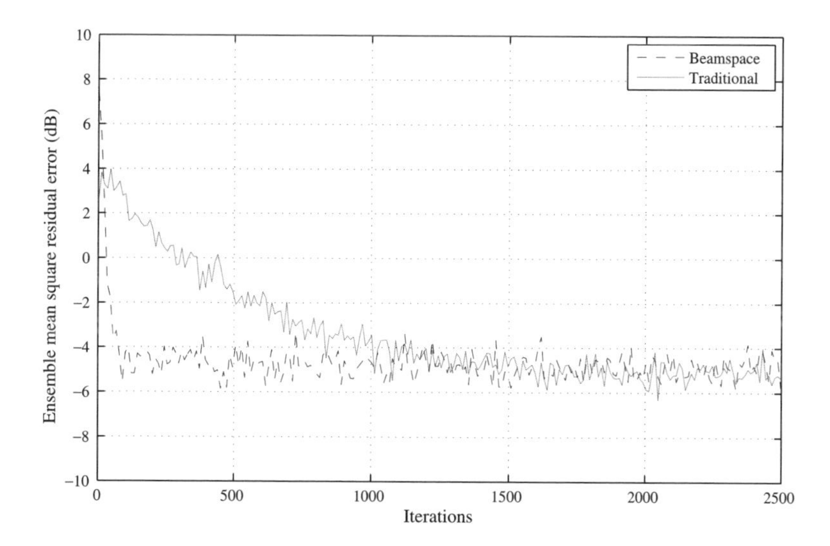

Figure 7.25 The learning curves for the beamspace adaptive beamformer and the GSC

The learning curves for these two beamformers are shown in Figure 7.25, where the solid line is for the proposed method and the dashed line is for the GSC. The beamspace method reaches its steady state in about 100 iterations, whereas it takes the GSC about 2000 iterations. The much faster convergence speed is due to the fact that the beamspace method only needs one adaptive weight for each of the FIB outputs and as a result the adaptive filter length of the beamspace method is much shorter than the GSC. In this case the adaptive filter length for the beamspace method is $5 - 1 = 4$ and for the GSC it is $19 \times (19 - 1) = 342$.

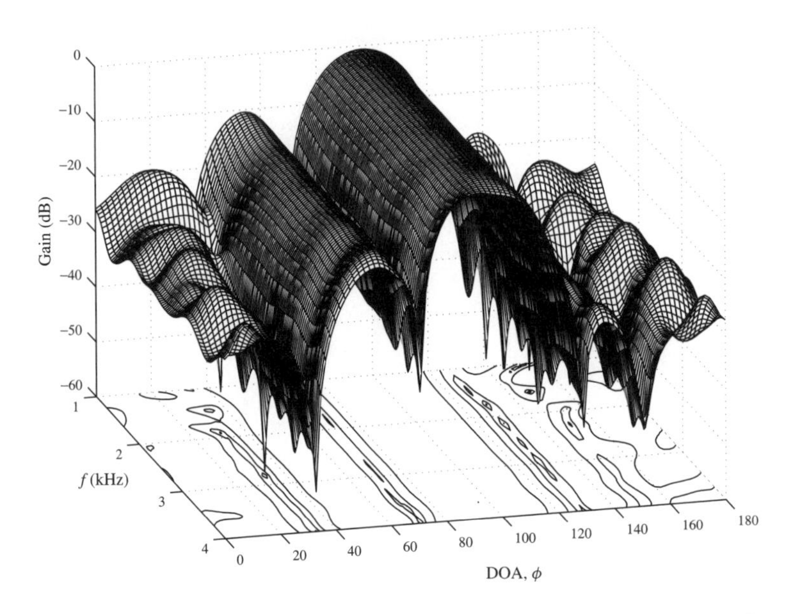

Figure 7.26 The resultant beam pattern of the proposed beamspace adaptive beamformer for the bandwidth of interest

As in the other beamspace adaptive arrays based on the FIB technique, the resultant beam pattern will be frequency invariant, which can be confirmed by examining the final response shown in Figure 7.26, where the nulls can be seen clearly in the three interference directions.

7.5 Summary

In this chapter we have introduced a different approach to wideband beamforming based on the recently proposed sensor delay-line system. A special property of the resultant wideband beamforming structure is that there is not any form of temporal processing required, such as tapped delay-lines or FIR/IIR filters. Therefore it can be considered as a wideband beamforming structure with spatial-only information.

There are two main advantages with this new approach: firstly, since only one coefficient is required for each received array signal, a simple analogue circuit can be employed for both adaptive and fixed beamforming if necessary, especially for the case with a very high signal frequency and bandwidth; secondly, it can provide a much wider spatial coverage for effective beamforming.

Most of the techniques developed for traditional TDL-based beamformers can be applied to this new structure directly and in this chapter we only studied several of them as examples to illustrate its performance, including frequency invariant beamforming in Section 7.2, reference signal based and LCMV beamforming in Section 7.3 and the beamspace adaptive beamforming in Section 7.4.

Further studies of the SDL-based structure are needed in the future to fully exploit its potential for wideband beamforming in various signal environments and applications.

8

Wideband Beamforming for Multipath Signals

In all of the studies so far, we have implicitly assumed that there is only one desired signal direction and no multipath propagation is present to the beamformer. However, in many practical scenarios, multipath propagation cannot be avoided or ignored. The multipath version of the desired signal can be considered as a coherent or correlated interference and it is well known that the performance of the conventional adaptive beamformers degrades severely due to the presence of such coherent or correlated interferences.

Various techniques have been developed to improve the performance of the beamformer for multipath signal reception, such as those based on convex optimization (Yu *et al.*, 2009a,b, 2010), and the classical spatial smoothing technique (Reddy *et al.*, 1987; Shan and Kailath, 1985; Unnikrishna Pillai and Kwon, 1989). However most of the efforts have been focused on the narrowband case and for wideband signals, although we can also apply the spatial smoothing technique as studied in Indukumar and Reddy (1990), a fixed beamformer is usually preferred due to its simplicity in implementation and a lack of effective methods for multipath adaptive wideband beamforming.

There is a major difference between a narrowband signal and a wideband signal for the multipath case. For narrowband signals, the multipath delayed signal is often assumed to be a scaled version of the original one up to a magnitude and phase change and the signal of interest could be completely cancelled by the multipath signal after adaptive beamforming; for the wideband case, normally any delayed version of the signal will only be partially correlated with the original one and in many cases after adaptive beamforming the desired signal will not be completely cancelled. Exploiting this property, we will present two solutions to the multipath problem based on the wideband beamspace adaptive beamforming structure (Liu, 2009d, 2010).

However, when a large number of multipath signals is present to the beamformer, we will have a generalized signal mixing problem independent of the array geometry, which leads to a MIMO (multiple input multiple output) system studied extensively in wireless communications. A brief introduction to MIMO systems will be provided from the viewpoint of beamforming at the end of this chapter.

Wideband Beamforming Wei Liu and Stephan Weiss

8.1 The Wideband Multipath Problem

Consider the sensor array shown in Figure 6.5 in Chapter 6. Now suppose $s_l(t), l = 0, \ldots, L-1$, are the L impinging plane-wave signals that are received at the *zero*th sensor. Then the signal received at the mth sensor will be $x_m(t) = \sum_{l=0}^{L-1} s_l(t - \tau_{m,l})$, where $\tau_{m,l}$ is the delay from the *zero*th to the mth sensor for the lth signal.

For convenience, we consider the discrete form of the signals. Then we have $s_l[n] = s_l(nT), l = 0, \ldots, L-1$, and $x_m[n] = x_m(nT), m = 0, \ldots, M-1$, where T is the sampling period.

As introduced in Chapter 7, wideband beamforming can be realized by sensor delay-lines. Without loss of generality, we will consider the TDL-based wideband beamforming structure shown in Figure 7.1 in that chapter and the two presented solutions can be extended to the SDL-based systems straightforwardly. For the TDL-based beamformer, its output $y[n]$ is given by:

$$y[n] = \mathbf{w}^H \mathbf{x} \tag{8.1}$$

where the coefficients vector and the input samples vector are defined as:

$$\mathbf{w} = \begin{bmatrix} \mathbf{w}_0^T & \mathbf{w}_1^T & \ldots & \mathbf{w}_{J-1}^T \end{bmatrix}^T \tag{8.2}$$

$$\mathbf{w}_j = \begin{bmatrix} w_{0,j} & w_{1,j} & \ldots & w_{M-1,j} \end{bmatrix}^T \tag{8.3}$$

$$\mathbf{x} = \begin{bmatrix} \mathbf{x}^T[n] & \mathbf{x}^T[n-1] & \ldots & \mathbf{x}^T[n-J+1] \end{bmatrix}^T \tag{8.4}$$

$$\mathbf{x}[n-j] = \begin{bmatrix} x_0[n-j] & x_1[n-j] & \ldots & x_{M-1}[n-j] \end{bmatrix}^T \tag{8.5}$$

Assume that $s_0(t)$ is the desired signal and $s_1(t)$ is a multi-path version of $s_0(t)$ given by $s_1(t) = \alpha s_0(t - \delta T)$, where α is a scalar and δT is the corresponding delay. (The discussion and solutions can be extended to the case with more than one multipath interferences in a straightforward way.) All of the remaining $L-2$ signals are uncorrelated interferences. For adaptive beamforming, as an example, we consider the classical LCMV beamformer studied in Section 2.2, for which the main direction θ_0 of the signal of interest is assumed to be known. The LCMV problem can be formulated as:

$$\min_{\mathbf{w}} \mathbf{w}^T \boldsymbol{R}_{xx} \mathbf{w} \qquad \text{subject to} \qquad \boldsymbol{C}^H \mathbf{w} = \mathbf{f} \tag{8.6}$$

where $\boldsymbol{R}_{xx}$ is the covariance matrix of the observed array data in $\mathbf{x}$, $\boldsymbol{C}$ is the constraint matrix and $\mathbf{f}$ the response vector. As shown in Section 2.2.2, its solution can be obtained by the method of Lagrange multipliers, given by:

$$\mathbf{w}_{opt} = \boldsymbol{R}_{xx}^{-1} \boldsymbol{C} (\boldsymbol{C}^H \boldsymbol{R}_{xx}^{-1} \boldsymbol{C})^{-1} \mathbf{f} \tag{8.7}$$

During the minimization of the output variance, the uncorrelated interferences will be suppressed effectively. However, for the correlated interference $s_1(t)$, it will try to cancel at least part of the desired signal in the output to reduce the overall output signal variance.

Suppose the impulse response of the beamformer to the desired signal $s_0(t)$ from the direction θ_0 is simply a delay T_1 and to $s_1(t)$ from θ_1 is given by $h_1(t)$, i.e. at the beamformer output, the component corresponding to $s_0(t)$ is $s_0(t - T_1)$ and that corresponding to $s_1(t)$ is $\alpha s_0(t-\delta T) * h_1(t)$, where $*$ represents the convolution operation. When $h_1(t)$ has a magnitude response $1/\alpha$ and a phase response $-(T_1-\delta T)\omega$, the desired signal will be cancelled completely.

To reduce the cancelation to the desired signal, in the next two sections we will present two solutions based on the frequency invariant beamforming network used in the previously studied wideband beamspace adaptive beamforming structure shown in Figures 5.43 and 6.6 (Liu, 2009d, 2010). The received array signals are first passed through the beamforming network, which transforms the wideband beamforming problem into a narrowband one and then a traditional narrowband adaptive beamformer is applied to the network output.

It will be shown that after this FIB network, the cancelation of the desired signal due to its multipath signals will be reduced greatly and a significantly improved output SINR can be achieved. Instead of applying the narrowband adaptive beamformer to the FIB outputs, we can also use an instantaneous BSS algorithm to extract the signal of interest, according to the statistical properties of the source signals, without knowing the main DOA angle of the signal of interest (Liu, 2010).

8.2 Approach Based on a Narrowband Beamformer

8.2.1 *Structure*

Assume the main direction θ_0 of the signal of interest is known. Then a solution to the multipath problem can be obtained based on the traditional beamforming idea by employing the frequency invariant beamforming technique (Liu, 2009d).

The structure is shown in Figure 8.1, where each of the blocks labelled as FIB_i, $i = 0, 1, \ldots, N-1$, represents a frequency invariant beamformer with a response $P_i(\theta)$. The beam FIB_0 is the main beam pointing to the main direction of the signal of interest and the remaining $N-1$ beams are auxiliary beams pointing to the other directions which have a zero response to the main direction of the desired signal. The output of FIB_0 is $b_0[n]$ and $y[n]$ is the final array output by combining the other FIB outputs $b_1[n], b_2[n], \ldots, b_{N-1}[n]$ with one adaptive coefficient w_i, $i = 1, \ldots, N-1$ for each of the them, given by:

$$y[n] = b_0[n] - \hat{\mathbf{w}}^H \mathbf{b}[n] \tag{8.8}$$

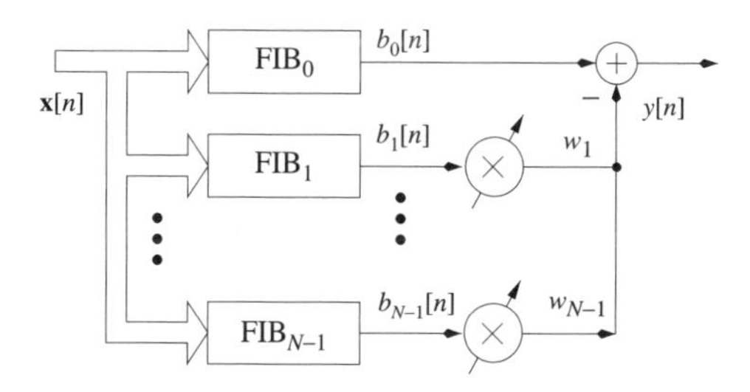

Figure 8.1 The beamforming structure based on frequency invariant transformations for multipath wideband signals with a following narrowband beamformer

where:

$$\mathbf{b}[n] = \left[b_1[n]\ b_2[n]\ \cdots\ b_{N-1}[n]\right]^{\mathrm{T}}$$
$$\hat{\mathbf{w}} = \left[w_1\ w_2\ \cdots\ w_{N-1}\right]^{\mathrm{H}} \tag{8.9}$$

The optimum weight vector $\hat{\mathbf{w}}_{opt}$ can be obtained by minimizing the mean square error $E\{|y[n]|^2\}$ between the main beam and the auxiliary beams, which is given by:

$$\hat{\mathbf{w}}_{opt} = \boldsymbol{R}_{bb}^{-1}\mathbf{p}_{bb} \tag{8.10}$$

where $\boldsymbol{R}_{bb} = E\{\mathbf{b}[n]\mathbf{b}[n]^H\}$ and $\mathbf{p}_{bb} = E\{\mathbf{b}[n]b_0^*[n]\}$. Alternatively, we can employ an adaptive algorithm to update the weight vector iteratively, such as the normalized LMS algorithm:

$$\hat{\mathbf{w}}[n+1] = \hat{\mathbf{w}}[n] + \frac{\mu}{\mathbf{b}[n]^H\mathbf{b}[n]} y^*[n]\mathbf{b}[n] \tag{8.11}$$

where μ is the stepsize.

Note here the responses $P_i(\theta), i = 0, 1, \ldots, N-1$, of the N FIBs are independent of frequency. Then their outputs $b_i[n], i = 0, 1, \ldots, N-1$ can be expressed as a weighted sum of the impinging signals, given by:

$$b_i[n] = \mathbf{p}_i\mathbf{s}[n] \tag{8.12}$$

with:

$$\mathbf{s}[n] = \left[s_0[n]\ s_1[n]\ \cdots\ s_{L-1}[n]\right]^{\mathrm{T}}$$
$$\mathbf{p}_i = \left[P_i(\theta_0)\ P_i(\theta_1)\ \cdots\ P_i(\theta_{L-1})\right] \tag{8.13}$$

Then the component $b_{i,l}$ corresponding to $s_l[n]$ in the output $b_i[n]$ of FIB$_i$ will be $b_{i,l} = P_i(\theta_l)s_l[n]$. For the desired signal $s_0[n]$ from the direction θ_0, since all of the other FIB$_i$, $i = 1, 2, \ldots, N-1$, have a zero response to the direction θ_0, i.e. $P_i(\theta_0) = 0$ for $i = 1, 2, \ldots, N-1$, the component $b_{0,0}$ corresponding to $s_0[n]$ in the output $b_0[n]$ of FIB$_0$ will be:

$$b_{0,0} = P_0(\theta_0)s_0[n] = P_0(\theta_0)s_0(nT) \tag{8.14}$$

and the signal component y_0 corresponding to $s_0[n]$ in the output $y[n]$ will be the same as in $b_0[n]$, namely:

$$y_0 = b_{0,0} = P_0(\theta_0)s_0(nT) \tag{8.15}$$

For the component y_1 in the output $y[n]$ corresponding to $s_1[n]$, the scaled delayed version (multipath) of $s_0[n]$, it is:

$$\begin{aligned} y_1 &= (P_0(\theta_1) - \sum_{i=1}^{N-1} P_i(\theta_1)w_i)s_1[n] \\ &= \alpha(P_0(\theta_1) - \sum_{i=1}^{N-1} P_i(\theta_1)w_i)s_0(nT - \delta T) \end{aligned} \tag{8.16}$$

Since $s_1[n] = s_0(nT - \delta T)$ is only partially correlated with $s_0[n] = s_0(nT)$ for the zero lag, no matter how to adjust the coefficients of the weight vector $\hat{\mathbf{w}}$, the desired signal will not be completely cancelled and the total output y_s corresponding to the signal of interest in the beamformer output $y[n]$ is given by:

$$\begin{aligned} y_s &= y_0 + y_1 \\ &= P_0(\theta_0)s_0(nT) + \alpha(P_0(\theta_1) - \sum_{i=1}^{N-1} P_i(\theta_1)w_i)s_0(nT - \delta T) \end{aligned} \tag{8.17}$$

Assuming the uncorrelated interferences have been suppressed sufficiently, the beamformer output $y[n]$ will include a simple filtered version of the desired signal $s_0(nT)$ plus noise and the original signal can be recovered by some appropriate deconvolution methods such as those proposed in Gillespie *et al.* (2001) and Haykin (2000b).

8.2.2 Simulations

The simulations are based on an equally spaced linear array with $N = 17$ sensors. For the FIB based beamformer, 5 frequency invariant beams are designed, each of them with dimensions of 17×80, and the desired responses for the design are given in Figure 6.8 in Chapter 6. Its performance is compared with those of the original LCMV beamformer and the spatial-smoothing based LCMV beamformer in term of the output SINR for different input signal to noise ratios (SNRs).

The original LCMV beamformer has dimensions of 17×80 with a zero-order derivative constraint. For the spatial-smoothing based LCMV beamformer, the sub-array size is 6 and in total there are $17 - 6 + 1 = 12$ sub-arrays, whose outputs are added together to form a new LCMV beamformer with dimensions 6×80. The desired signal comes from the broadside $\theta_0 = 0^\circ$ and three uncorrelated interfering signals with an SIR of -20 dB arrive from the directions $\theta = 60^\circ$, -25° and 50°, respectively. One correlated interfering signal, which is a delayed version of the desired signal by 20 samples and scaled by 0.5 in magnitude, arrives from the direction $\theta_1 = 30^\circ$. All of the source signals are bandlimited to the range $\Omega \in [0.4\pi\ 0.9\pi]$.

We change the input noise level from SNR = 0 dB to SNR = 20 dB and then draw the output SINR curve for each of the beamformers. The result is shown in Figure 8.2, where it can be seen that the FIB based beamformer can deal with the correlated interference problem effectively and its output SINR increases steadily with respect to a reducing input noise level.

On the other hand, the traditional LCMV beamformer has failed to pass through the signal of interest at its output and its output SINR stays at about 8 dB with a tiny fluctuation with respect to the input SNR level. The spatial-smoothing method can improve the output SINR of the traditional LCMV beamformer, but the improvement is limited and not as significant as the FIB based beamformer. To show the output SINR change during the adaptive process using the algorithm in Equation (8.11), a learning curve is provided in Figure 8.3 with a stepsize of $\mu = 0.005$ and an input SNR of 20 dB.

In the second set of simulations, we add one more multipath signal to the previous setting and it is a delayed version of the desired signal by 10 samples and scaled by 0.5

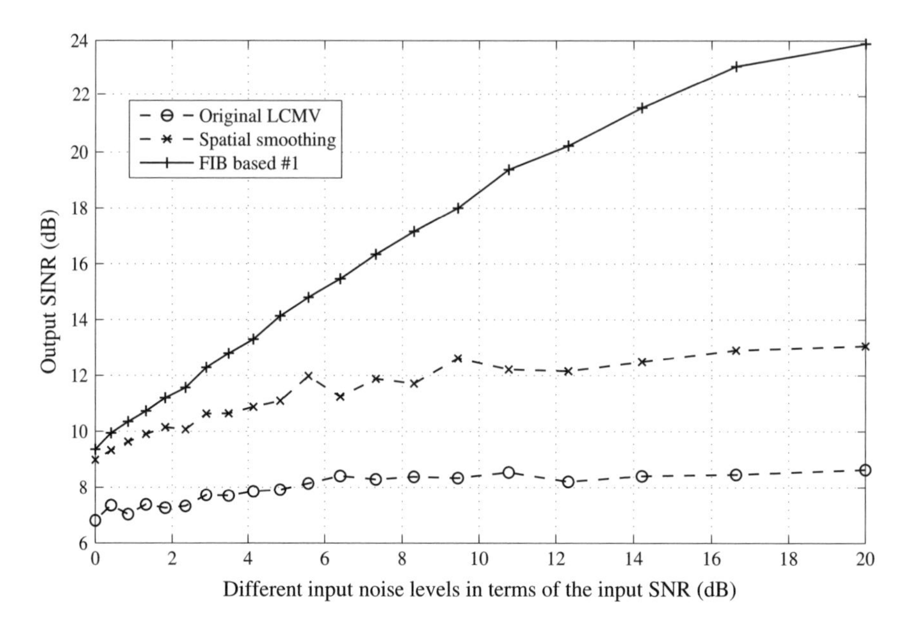

Figure 8.2 Output SINR of the original LCMV beamformer ('original LCMV'), the spatial-smoothing based LCMV beamformer ('spatial smoothing') and the FIB based beamformer in Figure 8.1 ('FIB based #1') for different values of input SNR in the first set of simulations

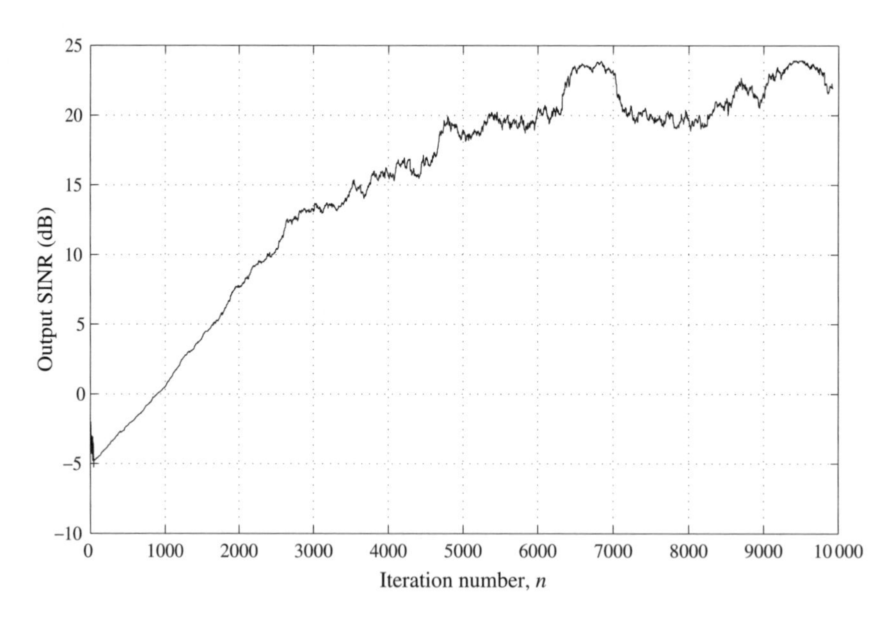

Figure 8.3 A learning curve for the output SINR of the FIB based beamformer shown in Figure 8.1 using the normalized LMS algorithm in Equation (8.11) (input SNR = 20 dB, $\mu = 0.005$)

in magnitude, arriving from the direction $\theta_1 = -70^\circ$. The result is shown in Figure 8.4, which is very similar to Figure 8.2.

Interestingly, the additional multipath signal has not degraded the beamformers's performance much as expected, especially for cases with lower input SNR values. For some of the input SNR values, the output SINR with two multipath signals is even a little bit higher than the one multipath signal case, which is shown in Figure 8.5, where we have

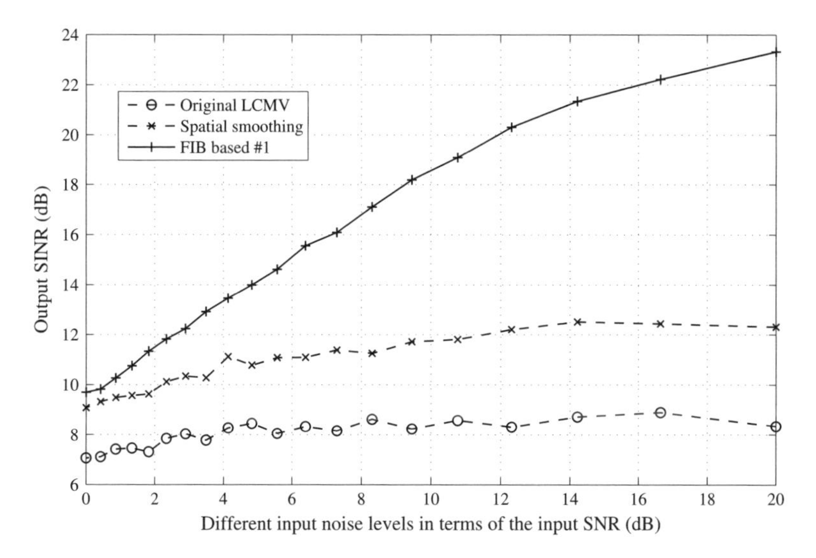

Figure 8.4 Output SINR of the three beamformers ('original LCMV', 'spatial smoothing' and 'FIB based #1') for different values of input SNR in the second set of simulations

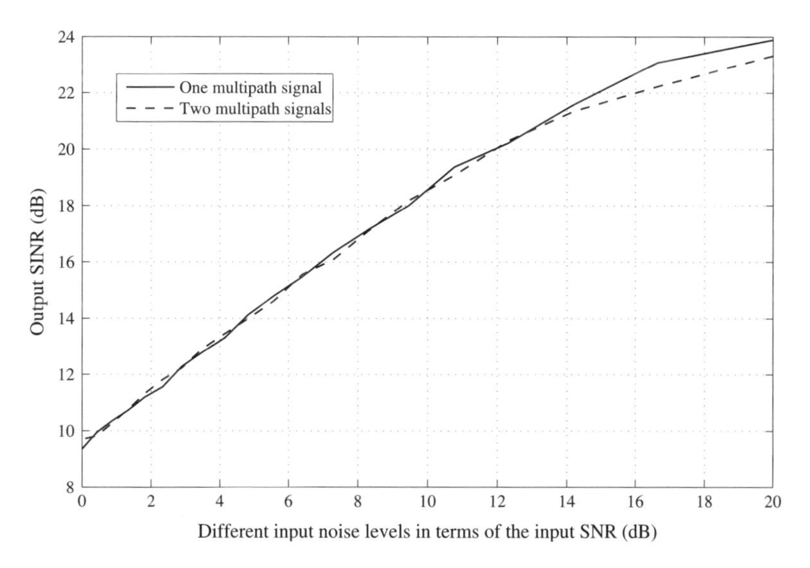

Figure 8.5 A comparison of the output SINR curves for the FIB based beamformer in the two sets of simulations

put the output SINR curves of the FIB based beamformer in the two set of simulations together for a comparison. A possible explanation for this phenomenon may be that the second multipath signal can add either constructively or destructively to the first multipath signal and their overall effect to the main desired signal will not change linearly with respect to the number of multipath signals. This is a complicated issue and further study is needed in the future.

8.3 Approach Based on Blind Source Separation

8.3.1 Structure

In the above discussion, we have assumed that we know the main DOA angle of the signal of interest, so that the beam FIB_0 can be designed to point to that direction with all of the remaining beams having a zero response to it. However, it is still possible to extract the signal of interest even if we do not have the DOA information for its main direction. The solution is to use an instantaneous BSS algorithm (Liu, 2010), as already studied in Section 6.3.

In Equation (8.12), each FIB output $b_i[n]$ is an instantaneous mixture of the original L source signals. Then we can apply a simple instantaneous BSE algorithm to the N outputs to extract the desired signal, as shown in Figure 8.6, where the desired signal is extracted by applying the set of coefficients $w_i, i = 0, 1, \ldots, N-1$. The output $y[n]$ is given by:

$$y[n] = \sum_{i=0}^{N-1} w_i b_i[n] = \tilde{\mathbf{w}}^T \tilde{\mathbf{b}}[n] \tag{8.18}$$

where:

$$\begin{aligned} \tilde{\mathbf{b}}[n] &= \left[b_0[n]\ b_1[n]\ \cdots\ b_{N-1}[n]\right]^{\mathrm{T}} \\ \tilde{\mathbf{w}} &= \left[w_0\ w_1\ \cdots\ w_{N-1}\right] \end{aligned} \tag{8.19}$$

As mentioned in Chapter 6, BSS/BSE algorithms are based on some assumptions on the statistical properties of the source signals. For example, if the original source signals

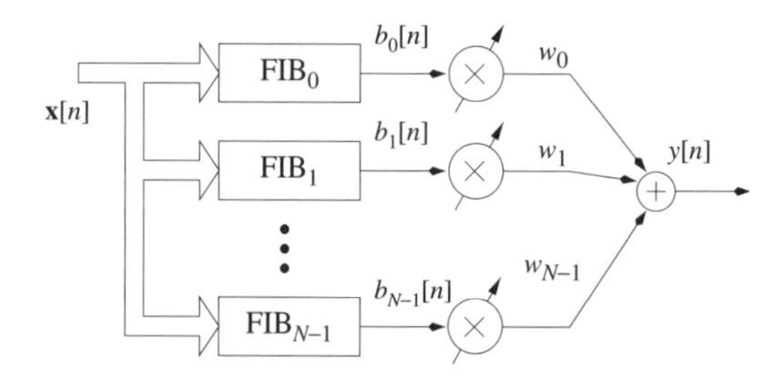

Figure 8.6 The beamforming structure based on frequency invariant transformations for multipath wideband signals with a following blind source separation/extraction operation

are independent of each other with at most one Gaussian signal, we can separate the original signals based on the principle that the separated signals are as independent with each other as possible.

For the multipath case when the source signal $s_1[n]$ is a delayed version of $s_0[n]$, we will not be able to separate them from each other. However since both of them are independent of the remaining signals, as long as the number of mixtures N is not smaller than the source signal number L, we will be able to separate them from the remaining signals by finding an appropriate linear combination $\tilde{\mathbf{w}}$ of the mixtures $\tilde{\mathbf{b}}[n]$. As a result, the extracted signal for $s_0[n]$ and $s_1[n]$ will be one of their mixtures, just like the beamforming case discussed in the previous section.

In the simulations part, the desired signal is sub-Gaussian with a negative kurtosis value and in this scenario we can minimize the normalized kurtosis value of the output $y[n]$. As given in Section 6.1.2, the resulting adaptive algorithm is:

$$\tilde{\mathbf{w}}[n+1] = \tilde{\mathbf{w}}[n] - \mu\phi(y[n])\tilde{\mathbf{b}}[n] \tag{8.20}$$

where:

$$\phi(y[n]) = \frac{m_2(y)[n]}{m_4(y)[n]}y^3[n] - y[n] \tag{8.21}$$

with:

$$m_q(y)[n] = (1-\beta)m_q(y)[n-1] + \beta|y[n]|^q, \quad q = 2, 4 \tag{8.22}$$

β is the forgetting factor for estimating $m_q(y)[n]$ and μ is the update stepsize.

8.3.2 *Simulations*

In order to use the normalized kurtosis based BSE algorithm, we have generated a sub-Gaussian signal with a normalized kurtosis value of about 0.7, which is the desired signal arriving from the broadside of the array. There are two uncorrelated interfering signals arriving from the directions $\theta = 60°$ and $-50°$, respectively, with the same power as the desired signal (SIR = 0 dB). All of the source signals are bandlimited to the normalized frequency range $\Omega \in [0.45\pi\ \pi]$. All the other parameters are the same as in the simulations of Section 8.2.

In the first set of simulations, we have one correlated multipath signal, which is a delayed version of the desired signal by 20 samples and scaled by 0.5 in magnitude, and arrives from the direction $\theta_1 = 30°$. The output SINRs of the original LCMV beamformer, the spatial-smoothing based LCMV beamformer and the FIB-based beamformer in Figure 8.6, are shown in Figure 8.7 with respect to different input noise levels. A learning curve to show the output SINR changes for the FIB based beamformer is given in Figure 8.8 with $\mu = 0.0008$, $\beta = 0.02$ and input SINR = 20 dB.

We then add one more multipath signal to the previous setting and it is a delayed version of the desired signal by 10 samples and scaled by 0.5 in magnitude, arriving from the direction $\theta_2 = -70°$. The result is shown in Figure 8.9.

We can see that based on the frequency invariant beamforming network, the simple instantaneous BSE algorithm can extract the desired signal successfully in both multipath

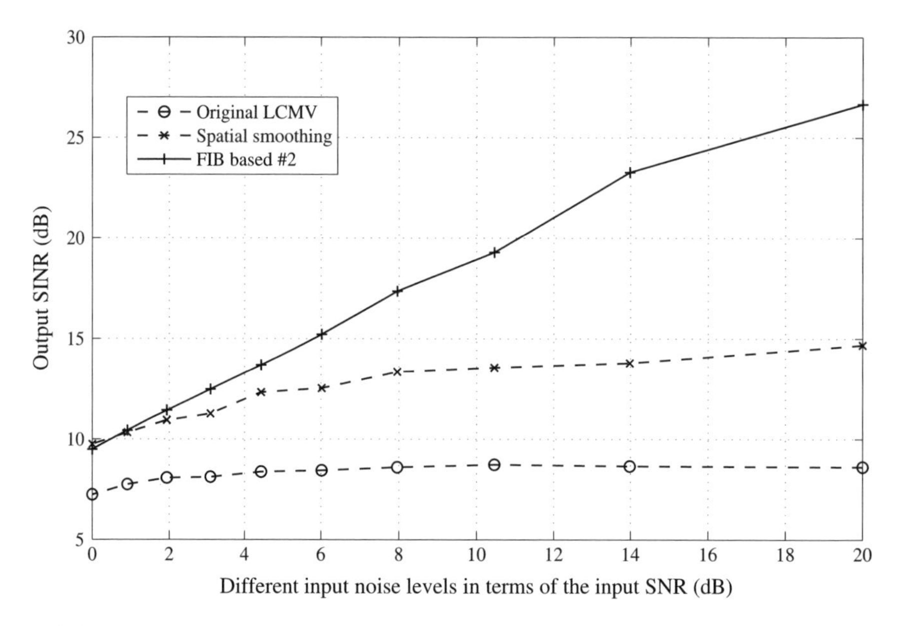

Figure 8.7 Output SINRs of the original LCMV beamformer ('original LCMV'), the spatial-smoothing based LCMV beamformer ('spatial smoothing') and the FIB based beamformer in Figure 8.6 ('FIB based #2'), for different values of input SNR in the first set of simulations

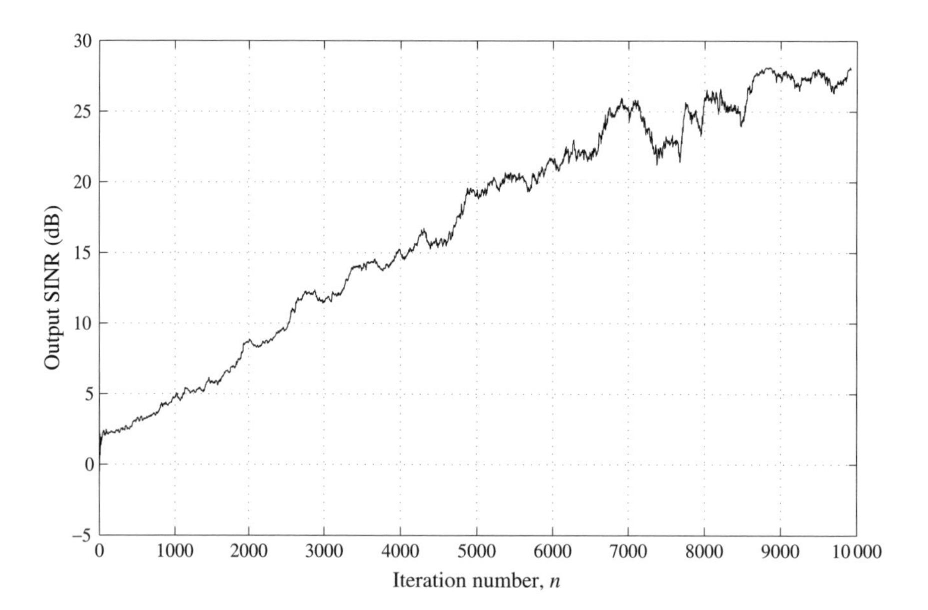

Figure 8.8 A learning curve for the output SINR of the FIB based beamformer shown in Figure 8.6 using the normalized kurtosis based algorithm in Equation (8.20) (input SNR $= 20\,\text{dB}$, $\mu = 0.0008$, $\beta = 0.02$)

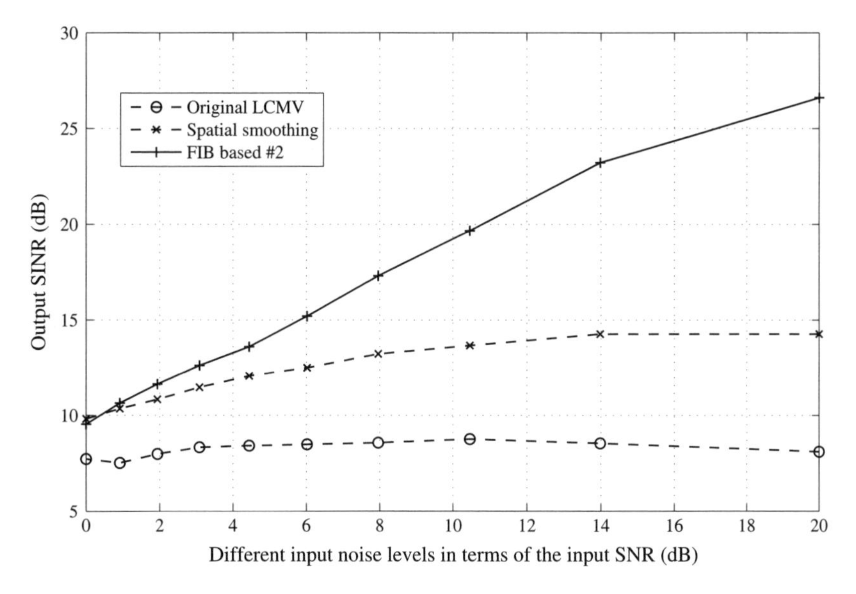

Figure 8.9 Output SINR of the three beamformers ('original LCMV', 'spatial smoothing' and 'FIB based #2'), for different values of input SNR in the second set of simulations

environments, significantly outperforming the spatial smoothing based beamformer. Again the results for both scenarios are very similar to each other and the additional multipath signal has not degraded their performance in a notable way.

8.4 MIMO System

In the previous sections, we have introduced a FIB-based approach to the wideband multipath beamforming problem. However, that approach is only effective when the number of multipath signals is relatively small compared to the number of array sensors. When we have a large number of multipath signals, the discussions there will not be valid any more since we are now facing a general signal mixing problem independent of the array geometry. In this scenario, the traditional array system has evolved into the one called 'MIMO' (Huang *et al.*, 2006).

The term MIMO is the acronym for 'multiple input multiple output' and MIMO systems have been undergoing extensive study in recent years due to their applications in wireless communications (Biglieri *et al.*, 2007; Oestges and Clerckx, 2007). Like traditional beamforming systems, they are also based on the use of multiple sensors/antennas for transmission and reception. It is a fast-developing area and its effective operation often involves many other techniques, such as space-time coding (Alamouti, 1998; Naguib *et al.*, 2000; Tarokh *et al.*, 1998; Telatar, 1999). The aim of this section is to give a brief introduction to this technology from the viewpoint of beamforming.

8.4.1 Evolution to a MIMO System

In this part, we discuss how a traditional beamforming system evolves into a MIMO one in a rich-multipath environment. For simplicity, we first consider a narrowband array system with M sensors in a multipath environment.

Suppose there are in total L narrowband source signals $s_l(t), l = 0, \ldots, L-1$, and each of them has Q_l multipath signals, which are denoted as $s_{l,q}(t) = \alpha_{l,q} s_l(t), q = 0, 1, \ldots, Q_l - 1$, at the *zero*th sensor, where $\alpha_{l,q}$ is a complex-valued constant representing the phase shift and amplitude attenuation due to multipath propagation. The multipath signal $s_{l,q}(t)$ arrives at the array from the direction $\theta_{l,q}$ with a propagation delay of $\tau_{l,q,m}$ from the *zero*th sensor to the mth sensor. Then the received array signal $x_m(t)$ at the mth sensor is given by:

$$\begin{aligned} x_m(t) &= \sum_{l=0}^{L-1} \sum_{q=0}^{Q_l-1} s_{l,q}(t) \mathrm{e}^{-j\omega\tau_{l,q,m}} + v_m(t) \\ &= \sum_{l=0}^{L-1} \sum_{q=0}^{Q_l-1} \alpha_{l,q} \mathrm{e}^{-j\omega\tau_{l,q,m}} s_l(t) + v_m(t) \end{aligned} \tag{8.23}$$

where ω is the signal frequency and $v_m(t)$ is the added noise for the mth sensor.

In vector form, we have:

$$\boldsymbol{x}(t) = \mathbf{A}\mathbf{s}(t) + \mathbf{v}(t) \tag{8.24}$$

with:

$$\begin{aligned} \mathbf{x}(t) &= \left[x_0(t), x_1(t), \cdots, x_{M-1}(t)\right] \\ \mathbf{s}(t) &= \left[s_0(t), s_1(t), \cdots, s_{L-1}(t)\right] \\ A &= \left[\mathbf{a}_0,\ \mathbf{a}_1,\ \cdots,\ \mathbf{a}_{L-1}\right] \end{aligned} \tag{8.25}$$

where $\mathbf{a}_l$ is the composite steering vector for the source signal $s_l(t)$:

$$\mathbf{a}_l = \sum_{q=0}^{Q_l-1} \alpha_{l,q} \mathbf{a}_{l,q} \tag{8.26}$$

with $\mathbf{a}_{l,q}$ being the original steering vector of the array for a signal arriving from the direction $\theta_{l,q}$, given by:

$$\mathbf{a}_{l,q} = \left[\mathrm{e}^{-j\omega\tau_{l,q,0}},\ \mathrm{e}^{-j\omega\tau_{l,q,1}},\ \ldots,\ \mathrm{e}^{-j\omega\tau_{l,q,M-1}}\right]^T \tag{8.27}$$

The structure of each steering vector is dependent on the array geometry and for the single-path case ($Q_l = 1$), the composite steering vector is then reduced to the original steering vector.

However, when a large number of multipath signals are observed by the array, the composite steering vector will become a general one, not dependent on the array geometry any more. In this case, even if we know all of the directions of the multipath signals, we still cannot decide the composite steering vector since we do not know the value of $\alpha_{l,q}$. Then the received array signals become general mixtures of the source signals and the traditional beamformer based on the angle information of the impinging signals will not work. With L source signals and M received array signals, we consider such a system as a general mutiple input, mutiple output system or a MIMO system.

For wideband signals, the input–output relationship described by Equation (8.24) will be frequency dependent and for each frequency we will have a different matrix $\boldsymbol{A}$. In the time domain it is a convolution operation between the channel response and the source signal, which leads to the following input–output relationship:

$$\mathbf{x}(t) = \int_{-\infty}^{+\infty} \mathbf{A}(\tau)\mathbf{s}(t-\tau)\mathrm{d}\tau + \mathbf{v}(t) \tag{8.28}$$

where $\boldsymbol{A}(\tau)$ is the matrix impulse response of the channel between the L inputs and the M outputs. Alternatively, we can reformulate it into:

$$\mathbf{x}(t) = \boldsymbol{A} * \mathbf{s}(t) + \mathbf{v}(t) \tag{8.29}$$

with $*$ denoting the convolution operation and $\boldsymbol{A}$ is now defined as:

$$\boldsymbol{A} = \begin{bmatrix} h_{0,0}(t) & h_{0,1}(t) & \dots & h_{0,L-1}(t) \\ h_{1,0}(t) & h_{1,1}(t) & \dots & h_{0,L-1}(t) \\ \vdots & \vdots & \ddots & \vdots \\ h_{M-1,0}(t) & h_{M-1,1}(t) & \dots & h_{M-1,L-1}(t) \end{bmatrix} \tag{8.30}$$

where $h_{m,l}(t)$ is the channel impulse response between the lth input (source signal) and the mth output (received array signal). Such a channel is called frequency-selective since the channel distortion introduced to the transmitted signal is different for different frequencies (Morgan, 2008).

In wireless communications, the L source signals can be considered as being transimitted by L antennas. Then we obtain the typical wireless MIMO system shown in Figure 8.10 with L transmit antennas and M receive antennas.

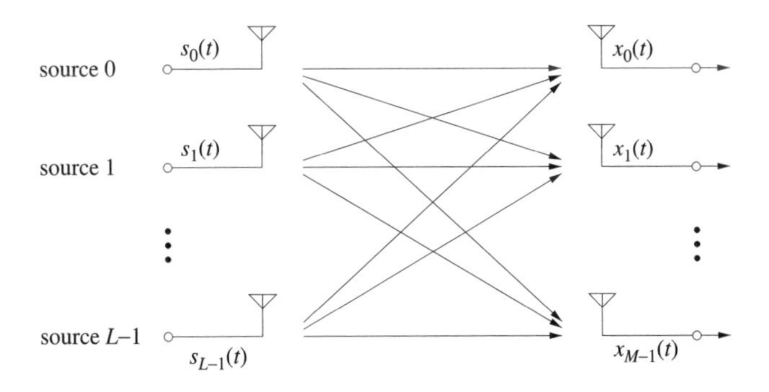

Figure 8.10 A structure for a wirelsss MIMO system with L input signals and M output signals

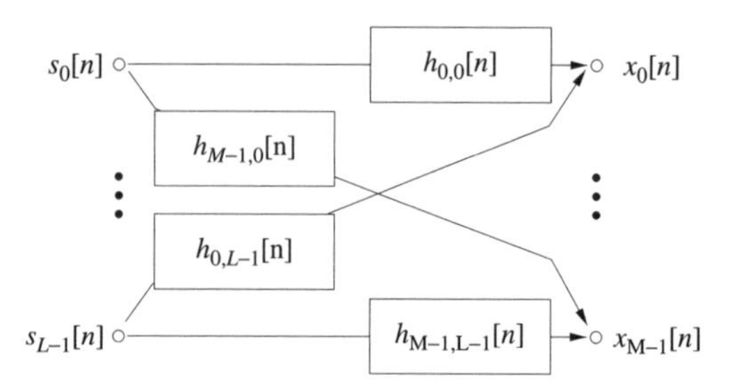

Figure 8.11 A structure for a wideband MIMO system with frequency-selective channels

8.4.2 MIMO Beamforming and Equalization

For a MIMO system, we can still perform 'beamforming' in a similar way as in the traditional settings and here we only consider the wideband case with frequency-selective channels, as shown in Figure 8.11 based on the input–output relationship in Equation (8.29), where we have used the discrete form for both the signals and the channel impulse responses.

For wideband beamforming, we need to employ TDLs or FIR/IIR filters with a length of J as shown in Figure 1.5 and discussed in Section 1.3, Chapter 1. Then the output of the beamformer based on the M received signals $x_m[n], m = 0, 1, \ldots, M-1$ is given by:

$$y[n] = \mathbf{g}^H \mathbf{x} \tag{8.31}$$

where the signal vector $\mathbf{x}$ and the weight vector $\mathbf{g}$ is defined as follows:

$$\mathbf{g} = \begin{bmatrix}\mathbf{g}_0^T & \mathbf{g}_1^T & \cdots & \mathbf{g}_{M-1}^T\end{bmatrix}^T \tag{8.32}$$

$$\mathbf{g}_m = \begin{bmatrix}w_{m,0} & w_{m,1} & \cdots & w_{m,J-1}\end{bmatrix}^T \tag{8.33}$$

$$\mathbf{x} = \begin{bmatrix}\mathbf{x}_0^T & \mathbf{x}_1^T & \cdots & \mathbf{x}_{M-1}^T\end{bmatrix}^T \tag{8.34}$$

$$\mathbf{x}_m = [x_m[n] \;\; x_m[n-1] \;\; \ldots \;\; x_m[n-J]]^T \tag{8.35}$$

Note here they have been arranged in a different order as in Equations (8.2–5).

If a reference signal or a training sequence is available, then we can form a reference signal based beamformer and the optimum weight vector can be obtained by minimizing the mean square error $E\{|e[n]|^2\}$ between the reference signal $r[n]$ and the beamformer output $y[n]$, as shown in Figure 2.1 in Section 2.1, Chapter 2.

If we know the composite steering vector of the desired signal, then we can form a traditional LCMV beamformer to reserve the signal of interest and suppress the interferences (Benesty *et al.*, 2007). Suppose the first signal $s_0(t)$ is the desired signal and its composite steering vector $\mathbf{a}_0$ in discrete form is given by:

$$\mathbf{a}_0 = [h_{0,0}[n],\; h_{1,0}[n], \ldots, h_{M-1,0}[n]]^T \tag{8.36}$$

where each of the element $h_{m,0}[n]$ of $\mathbf{a}_0$ represents a channel impulse response between $s_0(t)$ ($s_0[n]$ in discrete form) and the mth receive sensor. Suppose all of the channel impulse responses $h_{m,l}[n]$ are finite and the maximum length is P. Then the overall channel impulse response $f_0[n]$ from the source $s_0[n]$ to the beamformer output $y[n]$ is given by:

$$f_0[n] = \sum_{m=0}^{M-1} \sum_{j=0}^{J-1} w_{m,j} h_{m,0}[n-j] \tag{8.37}$$

Since the length of the mth TDL $\mathbf{g}_m$ is J and that of the filter $h_{m,0}[n-j]$ is P, the maximum length of the response filter $f_0[n]$ is $J+P-1$. In vector form, we have:

$$\begin{aligned}\mathbf{f}_0 &= \sum_{m=0}^{M-1} \boldsymbol{H}_{m,0}^H \mathbf{g}_m \\ &= \boldsymbol{H}_0^H \mathbf{g}\end{aligned} \tag{8.38}$$

where $\boldsymbol{H}_{m,0} \in \mathbf{C}^{J\times(J+P-1)}$ is defined as:

$$\boldsymbol{H}_{m,0}^* = \begin{bmatrix} h_{m,0}[0] & \dots & h_{m,0}[P-1] & 0 & 0 & \dots & 0 \\ 0 & h_{m,0}[0] & \dots & h_{m,0}[P-1] & 0 & \dots & 0 \\ \vdots & \vdots & \vdots & \vdots & \vdots & \vdots & \vdots \\ 0 & 0 & \dots & 0 & h_{m,0}[0] & \dots & h_{m,0}[P-1] \end{bmatrix} \tag{8.39}$$

and:

$$\begin{aligned}\boldsymbol{H}_0 &= \left[\boldsymbol{H}_{0,0}^H \; \boldsymbol{H}_{1,0}^H \; \cdots \; \boldsymbol{H}_{M-1,0}^H\right]^H \\ \mathbf{f}_0 &= \left[f_0[0] \; f_0[1] \; \dots \; f_0[J+P-1]\right]^T\end{aligned} \tag{8.40}$$

For the desired signal, we would like the beamformer to have a response vector $\mathbf{f}_0$ with only one non-zero value and the position of the non-zero value will represent the delay introduced to the desired signal at the beamformer output. We denote this desired response vector as $\mathbf{f}_{0,d}$. Then the LCMV problem can be formulated as:

$$\arg\min_{\mathbf{g}} \mathbf{g}^H \boldsymbol{R}_{xx} \mathbf{g} \qquad \text{subject to} \qquad \boldsymbol{H}_0^H \mathbf{g} = \mathbf{f}_{0,d} \tag{8.41}$$

where $\boldsymbol{R}_{xx}$ is the correlation matrix of $\mathbf{x}$ defined in Equation (8.34). As discussed in Chapter 2.2, its optimum solution is given by:

$$\mathbf{g}_{opt} = \boldsymbol{R}_{xx}^{-1} \boldsymbol{H}_0 (\boldsymbol{H}_0^H \boldsymbol{R}_{xx}^{-1} \boldsymbol{H})^{-1} \mathbf{f}_{0,d} \tag{8.42}$$

If we have more information about the MIMO channels, we can add further constraints to the formulation in the same way as in the traditional LCMV beamformer (Benesty *et al.*, 2007).

For a wireless MIMO system, when the channel information is unknown, the beamforming process of recovering one of the source signals from the received multiple outputs is also considered as channel equalization since the distortion introduced by the channel impulse response has been canceled/equalized by the beamformer weight vector $\mathbf{w}$.

We can perform channel equalization with a training sequence using the reference signal based beamformer (Fragouli *et al.*, 2003; Tong *et al.*, 2004). If the training sequence is not available, we can also employ convolutive blind source separation algorithms to recover the source signals since the multiple outputs in a MIMO system are convolutive mixtures of the multiple inputs, as illustrated in Equation (8.28). This leads to a blind wideband beamformer/equalizer (Inouye and Liu, 2002; Li and Liu, 1998), as discussed in Chapter 6, which is the third class of MIMO beamformers. Details about the general area of blind equalization can be found in Ding and Li (2001).

8.5 Summary

In this chapter, we have studied the wideband beamforming problem in different multipath environments.

For wideband signals, normally any delayed version will only be partially correlated with the original signal. Based on this observation, for the case with a small number of multipath signals, two solutions have been provided employing the wideband beamspace adaptive beamforming structure studied in previous chapters: one is the narrowband adaptive beamforming approach when the main direction of the desired signal is available and the other one is the instantaneous BSS/BSE approach when the DOA information of the signals is not available.

When a large number of multipath signals is present to the beamformer, we will have a generalized signal mixing problem independent of the array geometry and the original array system can be considered as a general multiple input multiple output system. A brief introduction to MIMO systems was then provided from the viewpoint of beamforming at the end of this chapter.

Appendix A: Matrix Approximation

The Frobenius norm of a $M \times N$ matrix $\boldsymbol{A}$ is defined as the square root of the sum of the squares of all its elements $a_{m,n}$ (Stewart, 1973, 1993):

$$||\boldsymbol{A}|| = \sqrt{\sum_{m,n=1,1}^{M,N} a_{m,n}^2} \tag{A.1}$$

By singular value decomposition, $\boldsymbol{A}$ can be decomposed into the product of three matrices as follows:

$$\boldsymbol{A} = \boldsymbol{U}\boldsymbol{\Sigma}\boldsymbol{V}^H = \boldsymbol{U}\begin{bmatrix} \boldsymbol{\Sigma}_k & \boldsymbol{0} \\ \boldsymbol{0} & \boldsymbol{0} \end{bmatrix}\boldsymbol{V}^H \tag{A.2}$$

where $\boldsymbol{U}$ and $\boldsymbol{V}$ are two unitary matrices and $\boldsymbol{\Sigma}_k$ is a $k \times k$ diagonal matrix containing the k ordered positive definite singular values $\sigma_1 \geq \sigma_2 \geq \cdots \geq \sigma_k > 0$. The variable k is the rank of $\boldsymbol{A}$ and represents the number of linearly independent columns in it. The Frobenius norm is invariant under unitary transformations. Then we have:

$$||\boldsymbol{A}||^2 = \sum_{m=1}^{k} \sigma_m^2 \tag{A.3}$$

Now we form a rank $r \leq k$ matrix $\hat{\boldsymbol{A}}$ by setting the values $\sigma_{r+1}, \ldots, \sigma_k$ in $\boldsymbol{\Sigma}_k$ to be zero, namely:

$$\hat{\boldsymbol{A}} = \boldsymbol{U}\hat{\boldsymbol{\Sigma}}\boldsymbol{V}^H = \boldsymbol{U}\begin{bmatrix} \boldsymbol{\Sigma}_r & \boldsymbol{0} \\ \boldsymbol{0} & \boldsymbol{0} \end{bmatrix}\boldsymbol{V}^H \tag{A.4}$$

which is the same way as forming the matrix '$\boldsymbol{C}_r$' in Section 2.3.1. Then the Frobenius norm ε_r of the error matrix $(\boldsymbol{A} - \hat{\boldsymbol{A}})$ is given by:

$$\begin{aligned} \varepsilon_r &= ||\boldsymbol{A} - \hat{\boldsymbol{A}}|| \\ &= ||\boldsymbol{U}(\boldsymbol{\Sigma} - \hat{\boldsymbol{\Sigma}})\boldsymbol{V}^H|| \\ &= ||\boldsymbol{\Sigma} - \hat{\boldsymbol{\Sigma}}|| \\ &= \sqrt{\sum_{m=r+1}^{k} \sigma_m^2} \end{aligned} \tag{A.5}$$

Wideband Beamforming Wei Liu and Stephan Weiss

For an arbitrary rank r matrix $\boldsymbol{B}$, we assume ε_B is the Frobenius norm of the error matrix $(\boldsymbol{A}-\boldsymbol{B})$. Then the theorem we want to prove is ε_r gives the minimum value for ε_B, i.e. $\hat{\boldsymbol{A}}$ is the best rank r approximation to $\boldsymbol{A}$ based on minimization of the error matrix' Frobenius norm $||\boldsymbol{A}-\boldsymbol{B}||$.

Here we follow the proof provided in Stewart (1973) (pp. 322–323). Assume $\boldsymbol{B}$ is already a matrix giving the minimum value of ε_B and its singular value decomposition is given by:

$$\boldsymbol{B} = \boldsymbol{U}_b\boldsymbol{\Sigma}_b\boldsymbol{V}_b^H = \boldsymbol{U}_b\begin{bmatrix}\hat{\boldsymbol{\Sigma}}_b & \boldsymbol{0}\\ \boldsymbol{0} & \boldsymbol{0}\end{bmatrix}\boldsymbol{V}_b^H \tag{A.6}$$

where $\hat{\boldsymbol{\Sigma}}_b$ is an $r \times r$ diagonal matrix holding its singular values. Now we define a new matrix $\boldsymbol{C}$ which is given by:

$$\boldsymbol{C} = \boldsymbol{U}_b^H\boldsymbol{A}\boldsymbol{V}_b = \begin{bmatrix}\boldsymbol{C}_{11} & \boldsymbol{C}_{12}\\ \boldsymbol{C}_{21} & \boldsymbol{C}_{22}\end{bmatrix} \tag{A.7}$$

Then we have:

$$\begin{aligned}\varepsilon_B &= ||\boldsymbol{A}-\boldsymbol{B}||\\ &= ||\boldsymbol{U}_b^H(\boldsymbol{A}-\boldsymbol{B})\boldsymbol{V}_b||\\ &= ||\boldsymbol{C}-\boldsymbol{\Sigma}_b||\\ &= ||\boldsymbol{C}_{11}-\hat{\boldsymbol{\Sigma}}_b|| + ||\boldsymbol{C}_{12}|| + ||\boldsymbol{C}_{21}|| + ||\boldsymbol{C}_{22}||\end{aligned} \tag{A.8}$$

Since $\boldsymbol{B}$ is already a matrix giving the minimum value of ε_B, we must have $\boldsymbol{C}_{12} = \boldsymbol{0}$. Otherwise, we will be able to construct a new rank r matrix $\hat{\boldsymbol{B}}$, given by:

$$\hat{\boldsymbol{B}} = \boldsymbol{U}_b\begin{bmatrix}\hat{\boldsymbol{\Sigma}}_b & \boldsymbol{C}_{12}\\ \boldsymbol{0} & \boldsymbol{0}\end{bmatrix}\boldsymbol{V}_b^H \tag{A.9}$$

so that the new Frobenius norm:

$$\begin{aligned}\varepsilon_{\hat{B}} &= ||\boldsymbol{A}-\hat{\boldsymbol{B}}||\\ &= ||\boldsymbol{U}_b^H(\boldsymbol{A}-\hat{\boldsymbol{B}})\boldsymbol{V}_b||\\ &= ||\boldsymbol{C}_{11}-\hat{\boldsymbol{\Sigma}}_b|| + ||\boldsymbol{C}_{21}|| + ||\boldsymbol{C}_{22}||\end{aligned} \tag{A.10}$$

will be smaller than ε_B, which contradicts the assumption that $\boldsymbol{B}$ gives the minimum value. In the same way, we have $\boldsymbol{C}_{21} = \boldsymbol{0}$ and $\boldsymbol{C}_{11} = \hat{\boldsymbol{\Sigma}}_b$. Then we have:

$$\boldsymbol{C} = \boldsymbol{U}_b^H\boldsymbol{A}\boldsymbol{V}_b = \begin{bmatrix}\hat{\boldsymbol{\Sigma}}_b & \boldsymbol{0}\\ \boldsymbol{0} & \boldsymbol{C}_{22}\end{bmatrix} \tag{A.11}$$

Since $\hat{\boldsymbol{\Sigma}}_b$ is diagonal, it consists of r singular values of $\boldsymbol{A}$. Equation (A.8) becomes:

$$\begin{aligned}\varepsilon_B &= ||\boldsymbol{C}-\boldsymbol{\Sigma}_b||\\ &= ||\boldsymbol{C}_{22}||\end{aligned} \tag{A.12}$$

Since both $\boldsymbol{U}_b$ and $\boldsymbol{V}_b$ are unitary matrices, we have:

$$||\boldsymbol{A}||^2 = ||\boldsymbol{C}||^2 = ||\hat{\boldsymbol{\Sigma}}_b||^2 + ||\boldsymbol{C}_{22}||^2 \tag{A.13}$$

Then:

$$\begin{aligned} ||\boldsymbol{C}_{22}||^2 &= ||\boldsymbol{A}||^2 - ||\hat{\boldsymbol{\Sigma}}_b||^2 \\ &= \sum_{m=1}^{k} \sigma_m^2 - ||\hat{\boldsymbol{\Sigma}}_b||^2 \end{aligned} \tag{A.14}$$

Obviously, when $\hat{\boldsymbol{\Sigma}}_b$ holds the r largest singular values $\sigma_1, \ldots, \sigma_r$ of the matrix $\boldsymbol{A}$, $||\boldsymbol{C}_{22}||^2$ and then ε_B reaches its minimum value:

$$\varepsilon_B^2 = |\boldsymbol{C}_{22}||^2 = \sum_{m=r+1}^{k} \sigma_m^2 = \varepsilon_r^2 \tag{A.15}$$

Therefore, we can draw the conclusion that $\hat{\boldsymbol{A}}$ is the best rank r approximation to $\boldsymbol{A}$ based on minimisation of the error matrix' Frobenius norm $||\boldsymbol{A} - \boldsymbol{B}||$.

Appendix B: Differentiation with Respect to a Vector

Differentiating a cost function with respect to a general complex-valued vector is often encountered in the area of array signal processing. In this appendix we give a brief explanation to the concept behind it.

Suppose the function $f(z)$ is a function of the complex variable z, with their real and imaginary parts, respectively, given by:

$$\begin{aligned} z &= x + jy \\ f(z) &= u + jv \end{aligned} \tag{B.1}$$

Then the complex differentiation $\mathrm{d}f/\mathrm{d}z$ at the point $z = z_0$ is defined as:

$$\begin{aligned} \frac{\mathrm{d}f}{\mathrm{d}z} &= \lim_{\Delta z \to 0} \frac{f(z_0 + \Delta z) - f(z_0)}{\Delta z} \\ &= \lim_{\Delta x, \Delta y \to 0} \frac{\Delta u + j\Delta v}{\Delta x + j\Delta y} \end{aligned} \tag{B.2}$$

For an analytic function, Δz in the above definition can approach to zero from arbitrary directions, which will always give the same derivative value at z_0. For example, we can set $\Delta x = 0$ and let $\Delta y \to 0$, or we can set $\Delta y = 0$ and let $\Delta x \to 0$. Then we have the following property:

$$\frac{\mathrm{d}f}{\mathrm{d}z} = \frac{\partial f}{\partial x} = -j\frac{\partial f}{\partial y} \tag{B.3}$$

which gives the definition for a general function $f(z)$ (not necessarily analytic) when both its real part u and its imaginary part v are differentiable (Remmert, 1991):

$$\frac{\mathrm{d}f}{\mathrm{d}z} = \frac{1}{2}\left(\frac{\partial f}{\partial x} - j\frac{\partial f}{\partial y}\right) \tag{B.4}$$

The derivative of $f(z)$ with respect to z^* is defined in a similar way as follows (Remmert, 1991):

$$\frac{\mathrm{d}f}{\mathrm{d}z^*} = \frac{1}{2}\left(\frac{\partial f}{\partial x} + j\frac{\partial f}{\partial y}\right) \tag{B.5}$$

Wideband Beamforming Wei Liu and Stephan Weiss

One important result with this definition is that a complex variable z is independent of its conjugate z^* since:

$$\frac{\mathrm{d}z}{\mathrm{d}z^*} = \frac{\mathrm{d}z^*}{\mathrm{d}z} = 0 \tag{B.6}$$

When the complex variable z is replaced by a complex-valued vector $\mathbf{w}$, given by:

$$\mathbf{w} = [w_0 \ w_1 \ \dots \ w_{M-1}]^T \tag{B.7}$$

where $w_m = x_m + jy_m$, $m = 0, 1, \dots, M-1$, we can define the derivative of $f(\mathbf{w})$ with respect to $\mathbf{w}$ by extending the result in Equation (B.4) directly, given by (Haykin, 1996):

$$\frac{\mathrm{d}f}{\mathrm{d}\mathbf{w}} = \frac{1}{2}\begin{bmatrix} \partial f/\partial x_0 - j\partial f/\partial y_0 \\ \partial f/\partial x_1 - j\partial f/\partial y_1 \\ \vdots \\ \partial f/\partial x_{M-1} - j\partial f/\partial y_{M-1} \end{bmatrix} \tag{B.8}$$

In the same way, we define $\mathrm{d}f/\mathrm{d}\mathbf{w}^*$ as in Haykin (1996):

$$\frac{\mathrm{d}f}{\mathrm{d}\mathbf{w}^*} = \frac{1}{2}\begin{bmatrix} \partial f/\partial x_0 + j\partial f/\partial y_0 \\ \partial f/\partial x_1 + j\partial f/\partial y_1 \\ \vdots \\ \partial f/\partial x_{M-1} + j\partial f/\partial y_{M-1} \end{bmatrix} \tag{B.9}$$

When $M = 1$, Equations (B.8) and (B.9) are reduced to Equations (B.4) and (B.5). Similarly $\mathbf{w}$ and $\mathbf{w}^*$ are also independent of each other.

As a simple example, consider the quadratic form $f(\mathbf{w}) = \mathbf{w}^H\mathbf{w}$ and we have:

$$\begin{aligned} \frac{\mathrm{d}f}{\mathrm{d}\mathbf{w}} &= \frac{\mathrm{d}(\mathbf{w}^H\mathbf{w})}{\mathrm{d}\mathbf{w}} = \mathbf{w}^* \\ \frac{\mathrm{d}f}{\mathrm{d}\mathbf{w}^*} &= \frac{\mathrm{d}(\mathbf{w}^H\mathbf{w})}{\mathrm{d}\mathbf{w}^*} = \mathbf{w} \end{aligned} \tag{B.10}$$

Appendix C: Genetic Algorithm

Genetic algorithms (GAs) are a class of optimization techniques based on the law of natural selection and have been successfully applied to many different areas, including the design of filters and filter banks with sum-of-powers-of-two (SOPOT) coefficients (Liu, 2000; Redmill *et al.*, 1997; Sriranganathan *et al.*, 1997). In this appendix, we will give a brief introduction and more details can be found in Houck *et al.* (1995), Man *et al.* (1997) and Tang *et al.* (1996).

C.1 The Principle

In a GA, it is assumed that the potential solution to any problem can be represented by a set of parameters, which are regarded as the genes of a chromosome. According to the optimization criterion, each chromosome is assigned a fitness value as an indication of its closeness to the final solution.

A GA consists of three consecutive processes: selection, genetic operation and replacement. The first group of chromosomes (the first generation) can be generated randomly or from some initial results provided by other optimization methods. In the selection process, the 'parents' are selected according to their fitness values and the chromosome with a larger fitness value has a greater chance to be selected. Then the genetic operation, including crossover and mutation, is applied to the chosen 'parents' to give birth to a new set of offspring chromosomes. Based on some replacement strategy, the chromosomes in the first group are then replaced by the newly generated ones.

The whole process is shown in Figure C.1 and is repeated until the desired chromosome is found.

According to Houck *et al.* (1995), there are six key issues for the successful operation of a GA: chromosome representation/encoding scheme, selection function, genetic operators, initial population creation, termination criteria and fitness evaluation function.

C.1.1 Chromosome Representation

Each chromosome represents a trial solution to the problem and is composed of a string of variables, i.e. genes. The variables can be binary digits (0 and 1), real numbers or other forms such as symbols $(A, B, C, \dots)$, matrices, etc, depending on the problem specifications. Bit-string representation is the most classic scheme (Holland, 1975) and is adopted in the design of SOPOT coefficients in Section 3.5, Chapter 3 since the SOPOT solutions are points in a discrete domain.

Wideband Beamforming Wei Liu and Stephan Weiss

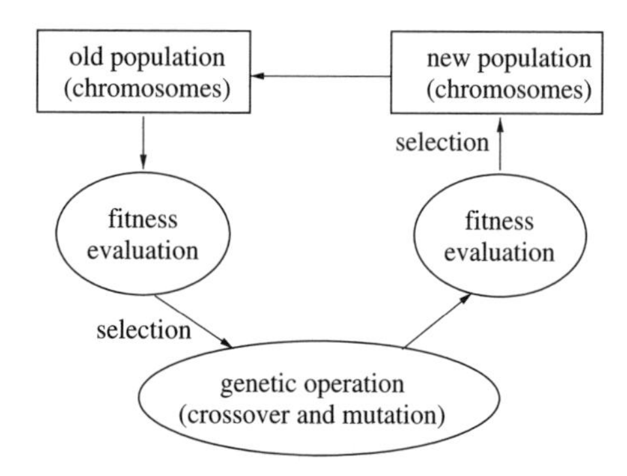

Figure C.1 The operation process of a general genetic algorithm

C.1.2 Parent Selection

The selection process is designed based on the 'survival-of-the-fittest' mechanism in nature, which means that chromosomes with a higher fitness value have a better chance to survive through being selected to produce offspring. It is often made probabilistic to reflect the intrinsic flexibility and freedom of this process in nature and many selection methods have been proposed, such as ranking, tournament and the proportionate scheme (Man *et al.*, 1997).

One example is the 'Roulette Wheel Selection' (Holland, 1975), where we first assign a probability of selection p_m to each chromosome $X_m[n]$ according to its fitness value $F(X_m[n])$, which is defined by:

$$p_m = \frac{F(X_m[n])}{\sum_{i=0}^{M-1} F(X_i[n])} \tag{C.1}$$

where $X_m[n]$ represents the mth chromosome for the nth generation and M is the population size; then a random number ε is generated between 0 and 1 and compared with the cumulative probability $\tilde{p}_m = \sum_{i=0}^{m} p_i$. If $\tilde{p}_{m-1} < \varepsilon \leq \tilde{p}_m$, then $X_m[n]$ is selected.

C.1.3 Genetic Operation

Genetic operation determines how the 'parents' produce their 'offspring' and in nature it refers to the recombination of different chromosomes after crossover and mutation. Because here all the information is represented by only one chromosome, the recombination is realized by crossover between the corresponding genes of the parental chromosomes, during which mutation might happen, introducing a sudden change of some genes in a chromosome.

For crossover, we should choose one or more crossover points in the parental chromosomes. These points can be generated from a number generator with a value between 1 and the length of the chromosome. Crossover can be further divided into a single-point crossover and a multi-point crossover.

For a single-point crossover, each of the two chromosomes is divided into two parts by a single randomly generated crossover point to form two new chromosomes, while for multipoint crossover, more than one crossover point are chosen randomly. One simple way of generating multi crossover points is to create a random binary string of 1s and 0s. The length of the string matches the number of genes in a chromosome. The changes in binary values within this string determine the positions of the crossover points.

Mutation introduces another random variation to the chromosome, involving a change of one, more or even all of its genes. It also occurs with a specified probability and the genes to mutate can be determined in a similar way as the determination of crossover points. Its range is determined by the specified optimization problem.

In the scenario where the solution is represented by binary digits, having determined the genes (variables) to mutate, we can randomly choose one or more bits in the variable to mutate and replace them by randomly generated binary values. Another commonly used standard mutation is flipping bits, where 1 is replaced by 0 and 0 replaced by 1, if the probability test used for mutation is passed.

C.1.4 Fitness Evaluation

Fitness evaluation is the starting point of the selection process and also the only link between a GA and the corresponding optimization problem. It assigns a value to each of the chromosomes according to their performance in the system. The range of the values varies from problem to problem and there are many different ways to map the performance to a proper fitness value.

To maintain uniformity, a linear normalization approach can be adopted, where the chromosomes are first ranked in descending order according to their performances; the best chromosome is assigned a value of α_0, and the worst the value α_{M-1}, with $\alpha_{M-1} < \alpha_0$; then the fitness value for the ith chromosome in the ordered list is given by:

$$\alpha_i = \alpha_0 - (i-1) \times \frac{(\alpha_0 - \alpha_{M-1})}{M} \tag{C.2}$$

C.1.5 Initialization

The initialization of a GA is to generate the first generation of chromosomes for the given problem. There are two commonly used methods in practice. One is to generate the initial chromosomes randomly, although most of them may be far away from the optimal solution, which will result in a larger number of iterations to converge to the optimum. If other methods are available to provide a good initial guess to the problem, then it is advantageous to include the guess results in the first generation. For example, in the design of multiplier-less filters and filter banks with SOPOT representation, optimal or sub-optimal real-valued results are available for some problems and these results can be used to aid a GA to search for the solution in the discrete domain.

C.1.6 Termination

We can end the optimization process when a pre-determined number of generations is reached or we can measure the deviation in the fitness values among the members of one generation and end the algorithm when it is smaller than some threshold value. In another way, the iteration stops once the best chromosome in one generation reaches a predefined fitness value. Further methods can be developed by combining these methods together in various ways.

C.2 Design Example in Section 3.5.2

In the blocking matrix design in Section 3.5.2, each of the coefficients $b_l[m]$, $m = 2, 3, \ldots, M-1$ is represented as:

$$b_{l,m} = \sum_{i=0}^{P[l,m]-1} a_i[l,m] \cdot 2^{L_i[l,m]} \qquad \text{with :}$$
$$a_i[l,m] \in \{-1, 1\}, \quad L_i[l,m] \in \{Q_1, Q_1+1, \ldots, Q_2-1, Q_2\} \tag{C.3}$$

Table C.1 SOPOT coefficients for the 16 × 8 blocking matrix $\tilde{\boldsymbol{B}}$

m	$\mathbf{b}_0$	$\mathbf{b}_1$	$\mathbf{b}_2$	$\mathbf{b}_3$
2	$2^{-2}+2^{-8}$	-2^{-9}	$-2^{-2}-2^{-4}+2^{-9}$	$-2^{-1}-2^{-7}$
3	$2^{-4}+2^{-5}-2^{-8}$	$2^{-2}+2^{-5}$	$-2^{-2}+2^{-7}-2^{-7}$	-2^{-7}
4	$-2^{-3}-2^{-6}$	$2^{-1}-2^{-5}+2^{-8}$	$2^{-2}+2^{-5}$	$2^{-1}+2^{-3}-2^{-6}$
5	$-2^{-1}+2^{-4}$	$2^{-1}-2^{-4}$	$2^{-1}+2^{-5}-2^{-9}$	$-2^{-3}+2^{-5}+2^{-8}$
6	$-2^{-1}-2^{-4}$	-2^{-7}	$2^{-4}+2^{-7}+2^{-9}$	$-2^{-1}-2^{-2}-2^{-8}$
7	$-2^{-1}+2^{-7}+2^{-8}$	$-2^{-1}+2^{-3}+2^{-8}$	$-2^{-1}-2^{-4}$	$2^{-3}+2^{-6}$
8	$-2^{-2}-2^{-5}-2^{-6}$	$-2^{-1}+2^{-6}+2^{-8}$	$-2^{-1}+2^{-7}$	$2^{-1}+2^{-3}-2^{-5}$
9	-2^{-6}	$-2^{-2}-2^{-8}-2^{-9}$	$2^{-2}-2^{-4}+2^{-6}$	$-2^{-3}-2^{-6}$
10	$2^{-2}-2^{-5}$	$2^{-5}-2^{-9}$	$2^{-1}+2^{-5}$	-2^{-1}
11	$2^{-2}+2^{-4}+2^{-8}$	$2^{-2}-2^{-8}-2^{-9}$	$2^{-2}-2^{-4}$	$2^{-3}+2^{-4}+2^{-8}$
12	$2^{-2}+2^{-6}+2^{-7}$	$2^{-2}+2^{-7}$	$-2^{-2}-2^{-6}-2^{-8}$	$2^{-1}-2^{-3}-2^{-5}$
13	$2^{-3}+2^{-5}$	2^{-5}	$-2^{-2}-2^{-9}$	$2^{-1}+2^{-3}-2^{-5}$
14	$2^{-6}+2^{7}$	-2^{-5}	$2^{-5}-2^{-9}$	$-2^{-3}-2^{-6}$
15	$-2^{-6}-2^{-7}$	$-2^{-2}+2^{-3}-2^{-9}$	$2^{-3}+2^{-8}$	$2^{-3}-2^{-7}$

m	$\mathbf{b}_4$	$\mathbf{b}_5$	$\mathbf{b}_6$	$\mathbf{b}_7$
2	2^{-2}	$2^{-2}-2^{-4}$	$2^{-1}-2^{-4}$	$2^{-3}+2^{-4}+2^{-8}$
3	$-2^{-1}-2^{-7}$	$2^{-3}+2^{-5}+2^{-6}$	$-2^{-2}-2^{-5}-2^{-6}$	$-2^{-2}+2^{-7}$
4	$2^{-3}+2^{-5}+2^{-6}$	$-2^{-1}-2^{-3}-2^{-5}$	$2^{-3}-2^{-6}$	$2^{-2}-2^{-6}$
5	$2^{-1}+2^{-6}$	$2^{-1}+2^{-2}-2^{-7}$	$2^{-2}-2^{-5}-2^{-6}$	$-2^{-2}-2^{-5}-2^{-7}$
6	$-2^{-1}-2^{-5}-2^{8}$	$-2^{-2}-2^{-4}-2^{-6}$	$-2^{-1}-2^{-3}$	$2^{-2}+2^{-6}+2^{-8}$
7	-2^{-5}	$-2^{-2}-2^{-7}$	$2^{-1}+2^{-4}+2^{-4}$	$-2^{-2}-2^{-4}+2^{-6}$
8	$2^{-1}+2^{-4}$	$2^{-1}+2^{-4}$	$-2^{-1}-2^{-2}+2^{-6}$	$2^{-2}+2^{-8}$
9	$-2^{-2}-2^{-4}+2^{-8}$	$-2^{-1}-2^{-4}-2^{-8}$	$2^{-1}+2^{-5}$	$-2^{-5}-2^{-6}$
10	-2^{-2}	$2^{-2}-2^{-5}+2^{-8}$	$2^{-4}+2^{-7}$	$2^{-2}+2^{-9}$
11	$2^{-1}-2^{-3}+2^{-5}$	$2^{-2}-2^{-4}-2^{-6}$	$-2^{-3}-2^{-4}+2^{-6}$	$-2^{-3}-2^{-7}-2^{-9}$
12	$-2^{-3}-2^{-7}$	$-2^{-2}-2^{-6}$	$2^{-1}-2^{-6}-2^{-7}$	$2^{-3}+2^{-6}-2^{-8}$
13	$-2^{-2}+2^{-4}-2^{-7}$	$2^{-3}+2^{-5}+2^{-6}$	$-2^{-1}+2^{-5}-2^{-7}$	$-2^{-3}+2^{-8}$
14	$2^{-4}+2^{6}$	-2^{-8}	$2^{-2}+2^{-4}+2^{-7}$	$2^{-3}-2^{-5}-2^{-7}$
15	$2^{-5}-2^{-7}$	-2^{-6}	$-2^{-2}+2^{-5}$	-2^{-5}

where $P[l, m]$ is a limit for the number of SOPOT terms and Q_1 and Q_2 are integers determined by the range of the corresponding variable. Normally, $P[l, m]$ is limited to a small number. Thus, the whole process will be kept in the discrete domain and good results in a SOPOT form can be obtained.

Any of the mutation and crossover techniques introduced in the previous section can be employed. In this design, we use the single-point crossover and randomly create a number to decide the crossover point. A mutation is set to occur with a probability of 0.5 in a chromosome, whereby within the chromosome each gene has a uniform probability for being affected by this mutation. The stopband attenuation is assigned to each of the chromosomes as its fitness value and the 'Roulette Wheel Selection' scheme is employed to select the appropriate parents. The initial population is generated randomly and the whole optimization process is terminated when a specified maximum number of generations is reached. Table C.1 gives the GA design results in a SOPOT notation.

Bibliography

Abramowitz M and Stegun IA (Eds) 1970 *Handbook of Mathematical Functions*. Dover Publications, New York, NY, USA.

Affes S and Grenier Y 1997 A signal subspace tracking algorithm for microphone array processing of speech. *IEEE Transactions on Speech and Audio Processing* **5**(5), 425–437.

Affes S, Gazor S and Grenier Y 1996 An algorithm for multisource beamforming and multitarget tracking. *IEEE Transactions on Signal Processing* **44**(6), 1512–1522.

Ahmed KM and Evans RJ 1983 Broadband adaptive array processing. *IEE Proceedings F: Communications, Radar and Signal Processing* **130**(5), 432–440.

Ahmed KM and Evans RJ 1984 An adaptive array processor with robustness and broadband capabilities. *IEEE Transactions on Antennas and Propagation* **AP-32**(9), 944–950.

Akansu A and Haddad R 1992 *Multiresolution Signal Decomposition: Transforms, Subbands and Wavelets*. Academic Press, Boston, MA, USA.

Alamouti SM 1998 A simple transmit diversity technique for wireless communications. *IEEE Journal on Selected Areas in Communications* **16**(8), 1451–1458.

Allen B and Ghavami M 2005 *Adaptive Array Systems, Fundamentals and Applications*. John Wiley & Sons, Ltd, Chichester, UK.

Amari S 1998 Natural gradient works efficiently in learning. *Neural Computation* **10**(2), 251–276.

An J and Champagne B 1994 GSC realisations using the two-dimensional transform LMS algorithm. *IEE Proceedings: Radar, Sonar and Navigation* **141**(5), 270–278.

Applebaum S and Chapman D 1976 Adaptive arrays with main beam constraints. *IEEE Transactions on Antennas and Propagation* **AP-24**(5), 650–662.

Bamberger RH and Smith MJT 1992 A Filter Bank for the Directional Decomposition of Images: Theory and Design. *IEEE Transactions on Signal Processing* **40**(4), 882–893.

Barros AK and Cichocki A 2001 Extraction of specific signals with temporal structure. *Neural Computation* **13**(9), 1995–2003.

Bellanger M, Bonnerot G and Coudreuse M 1976 Digital Filtering by Polyphase Network: Application to Sample Rate Alteration and Filter Banks. *IEEE Transactions on Acoustics, Speech and Signal Processing* **ASSP-24**(4), 109–114.

Belouchrani A, Abed-Meraim K, Cardoso JF and Moulines 1997 A blind source separation technique using second-order statistics. *IEEE Transactions on Signal Processing* **45**(2), 434–444.

Benesty J and Morgan D 2000 Frequency-domain adaptive filtering revisited, generalization to the multi-channel case, and application to acoustic echo cancellation *Proceedings of the IEEE International Conference on Acoustics, Speech and Signal Processing*, Vol. 2, pp. 789–792, Istanbul, Turkey.

Benesty J, Chen JD, Huang YT and Dmochowski J 2007 On microphone-array beamforming from a MIMO acoustic signal processing perspective. *IEEE Transactions on Acoustics, Speech and Language Processing* **15**(3), 1053–1065.

Biglieri E, Calderbank R, Constantinides A, Goldsmith A, Paulraj A and Poor HV 2007 *MIMO Wireless Communications*. Cambridge University Press, Cambridge, UK.

Wideband Beamforming Wei Liu and Stephan Weiss

Bitzer J, Simmer KU and Kammeyer KD 1999 Theoretical noise reduction limits of the generalized sidelobe canceller (GSC) for speech enhancement *Proceedings of the IEEE International Conference on Acoustics, Speech and Signal Processing*, Vol. 5, pp. 2965–2968, Phoenix, AZ, USA.

Björck A 1996 *Numerical Methods for Least Squares Problems*. Society for Industrial and Applied Mathematics (SIAM), Philadelphia, PA, USA.

Bölcskei H, Hlawatsch F and Feichtinger HG 1995 On the Equivalence of DFT Filter Banks and the Gabor Expansion *SPIE Proceedings – Wavelet Applications in Signal and Image Processing III*, Vol. 2569, pp. 128–139, San Diego, CA, USA.

Boyd S and Vandenberghe L 2004 *Convex Optimization*. Cambridge University Press, Cambridge, UK.

Brandstein MS and Ward D (Eds) 2001 *Microphone Arrays: Signal Processing Techniques and Applications*. Springer-Verlag, Berlin, Germany.

Breed B and Strauss J 2002 A short proof of the equivalence of lcmv and gsc beamforming. *IEEE Signal Processing Letters* **9**(6), 168–169.

Buckley K and Griffiths L 1986 Eigenstructure based broadband source location estimation *Proceedings of the IEEE International Conference on Acoustics, Speech and Signal Processing*, Vol. 11, pp. 1869–1872, Los Angeles, CA, USA.

Buckley KM 1986 Broad-band beamforming and the generalized sidelobe canceller. *IEEE Transactions on Acoustics, Speech and Signal Processing* **ASSP-34**(5), 1322–1323.

Buckley KM 1987 Spatial/Spectral Filtering with Linearly Constrained Minimum Variance Beamformers. *IEEE Transactions on Acoustics, Speech and Signal Processing* **ASSP-35**(3), 249–266.

Buckley KM and Griffith LJ 1986 An adaptive generalized sidelobe canceller with derivative constraints. *IEEE Transactions on Antennas and Propagation* **34**(3), 311–319.

Burrus CS and Parks TW 1985 *DFT/FFT and Convolution Algorithms: Theory and Implementation*. John Wiley & Sons, Inc., New York, NY, USA.

Sureau JC and Keeping K 1982 Sidelobe control in cylindrical arrays. *IEEE Transactions on Antennas and Propagation* **30**(5), 1027–1031.

Capon J 1969 High-resolution Frequency-wavenumber Spectrum Analysis. *Proceedings of the IEEE* **57**(8), 1408–1418.

Cardoso J and Souloumiac A 1993 Blind beamforming for non-Gaussian signals. *IEE Proceedings F: Radar and Signal Processing* **140**(6), 362–370.

Chan SC and Chen HH 2005 Theory and design of uniform concentric circular arrays with frequency invariant characteristics *Proceedings of the IEEE International Conference on Acoustics, Speech and Signal Processing*, Vol. 4, pp. 805–808, Philadelphia, PA, USA.

Chan SC and Chen HH 2006 Theory and design of uniform concentric spherical arrays with frequency invariant characteristics *Proceedings of the IEEE International Conference on Acoustics, Speech and Signal Processing*, Vol. IV, pp. 1057–1060, Toulouse, France.

Chan SC and Chen HH 2007 Uniform concentric circular arrays with frequency-invariant characteristics – theory, design, adaptive beamforming and DOA estimation. *IEEE Transactions on Signal Processing* **55**(1), 165–177.

Chan SC and Pun KS 2002 On the design of digital broadband beamformer for uniform circular array with frequency invariant characteristics *Proceedings of the IEEE International Symposium on Circuits and Systems*, Vol. 1, pp. 693–696, Phoenix, AZ, USA.

Chan SC, Liu W and Ho KL 2001 Multiplier-less perfect reconstruction modulated filter banks with sun-of-powers-of-two coefficients. *IEEE Signal Processing Letters* **8**(6), 163–166.

Chan SC, Pun KS and Ho KL 2002 Efficient implementation of wideband multibeam forming network using SOPOT coefficients and multiplier block *Proceedings of the International Conference on Digital Signal Processing*, Vol. 1, pp. 243–246, Santorini, Greece.

Chapman DJ 1976 Partial Adaptivity for Large Arrays. *IEEE Transactions on Antennas and Propagation* **24**(9), 685–696.

Chau E, Sheikhzadeh H, Brennan R and Schneider T 2002 A subband beamformer on an ultra low-power miniature dsp platform *Proceedings of the IEEE International Conference on Acoustics, Speech and Signal Processing*, Vol. 3, pp. 2953–2956, Orlando, FL, USA.

Chen HH and Chan SC 2007 Adaptive beamforming and DOA estimation using uniform concentric spherical arrays with frequency invariant characteristics. *Journal of VLSI Signal Processing* **46**(1), 15–34.

Chen HH, Chan SC and Ho KL 2007 Adaptive beamforming using frequency invariant uniform concentric circular arrays. *IEEE Transactions on Circuits and Systems I: Regular Papers* **54**(9), 1938–1949.

Chen T 1993 Unified eigenfilter approach: with applications to spectral/spatial filtering *Proceedings of the IEEE International Symposium on Circuits and Systems*, Vol. 1, pp. 331–334, Chicago, IL, USA.

Chen Y and Fang H 1992 Frequency-domain implementation of griffiths-jim adaptive beamformer. *Journal of the Acoustical Society of America* **91**(6), 3354–3366.

Chou T 1995 Frequency-independent beamformer with low response error *Proceedings of the IEEE International Conference on Acoustics, Speech and Signal Processing*, Vol. 5, pp. 2995–2998, Detroit, MI, USA.

Cichocki A and Amari S 2003 *Adaptive Blind Signal and Image Processing*. John Wiley & Sons, Inc., New York, NY, USA.

Cichocki A and Thawonmas R 2000 On-line algorithm for blind signal extraction of arbitrarily distributed, but temporally correlated sources using second order statistics. *Neural Processing Letters* **12**(1), 91–98.

Cichocki A, Thawonmas R and Amari S 1997 Sequential blind signal extraction in order specified by stochastics properties. *IEE Electronics Letters* **33**(1), 64–65.

Claesson I and Nordholm S 1992 A Spatial Filtering Approach to Robust Adaptive Beamforming. *IEEE Transactions on Antennas and Propagation* **40**(9), 1093–1096.

Cohen I 2003 Analysis of two-channel generalized sidelobe canceller (GSC) with post-filtering. *IEEE Transactions on Speech and Audio Processing* **11**(6), 684–699.

Compton, Jr RT 1988a The relationship between tapped delay-line and FFT processing in adaptive arrays. *IEEE Transactions on Antennas and Propagation* **36**(1), 15–26.

Compton RT 1988b *Adaptive Antennas: Concepts and Performance*. Prentice-Hall, Englewood Cliffs, NJ, USA.

Compton RT 1988c The bandwidth performance of a two-element adaptive array with tapped delay-line processing. *IEEE Transactions on Antennas and Propagation* **36**(1), 4–14.

Cox H, Zeskind RM and Owen MM 1987 Robust adaptive beamforming. *IEEE Transactions on Acoustics, Speech and Signal Processing* **ASSP-35**(10), 1365–1376.

Crawford FS 1968 *Waves* Berkeley Physics Courses, Vol. 3. McGraw-Hill, New York, NY, USA.

Crochiere RE 1977 On the design of subband coders for low bit rate speech communications. *The Bell System Technical Journal* **56**, 747–771.

Crochiere RE 1981 Subband coding. *The Bell System Technical Journal* **60**, 1633–1654.

Crochiere RE and Rabiner LR 1983 *Multirate Digital Signal Processing*. Prentice-Hall, Englewood Cliffs, NJ, USA.

Crochiere RE, Webber SA and Flanagan JL 1976 Digital coding of speech in subbands. *The Bell System and Technical Journal* **55**(8), 1069–1085.

Cruces-Alvarez SA, Cichocki A and Amari S 2004 From blind signal extraction to blind instantaneous signal separation: Criteria, algorithms and stability. *IEEE Transactions on Neural Networks* **15**(4), 859–873.

Cvetković Z and Vetterli M 1998 Tight Weyl–Heisenberg Frames in $l^2(\mathbb{Z})$. *IEEE Transactions on Signal Processing* **46**(5), 1256–1259.

Davies DEN 1983 Circular arrays, in *The Handbook of Antenna Design* (Rudge AW, Milne K, Olver AD and Knight P (Eds)), Vol. 2 Chapter 12. Peter Peregrinus Ltd, London, UK.

De Haan J, Grbic N and Claesson I 2002 Design and evaluation of nonuniform dft filter banks in subband microphone arrays *Proceedings of the IEEE International Conference on Acoustics, Speech and Signal Processing*, Vol. 2, pp. 1173–1176, Orlando, FL, USA.

Delfosse N and Loubaton P 1995 Adaptive blind separation of independent sources: A deflation approach. *Signal Processing* **95**(1), 59–83.

Di Claudio ED 2005 Asymptotically perfect wideband focusing of multiring circular arrays. *IEEE Transactions on Signal Processing* **53**(10), 3661–3673.

Ding Z and Li Y 2001 *Blind Equalisation and Identification* Signal Processing and Communications. CRC Press, New York, NY, USA.

Doclo S and Moonen M 2003a Design of broadband beamformers robust against gain and phase errors in the microphone array characteristics. *IEEE Transactions on Signal Processing* **51**(10), 2511–2526.

Doclo S and Moonen M 2003b Design of far-field and near-field broadband beamformers using eigenfilters. *Signal Processing* **83**(12), 2641–2673.

Dogan M and Mendel JM 1994 Cumulant-based blind optimum beamforming. *IEEE Transactions on Aerospace and Electronic Systems* **30**(3), 722–741.

Doles JH III and Benedict FD 1988 Broad-band array design using the asymptotic theory of unequally spaced arrays. *IEEE Transactions on Antennas and Propagation* **36**(1), 27–33.

Douglas SC 2002 Blind signal separation and blind deconvolution, in *The Handbook of Neural Network Signal Processing* (Hu YH and Hwang JN (Eds)), Chapter 7. CRC Press, Boca Raton, FL, USA.

Douglas SC and Sun X 2003 Convolutive blind separation of speech mixtures using the natural gradient. *Speech Communication* **39**(1), 65–78.

Duan H, Ng BP, See CM and Fang J 2008 Applications of the SRV constraint in broadband pattern synthesis. *Signal Processing* **88**(4), 1035–1045.

El-Keyi A, Kirubarajan T and Gershman A 2005 Wideband robust beamforming based on worst-case performance optimization *Proceedings of the IEEE Workshop on Statistical Signal Processing*, pp. 265–270, Bordeaux, France.

Er M 1993 On the limiting solution of quadratically constrained broad-band beam formers. *IEEE Transactions on Signal Processing* **43**(1), 418–419.

Er M 1994 A new approach to robust beamforming in the presence of steering vector errors. *IEEE Transactions on Signal Processing* **42**(7), 1826–1829.

Er M and Cantoni A 1983 Derivative constraints for broadband element space antenna array processors. *IEEE Transactions on Antennas and Propagation* **ASSP-31**(6), 1378–1393.

Er M and Cantoni A 1985 A new approach to the design of broad-band element space antenna array processors. *IEEE Journal of Oceanic Engineering* **OE-10**(3), 231–240.

Er M and Cantoni A 1986a A new set of linear constraints for broad-band time domain element space processors. *IEEE Transactions on Antennas and Propagation* **AP-34**(3), 320–329.

Er M and Cantoni A 1986b A new set of linear constraints for broad-band time domain element space processors. *IEEE Transactions on Acoustics, Speech and Signal Processing* **ASSP-34**(6), 1376–1379.

Er M and Cantoni A 1990 An alternative implementation of quadratically constrained broadband beamformers. *Signal Processing* **21**(2), 117–127.

Er M and Ng BP 1990 On derivative constrained broad-band beamforming. *IEEE Transactions on Acoustics, Speech and Signal Processing* **38**(3), 551–552.

Fenn AJ 1985 Maximizing jammer effectiveness for evaluating the performance of adaptive nulling array antennas. *IEEE Transactions on Antennas and Propagation* **AP-33**(10), 1131–1142.

Ferrara ER 1985 Frequency-domain adaptive filtering, in *Adaptive Filters* (Cowan CFN and Grant PM (Eds)), Chapter 6, pp. 145–179. Prentice-Hall, Englewood Cliffs, NJ, USA.

Fletcher R 2000 *Practical Methods of Optimization* 2nd Edn. John Wiley & Sons, Inc., New York, NY, USA.

Fliege NJ 1993 *Multiraten-Signalverarbeitung*. BG Teubner, Stuttgart, Germany.

Fliege NJ 1994 *Multirate Digital Signal Processing: Multirate Systems, Filter Banks, Wavelets*. John Wiley & Sons, Ltd, Chichester, UK.

Fourikis N 2000 *Advanced Array Systems, Applications and RF Technologies*. Academic Press, London, UK.

Fragouli C, Al-Dhahir N and Turin W 2003 Training-based channel estimation for multiple antenna broadband transmissions. *IEEE Transactions on Wireless Communications* **2**(2), 384–391.

Freese H, Sperry B and Votow K 2003 Sub-aperture beamspace adaptive processing *Proceedings of the Workshop on Adaptive Sensor Array Processing*, Lexington, MA, USA.

Friedlander B and Weiss AJ 1993 Direction finding for wide-band signals using an interpolated array. *IEEE Transactions on Signal Processing* **41**(4), 1618–1634.

Frost, III OL 1972 An algorithm for linearly constrained adaptive array processing. *Proceedings of the IEEE* **60**(8), 926–935.

Fudge GL and Linebarger DA 1994 A calibrated generalized sidelobe canceller for wideband beamforming. *IEEE Transactions on Signal Processing* **22**(10), 2871–2875.

Gabriel WF 1976 Adaptive arrays – an introduction. *Proceedings of the IEEE* **64**(2), 239–272.

Gabriel WF 1980 Spectral analysis and adaptive array superresolution techniques. *Proceedings of the IEEE* **68**(6), 654–666.

Galand CR and Nussbaumer HJ 1984 New quadrature mirror filter structures. *IEEE Transactions on Acoustics, Speech and Signal Processing* **ASSP-32**(3), 522–531.

Georgiev P, Theis F and Cichocki A 2005 Sparse component analysis and blind source separation of underdetermined mixtures. *IEEE Transactions on Neural Networks* **16**(4), 992–996.

Ghavami M 2002 Wideband smart antenna theory using rectangular array structures. *IEEE Transactions on Signal Processing* **50**(9), 2143–2151.

Gillespie BW, Malvar HS and Florencio DAF 2001 Speech dereverberation via maximum-kurtosis subband adaptive filtering *Proceedings of the IEEE International Conference on Acoustics, Speech and Signal Processing*, Vol. 6, pp. 3701–3704, Salt Lake City, UT, USA.

Gilloire A 1987 Experiments with subband acoustic echo cancelers for teleconferencing *Proceedings of the IEEE International Conference on Acoustics, Speech and Signal Processing*, pp. 2141–2144, Dallas, TX, USA.

Gilloire A and Vetterli M 1988 Adaptive filtering in subbands *Proceedings of the IEEE International Conference on Acoustics, Speech and Signal Processing*, pp. 1572–1575, New York, NY, USA.

Gilloire A and Vetterli M 1992 Adaptive filtering in subbands with critical sampling: Analysis, experiments and applications to acoustic echo cancelation. *IEEE Transactions on Signal Processing* **SP-40**(8), 1862–1875.

Girod B, Hartung F and Horn U 1995 Subband image coding, in *Subband and Wavelet Transforms: Design and Applications* (Smith MJT and Akansu AN (Eds)), Chapter 7. Kluwer Academic Publishers, Boston, MA, USA.

Gitlin R and Magee FJ 1977 Self-orthogonalizing adaptive equalization algorithms. *IEEE Transactions on Communications* **COM-25**(7), 666–672.

Goldstein J, Ingram M, Holder E and Smith R 1992 A linearly constrained wideband adaptive array antenna with orthogonal filter structure *Proceedings of the IEEE International Symposium on Antennas and Propagation*, pp. 612–615, Chicago, IL, USA.

Goldstein JS and Reed IS 1997 Theory of partially adaptive radar. *IEEE Transactions on Aerospace and Electronic Systems* **33**(4), 1309–1325.

Golub GH and Van Loan CF 1996 *Matrix Computations*, 3rd Edn. Johns Hopkins University Press, Baltimore, MA, USA.

Gönen, E. and Mendel, J. M. 1997 Applications of cumulants to array processing. III. blind beamforming for coherent signals. *IEEE Transactions on Signal Processing* **45**(9), 2252–2264.

Goodwin MM and Elko GW 1993 Constant beamwidth beamforming *Proceedings of the IEEE International Conference on Acoustics, Speech and Signal Processing*, Vol. 1, pp. 169–172, Minneapolis, MN, USA.

Griffiths LJ and Jim CW 1982 An alternative approach to linearly constrained adaptive beamforming. *IEEE Transactions on Antennas and Propagation* **30**(1), 27–34.

Harteneck M and Stewart RW 1997 Filterbank Design for Oversampled Filter Banks Without Aliasing in the Subbands. *IEE Electronics Letters* **38**(18), 1538–1539.

Harteneck M, Páez-Borrallo JM and Stewart RW 1998 An Oversampled Subband Adaptive Filter Without Cross Adaptive Filters. *Signal Processing* **64**(1), 93–101.

Harteneck M, Weiss S and Stewart R 1999 Design of near perfect reconstruction oversampled filter banks for subband adaptive filters. *IEEE Transactions on Circuits and Systems – II: Analog and Digital Signal Processing* **46**(8), 1081–1085.

Haykin S 1985 *Array Signal Processing*, Prentice-Hall Signal Processing Series. Prentice-Hall, Englewood Cliffs, NJ, USA.

Haykin S 1996 *Adaptive Filter Theory*, 3rd Edn. Prentice-Hall, Englewood Cliffs, NJ, USA.

Haykin S (Ed.) 2000a *Unsupervised Adaptive Filtering*, Vol. 1: *Blind Source Separation*. John Wiley & Sons, Inc., New York, NY, USA.

Haykin S 2000b *Unsupervised Adaptive Filtering*, Vol. 2: *Blind Deconvolution*. John Wiley & Sons, Inc., New York, NY, USA.

Hixson EL and Au KT 1970 Wide-bandwidth constant beamwidth acoustic array. *Journal of the Acoustic Society of America* **48**(1), 117.

Holland J 1975 *Adaptation in Natural and Artificial Systems*. The University of Michigan Press, Ann Arbor, MI, USA.

Horn RA and Johnson CR 1985 *Matrix Analysis*. Cambridge University Press, Cambridge, UK.

Hoshuyama O, Sugiyama A and Hirano A 1997 A robust generalized sidelobe canceller with a blocking matrix using leaky adaptive filters. *Electronics and Communications in Japan (Part III: Fundamental Electronic Science)* **80**(8), 56–65.

Hoshuyama O, Sugiyama A and Hirano A 1999a A robust adaptive beamformer for microphone arrays with a blocking matrix using constrained adaptive filters. *IEEE Transaction on Signal Processing* **47**(10), 2677–2684.

Hoshuyama O, Sugiyama A and Hirano A 1999b A robust adaptive beamformer with a blocking matrix using coefficients constrained adaptive filters. *IEICE Transactions. Fundamentals of Electronics, Communications and Computer Sciences* **E82-A**(4), 640–647.

Houck C, Joines J and Kay M 1995 A genetic algorithm for function optimization: A MATLAB© implementation. Technical report, North Carolina State University, Rayleigh, NC, USA.

Huang X, Wu HC and Principe JE 2007 Robust blind beamforming algorithm using joint multiple matrix diagonalization. *IEEE Sensors Journal* **7**(1), 130–136.

Huang Y, Benesty J and Chen J 2006 *Acoustic MIMO Signal Processing* Signals and Communication Technology. Springer-Verlag, New York, NY, USA.

Huarng K and Yeh C 1992 Performance Analysis of Derivative Constraint Adaptive Arrays with Pointing Errors. *IEEE Transactions on Antennas and Propagation* **40**(8), 975–981.

Hudson JE 1981 *Adaptive Array Principles*, IEE Electromagnetic Waves. The Institution of Electrical Engineers, London, UK.

Hung HS and Mao CY 1994 Robust coherent signal-subspace processing for directions-of-arrival estimation of wideband sources. *IEE Proceedings – Radar, Sonar and Navigation* **141**(5), 256–262.

Hyvarinen A and Oja E 1997 A fast fixed-point algorithm for independent component analysis. *Neural Computation* **9**(7), 1483–1492.

Hyvarinen A, Karhunen J and Oja E 2001 *Independent Component Analysis*. John Wiley & Sons, Inc., New York, NY, USA.

Indukumar KC and Reddy VU 1990 Broad-band DOA estimation and beamforming in multipath environment *Proceedings of the IEEE International Radar Conference*, pp. 532–537, Arlington, VA, USA.

Inouye Y and Liu RW 2002 A system-theoretic foundation for blind equalization of an FIR MIMO channel system. *IEEE Transactions on Circuits and Systems – I: Fundamental Theory and Applications* **49**(4), 425–436.

Jablon NK 1986 Steady state analysis of the generalized sidelobe canceler by adaptive noise canceling techniques. *IEEE Transactions on Antennas and Propagation* **AP-34**(3), 330–337.

Jafari M, Alty S and Chambers J 2004 New natural gradient algorithm for cyclostationary sources. *IEE Proceedings – Vision, Image and Signal Processing* **151**(1), 62–68.

Jafari M, Alty S and Chambers J 2006 Sequential blind source separation based exclusively on second order statistics developed for a class of periodic signals. *IEEE Transactions on Signal Processing* **54**(3), 1028–1040.

Jim CW 1977 A comparison of two LMS constrained optimal array structures. *Proceedings of the IEEE* **65**(12), 1730–1731.

Johnson CR, Schniter P, Endres TJ, Behm JD, Brown DR and Casas RA 1998 Blind equalization using the constant modulus criterion: A review. *Proceedings of the IEEE* **86**(10), 1927–1950.

Johnson DH 1982 The application of spectral estimation methods to bearing estimation problems. *Proceedings of the IEEE* **70**(9), 1018–1028.

Johnson DH and Dudgeon DE 1993 *Array Signal Processing: Concepts and Techniques*, Signal Processing Series. Prentice-Hall, Englewood Cliffs, NJ, USA.

Kajala M and Hämäläinen M 1999 Broadband beamforming optimization for speech enhancement in noisy environments *Proceedings of the IEEE Workshop on Applications of Signal Processing to Audio and Acoustics*, pp. 19–22, New York, NY, USA.

Kamp Y and Thiran J 1975 Chebyshev approximation for two-dimensional nonrecursive digital filters. *IEEE Transactions on Circuits and Systems* **CAS-22**(3), 208–218.

Kaneda Y and Ohga J 1986 Adaptive microphone-array system for noise reduction. *IEEE Transactions on Acoustics, Speech and Signal Processing* **ASSP-34**(6), 1391–1400.

Kazanci OR and Krolik JL 2005 Beamspace adaptive channel compensation for sensor arrays with faulty elements *Proceedings of the Asilomar Conference on Signals, Systems and Computers*, pp. 1316–1319, Pacific Grove, CA, USA.

Kellermann W 1988 Analysis and Design of Multirate Systems for Cancellation of Acoustical Echoes *Proceedings of the IEEE International Conference on Acoustics, Speech and Signal Processing*, Vol. 5, pp. 2570–2573, New York, NY, USA.

Kellermann W and Buchner H 2003 Wideband algorithms versus narrowband algorithms for adaptive filtering in the dft domain *Proceedings of the Asilomar Conference on Signals, Systems and Computers*, pp. 81–84, Pacific Grove, CA, USA.

Khalab JM and Ibrahim MK 1994 Novel Multirate Adaptive Beamforming Technique. *IEE Electronics Letters* **30**(15), 1194–1195.

Khalab JM and Woolfson MS 1994 Efficient Multirate Adaptive Beamforming Technique. *IEE Electronics Letters* **30**(25), 2102–2103.

Kikuma N and Takao K 1989 Broadband and robust adaptive antenna under correlation constraint. *IEE Proceedings, Part H* **136**(2), 85–89.

Kim SJ and Jones JD 1991 Optimal design of piezoactuators for active noise and vibration control. *AIAA journal* **29**(12), 2047–2053.

Koilpillai RD and Vaidyanathan PP 1992 Cosine-modulated FIR Filter Banks Satisfying Perfect Reconstruction. *IEEE Transactions on Signal Processing* **40**(4), 770–783.

Korompis D, Yao K and Lorenzelli F 1994 Broadband maximum energy array with user imposed spatial and frequency constraints *Proceedings of the IEEE International Conference on Acoustics, Speech and Signal Processing*, Vol. 4, pp. 529–532, Adelaide, SA, Australia.

Krim H and Viberg M 1996 Two decades of array signal processing research: the parametric approach. *IEEE Signal Processing Magazine* **13**(4), 67–94.

Krolik J and Swingler D 1990 Focused wide-band array processing by spatial resampling. *IEEE Transactions on Acoustics, Speech and Signal Processing* **38**(2), 356–360.

Lau BK, Leung YH, Teo KL and Steeram V 1999 Minimax filters for microphone arrays. *IEEE Transactions on Circuits and Systems – II: Analog and Digital Signal Processing* **46**(12), 1522–1524.

Lawson CL and Hanson RJ 1974 *Solving Least Squares Problems*, Automatic Computation. Prentice-Hall, Englewood Cliffs, NJ, USA.

Lebret H and Boyd S 1997 Antenna array pattern synthesis via convex optimization. *IEEE Transactions on Signal Processing* **45**(3), 526–532.

Lee J and Un CK 1986 Performance of Transform-domain LMS Adaptive Digital Filters. *IEEE Transactions on Acoustics, Speech and Signal Processing* **34**(3), 499–510.

Lee TS 1994 Efficient wideband source localization using beamforming invariance technique. *IEEE Transactions on Signal Processing* **42**(6), 1376–1387.

Lee TS and Tsai TC 2001 A beamspace-time interference cancelling cdma receiver for sectored communications in a multipath environment. *IEEE Journal on Selected Areas in Communications* **19**(7), 1374–1384.

Lee YP and Freese H 2005 Subarray beam-space adaptive beamforming for a dynamic long towed-array *Proceedings of the Workshop on Adaptive Sensor Array Processing*, Lexington, MA, USA.

Li J and Stoica P (Eds) 2005 *Robust Adaptive Beamforming*. John Wiley & Sons, Inc., Hoboken, NJ, USA.

Li Y and Liu KJ 1998 Adaptive blind source separation and equalization for multiple-input/multiple-output systems. *IEEE Transactions on Information Theory* **44**(7), 2864–2876.

Li YQ and Wang J 2002 Sequential blind extraction of instantaneously mixed sources. *IEEE Transactions on Signal Processing* **50**(5), 997–1006.

Lim Y and Parker S 1983 FIR Filter Design over a Discrete Power-of-two Coefficient Space. *IEEE Transactions on Acoustics, Speech and Signal Processing* **ASSP-31**(3), 583–591.

Lim Y, Yang R, Li D and Song J 1983 Discrete Coefficient FIR Digital Filter Design Based upon LMS Criteria. *IEEE Transactions on Circuits and Systems* **CAS-30**(10), 723–739.

Lin N, Liu W and Langley R 2007 Performance analysis of a broadband beamforming structure without tapped delay-lines *Proceedings of the International Conference on Digital Signal Processing*, pp. 583–586, Cardiff, UK.

Lin N, Liu W and Langley R 2008 Performance analysis of a two-element linearly constrained minimum variance beamformer with sensor delay-line processing *Proceedings of the IEEE International Conference on Acoustics, Speech and Signal Processing*, pp. 2601–2604, Las Vegas, NV, USA.

Lin N, Liu W and Langley RJ 2010 Performance analysis of an adaptive broadband beamformer based on a two-element linear array with sensor delay-line processing. *Signal Processing* **90**(1), 269–281.

Liu W 2000 *On the Design of Multiplier-less Perfect Reconstruction Filter Banks Using Genetic Algorithm and Sum-of-powers-of-two Representation. M.Phil. Thesis*, Department of Electrical and Electronic Engineering, University of Hong Kong, Hong Kong.

Liu W 2003 *Digital Beamforming Employing Subband Techniques. Ph.D. Thesis*, School of Electronics and Computer Science, University of Southampton, Southampton, UK.

Liu W 2007 Adaptive broadband beamforming with spatial-only information *Proceedings of the International Conference on Digital Signal Processing*, pp. 575–578, Cardiff, UK.

Liu W 2009a Adaptive wideband beamforming with sensor delay-lines. *Signal Processing* **89**(5), 876–882.

Liu W 2009b Blind adaptive beamforming for wideband circular arrays *Proceedings of the IEEE International Conference on Acoustics, Speech and Signal Processing*, pp. 2029–2032, Taipei, Taiwan.

Liu W 2009c Design of a rectangular frequency invariant beamformer with a full azimuth angle coverage *Proceedings of the European Signal Processing Conference*, pp. 579–582, Glasgow, UK.

Liu W 2009d A novel approach to adaptive beamforming for multi-path broadband signals *Proceedings of the IEEE Workshop on Statistical Signal Processing*, pp. 197–200, Cardiff, UK.

Liu W 2010 Blind Beamforming for Multi-path Wideband Signals Based on Frequency Invariant Transformation *Proceedings of the International Symposium on Communications, Control and Signal Processing*, Limassol, Cyprus.

Liu W and Langley R 2007 Adaptive wideband beamforming with combined spatial/temporal subband decomposition *Proceedings of the Progress in Electromagnetics Research Symposium*, pp. 1386–1391, Beijing, China.

Liu W and Langley RJ 2009 An adaptive wideband beamforming structure with combined subband decomposition. *IEEE Transactions on Antennas and Propagation* **89**(7), 913–920.

Liu W and Mandic DP 2005 Semi-blind source separation for convolutive mixtures based on frequency invariant transformation *Proceedings of the IEEE International Conference on Acoustics, Speech and Signal Processing*, Vol. 5, pp. 285–288, Philadelphia, PA, USA.

Liu W and Mandic DP 2006 A normalised kurtosis based algorithm for blind source extraction from noisy measurements. *Signal Processing* **86**(7), 1580–1585.

Liu W and Weiss S 2004a Frequency invariant beamforming in subbands *Proceedings of the Asilomar Conference on Signals, Systems and Computers*, Vol. 2, pp. 1968–1972, Pacific Grove, CA, USA.

Liu W and Weiss S 2004b A new class of broadband arrays with frequency invariant beam patterns *Proceedings of the IEEE International Conference on Acoustics, Speech and Signal Processing*, Vol. 2, pp. 185–188, Montreal, QU, Canada.

Liu W and Weiss S 2005 Design of frequency-invariant beamformers employing multi-dimensional fourier transforms *Proceedings of the Fourth International Workshop on Multidimensional (nD) Systems (NDS)*, pp. 19–23, Wuppertal, Germany.

Liu W and Weiss S 2008a Broadband beamspace adaptive beamforming with spatial-only information *Proceedings of the IEEE Workshop on Sensor Array and Multichannel Signal Processing*, pp. 330–334, Darmstadt, Germany.

Liu W and Weiss S 2008b Design of frequency invarint beamformers for broadband arrays. *IEEE Transactions on Signal Processing* **56**(2), 855–860.

Liu W and Weiss S 2008c Off-broadside main beam design for frequency invariant beamformers *Proceedings of the International ITG/IEEE Workshop on Smart Antennas*, pp. 190–194, Darmstadt, Germany.

Liu W and Weiss S 2009a Beam steering for wideband arrays. *Signal Processing* **89**(5), 941–945.

Liu W and Weiss S 2009b Off-broadside main beam design and subband implementation for a class of frequency invariant beamformers. *Signal Processing* **89**(5), 913–920.

Liu W, Chan SC and Ho KL 2000a Low-delay perfect reconstruction two-channel fir/iir filter banks and wavelet bases with sopot coefficients *Proceedings of the IEEE International Conference on Acoustics, Speech and Signal Processing*, Vol. 1, pp. 109–112, Istanbul, Turkey.

Liu W, Chan SC and Ho KL 2000b Multiplier-less low-delay FIR and IIR wavelet filter banks with SOPOT coefficients *Proceedings of the European Signal Processing Conference*, Vol. 3, p. 289, Tampere, Finland.

Liu W, Mandic DP and Cichocki A 2006 Blind second-order source extraction of instantaneous noisy mixtures. *IEEE Transactions on Circuits and Systems II: Express Briefs* **53**(9), 931–935.

Liu W, Mandic DP and Cichocki A 2007a Analysis and online realization of the CCA approach for blind source separation. *IEEE Transactions on Neural Networks* **18**(5), 1505–1510.

Liu W, Mandic DP and Cichocki A 2007b Blind source extraction based on a linear predictor. *IET Signal Processing* **1**(1), 29–34.

Liu W, Mandic DP and Cichocki A 2008a Blind source separation based on generalised canonical correlation analysis and its adaptive realization *Proceedings of the International Congress on Image and Signal Processing*, Vol. 5, pp. 417–421, Hainan, China.

Liu W, Mandic DP and Cichocki A 2008b A dual-linear predictor approach to blind source extraction for noisy mixtures *Proceedings of the IEEE Workshop on Sensor Array and Multichannel Signal Processing*, pp. 515–519, Darmstadt, Germany.

Liu W, McLernon D and Ghogho M 2009 Design of frequency invariant beamformer without temporal filtering. *IEEE Transactions on Signal Processing* **57**(2), 798–802.

Liu W, Weiss S and Hanzo L 2001a A novel method for partially adaptive broadband beamforming *Proceedings of the IEEE Workshop on Signal Processing Systems*, pp. 361–372, Antwerp, Belgium.

Liu W, Weiss S and Hanzo L 2001b Subband adaptive generalized sidelobe canceller for broadband beamforming *Proceedings of the IEEE Workshop on Statistical Signal Processing*, pp. 591–594, Singapore.

Liu W, Weiss S and Hanzo L 2002a Low-complexity frequency-domain GSC for broadband beamforming *Proceedings of the International Conference on Signal Processing*, Vol. 1, pp. 386–389, Beijing, China.

Liu W, Weiss S and Hanzo L 2002b Partially adaptive broadband beamforming with a subband-selective tranformation matrix *Proceedings of the IEEE Workshop on Sensor Array and Multichannel Signal Processing*, pp. 43–47, Arlington, VA, USA.

Liu W, Weiss S and Hanzo L 2002c Subband-selective partially adaptive broadband beamforming with cosine-modulated blocking matrix *Proceedings of the IEEE International Conference on Acoustics, Speech and Signal Processing*, Vol. 3, pp. 2913–2916, Orlando, FL, USA.

Liu W, Weiss S and Hanzo L 2003a A novel method for partially adaptive broadband beamforming *The Journal of VLSI Signal Processing – Systems for Signal, Image and Video Technology*, **33**(3), 337–344.

Liu W, Weiss S and Hanzo L 2003b An subband-selective transform-domain GSC with low computational complexity *Proceedings of the Postgraduate Research Conference in Electronics, Photonics, Communications and Software*, pp. 31–32, Exeter University, Exeter, UK.

Liu W, Weiss S and Hanzo L 2004a A subband-selective broadband GSC with cosine-modulated blocking matrix. *IEEE Transactions on Antennas and Propagation* **52**(3), 813–820.

Liu W, Weiss S and Hanzo L 2005 A generalised sidelobe canceller employing two-dimensional frequency invariant filters. *IEEE Transactions on Antennas and Propagation* **53**(7), 2339–2343.

Liu W, Weiss S and Koh CL 2004b Constrained adaptive broadband beamforming algorithm in frequency domain *Proceedings of the IEEE Workshop on Sensor Array and Multichannel Signal Processing*, pp. 94–98, Barcelona, Spain.

Liu W, Weiss S, McWhirter JG and Proudler IK 2007c Frequency invariant beamforming for two-dimensional and three-dimensional arrays. *Signal Processing* **87**(11), 2535–2543.

Liu W, Wu R and Langley R 2007d Design and analysis of broadband beamspace adaptive arrays. *IEEE Transactions on Antennas and Propagation* **55**(12), 3413–3420.

Lofberg J 2004 YALMIP: a toolbox for modeling and optimization in MATLAB© *Proceedings of the IEEE International Symposium on Computer Aided Control Systems Design*, pp. 284–289, Taipei, Taiwan.

Lorenz RG and Boyd SP 2005 Robust minimum variance beamforming. *IEEE Transactions on Signal Processing* **53**(5), 1684–1696.

Lorenzelli F, Wang A, Korompis D, Hudson R and Yao K 1996 Subband Processing for Broadband Microphone Arrays. *Journal of VLSI Signal Processing Systems for Signal, Image and Video Technology* **14**(1), 43–55.

Lu Y and Morris JM 1999 Gabor Expansion for Adaptive Echo Cancellation. *IEEE Signal Processing Magazine* **16**(2), 68–80.

Makhoul J 1981 On the eigenvectors of symmetric toeplitz matrices. *IEEE Transactions on Acoustics, Speech and Signal Processing* **29**(4), 868–872.

Malouche Z and Macchi O 1998 Adaptive unsupervised extraction of one component of a linear mixture with a single neuron. *IEEE Transactions on Neural Networks* **9**(1), 123–138.

Malvar HS 1992 Extended Lapped Transforms: Properties, Applications and Fast Algorithms. *IEEE Transactions on Signal Processing* **40**(11), 2703–2714.

Man KF, Tang KS, Wong S and Halang W 1997 *Genetic Algorithms for Control and Signal Processing*. Springer-Verlag, London, UK.

Mani VV and Base R 2008 Smart antenna design for beamforming of UWB signals in Gaussian noise *Proceedings of the International ITG/IEEE Workshop on Smart Antennas*, pp. 311–316, Darmstadt, Germany.

Mat 2001 *MATLAB© 6.1*. The Math Works, Inc., Natick, MA, USA.

Mathews CP and Zoltowski MD 1994 Eigenstructure techniques for 2-D angle estimation with uniform circular arrays. *IEEE Transactions on Signal Processing* **42**(9), 2395–2407.

Matsuoka K and Nakashima S 2001 Minimal distortion principle for blind source separation *Proceedings of the Third International Symposium on Independent Component Analysis and Blind Signal Separation*, pp. 722–727, San Diego, CA, USA.

Matsuoka K, Ohya M and Kawamoto M 1995 A neural net for blind separation of nonstationary signal sources. *Neural Networks* **8**(3), 411–419.

Mayhan JT, Simmons AJ and Cummings WC 1981 Wide-band adaptive antenna nulling using tapped delay lines. *IEEE Transactions on Antennas and Propagation* **AP-29**(6), 923–936.

Mintzer F 1985 Filter for Distortion-free Two-band Multirate Filter Banks. *IEEE Transactions on Acoustics, Speech and Signal Processing* **ASSP-33**(3), 626–630.

Molgedey L and Schuster HG 1994 Separation of a mixture of independent signals using time delayed correlations. *Physics Review Letters* **72**(23), 3634–3637.

Monzingo RA and Miller TW 2004 *Introduction to Adaptive Arrays*. SciTech Publishing Inc., Raleigh, NC, USA.

Morgan DR 2008 Analysis and realization of an exponentially-decaying impulse response model for frequency-selective fading channels. *IEEE Signal Processing Letters* **15**, 441–444.

Naguib AF, Seshadri N and Calderbank A 2000 Increasing data rate over wireless channels: Space time coding and signal processing for high data rate wireless communications. *IEEE Signal Processing Magazine* **17**(3), 76–92.

Neo W and Farhang-Boroujeny B 2002 Robust microphone arrays using subband adaptive filters. *IEE Proceedings – Vision, Image and Signal Processing* **149**(1), 17–25.

Nesterov Y and Nemirovskii A 1994 *Interior-Point Polynomial Algorithms in Convex Programming*, Vol. 13, *Studies in Applied Mathematics*. SIAM, Philadelphia, PA, USA.

Neuvo Y, Cheng-Yu D and Mitra SK 1984 Interpolated Finite Impulse Response Filters. *IEEE Transactions on Acoustics, Speech and Signal Processing* **32**(6), 563–570.

Nguyen TQ 1993 Design of arbitrary fir digital filters using the eigenfilter method. *IEEE Transactions on Signal Processing* **41**(3), 1128–1139.

Nordebo S, Claesson I and Nordholm S 1994 Weighted chebyshev approximation for the design of broadband-beamformers using quadratic programming. *IEEE Signal Processing Letters* **1**(7), 103–105.

Nordholm S, Claesson I and Eriksson P 1992 The broad-band Wiener solution for Griffiths–Jim beamformers. *IEEE Transactions on Signal Processing* **40**(2), 474–478.

Nordholm S, Rehbock V, Tee K and Nordebo S 1998 Chebyshev optimization for the design of broadband beamformers in the near field. *IEEE Transactions on Circuits and Systems II: Analog and Digital Signal Processing* **45**(1), 141–143.

Oestges C and Clerckx B 2007 *MIMO Wireless Communications: From Real-World Propagation to Space-Time Code Design*. Academic Press, London, UK.

Olen CA and Compton RT 1990 A numerical pattern synthesis algorithm for arrays. *IEEE Transactions on Antennas and Propagation* **38**(10), 1666–1676.

Oppenheim AV and Schafer RW 1975 *Digital Signal Processing*. Prentice-Hall, Englewood Cliffs, NJ, USA.

Owsley NL 1985 Sonar array processing, in *Array Signal Processing* (Haykin S (Ed.)), Chapter 3, pp. 115–184. Prentice-Hall, Englewood Cliffs, NJ, USA.

Parra L and Spence C 2000 Convolutive blind separation of non-stationary sources. *IEEE Transactions on Speech and Audio Processing* **8**(3), 320–327.

Parra LC 2006 Steerable frequency-invariant beamforming for arbitrary arrays. *Journal of the Acoustic Society of America* **119**(6), 3839–3847.

Pei SC and Shyu JJ 1990 2-d fir eigenfilters: A least-squares approach. *IEEE Transactions on Circuits and Systems* **37**(1), 24–34.

Pei SC and Shyu JJ 1993 Complex eigenfilter design of arbitrary complex coefficient fir digital filters. *IEEE Transactions on Circuits and Systems II: Analog and Digital Signal Processing* **40**(1), 32–40.

Pei SC and Tseng CC 2001 A new eigenfilter based on total least squares error criterion. *IEEE Transactions on Circuits and Systems I: Regular Papers* **48**(6), 699–709.

Pei SC, Tseng CC and Yang WS 1998 Fir filter designs with linear constraints using the eigenfilter approach. *IEEE Transactions on Circuits and Systems – II: Analog and Digital Signal Processing* **45**(2), 232–237.

Pham DT and Cardoso JF 2001 Blind separation of instantaneous mixtures of nonstationary sources. *IEEE Transactions on Signal Processing* **49**(9), 1837–1848.

Pridham RG and Mucci RA 1978 A novel approach to digital beamforming. *Journal of the Acoustic Society of America* **63**(2), 425–434.

Pridham RG and Mucci RA 1979 Digital interpolation beamforming for low-pass and bandpass signals. *Proceedings of the IEEE* **67**(6), 904–919.

Reddy VU, Paulraj A and Kailath T 1987 Performance analysis of the optimum beamformer in the presence of correlated sources and its behavior under spatial smoothing. *IEEE Transactions on Acoustics, Speech and Signal Processing* **35**(7), 927–936.

Redmill DW, Bull DR and Martin RR 1997 Design of Multiplier Free Linear Phase Perfect Reconstruction Filter Banks Using Transformations and Genetic Algorithms *Proceedings of the Sixth International Conference on Image Processing and Its Applications*, Vol. 2, pp. 766–770, Dublin, Ireland.

Remmert R 1991 *Theory of Complex Functions* Graduate Texts in Mathematics. Springer-Verlag, New York, NY, USA.

Rodgers WE and Compton, Jr RT 1979 Adaptive array bandwidth with tapped delay-line processing. *IEEE Transactions on Areospace Electronic Systems* **AES-15**(1), 21–27.

Roy R and Kailath T 1989 ESPRIT-estimation of signal parameters via rotational invariancetechniques. *IEEE Transactions on Acoustics, Speech and Signal Processing* **37**(7), 984–995.

Rübsamen M and Gershman AB 2008 Robust presteered broadband beamforming based on worst-case performance optimization *Proceedings of the IEEE Workshop on Sensor Array and Multichannel Signal Processing*, pp. 340–344, Darmstadt, Germany.

Salmon N, Beale J, Parkinson J, Hayward S, Hall P, Macpherson R, Lewis R and Harvey A 2007 Digital beam-forming for passive millimetre wave security imaging *Proceedings of the European Conference on Antennas and Propagation*, pp. 1–11, Edinburgh, UK.

Sawada H, Mukai R, Araki S and Makino S 2004 Convolutive blind source separation for more than two sources in the frequency domain *Proceedings of the IEEE International Conference on Acoustics, Speech and Signal Processing*, Vol. III, pp. 885–888, Montreal, QA, Canada.

Schafer RW and Rabiner LR 1973 A digital signal processing approach to interpolation. *Proceedings of the IEEE* **61**(12), 692–720.

Schmidt R 1986 Multiple emitter location and signal parameter estimation. *IEEE Transactions on Antennas and Propagation* **34**(3), 276–280.

Scholnik DP and Coleman JO 2000a Formulating wideband array-pattern optimizations *Proceedings of the IEEE International Conference on Phased Array Systems and Technology*, pp. 489–492, Dana Point, CA, USA.

Scholnik DP and Coleman JO 2000b Optimal design of wideband array patterns *Proceedings of the IEEE International Radar Conference*, pp. 172–177, Washington, DC, USA.

Scholnik DP and Coleman JO 2001 Superdirectivity and SNR constraints in wideband array-pattern design *Proceedings of the IEEE International Radar Conference*, pp. 181–186, Atlanta, GA, USA.

Schönle M, Fliege NJ and Zölzer U 1993 Parametric Approximation of Room Impulse Responses by Multirate Systems *Proceedings of the IEEE International Conference on Acoustics, Speech and Signal Processing*, Vol. I, pp. 153–156, Minneapolis, MN, USA.

Sekiguchi T and Karasawa Y 2000 Wideband beamspace adaptive array utilizing FIR fan filters for multibeam forming. *IEEE Transactions on Signal Processing* **48**(1), 277–284.

Shan TJ and Kailath T 1985 Adaptive beamforming for coherent signals and interference. *IEEE Transactions on Acoustics, Speech and Signal Processing* **33**(3), 527–536.

Sheinvald J 1998 On blind beamforming for multiple non-Gaussian signals and the constant-modulus algorithm. *IEEE Transactions on Signal Processing* **46**(7), 1878–1885.

Shynk JJ 1992 Frequency-domain and multirate adaptive filtering. *IEEE Signal Processing Magazine* **9**(1), 14–37.

Simanapalli S and Kaveh M 1994 Broadband focusing for partially adaptive beamforming. *IEEE Transactions on Aerospace and Electronic Systems* **30**(1), 68–80.

Skidmore ID and Proudler IK 2001 KAGE: a New Fast RLS Algorithm *Proceedings of the IEEE International Conference on Acoustics, Speech and Signal Processing*, Vol. 6, pp. 3773–3776, Salt Lake City, UT, USA.

Slavakis K and Yamada I 2007 Robust wideband beamforming by the hybrid steepest descent method. *IEEE Transactions on Signal Processing* **55**(9), 4511–4522.

Smaragdis P 1998 Blind separation of convolved mixtures in the frequency domain. *Nerocomputing* **22**(1), 21–34.

Smith MJT and Barnwell III TP 1986 Exact reconstruction techniques for tree-structured sub-band coders. *IEEE Transactions on Acoustics, Speech and Signal Processing* **34**(3), 434–441.

Smith R 1970 Constant beamwidth receiving arrays for broad band sonar systems. *Acustica* **23**, 21–26.

Somayazulu V, Mitra S and Shynk J 1989 Adaptive Line Enhancement Using Multirate Techniques *Proceedings of the IEEE International Conference on Acoustics, Speech and Signal Processing*, Vol. 2, pp. 928–931, Glasgow, UK.

Sriranganathan S, Bull DR and Redmill RW 1997 The Design of Low Complexity Two-channel Lattice-structure Perfect-reconstruction Filter Banks Using Genetic Algorithms *Proceedings of the IEEE International Symposium on Circuits and Systems*, Vol. 4, pp. 2393–2396, Hong Kong.

Steffen P, Heller PN, Gopinath RA and Burrus CS 1993 Theory of Regular M-Band Wavelet Bases. *IEEE Transactions on Signal Processing* **SP-41**(12), 3497–3511.

Stewart GW 1973 *Introduction to Matrix Computations*. Academic Press, New York, NY, USA.

Stewart GW 1993 On the early history of the singular value decomposition. *SIAM Review* **35**(4), 551–566.

Steyskal H 1983 Wide-band nulling performance versus number of pattern constraints for an array antenna. *IEEE Transactions on Antennas and Propagation* **AP-31**(1), 159–163.

Strang G 1980 *Linear Algebra and Its Applications*, 2nd Edn. Academic Press, New York, NY, USA.

Strang G and Nguyen T 1996 *Wavelets and Filter Banks*. Wellesley-Cambridge Press, Wellesley, MA, USA.

Sturm JF 1999 Using SeDuMi 1.02, a MATLAB© toolbox for optimization over symmetric cones. *Optimization Methods and Software* **11**(1), 625–653.

Takao K and Komiyama K 1980 An adaptive antenna for rejection of wide-band interference. *IEEE Transactions on Aerospace and Electronic Systems* **AES-16**(7), 452–459.

Takao K and Uchida K 1989 Beamspace partially adaptive antenna. *IEE Proceddings H, Microwaves, Antennas and Propagation* **136**(6), 439–444.

Tang KS, Man KF, Kwong S and He Q 1996 Genetic Algorithms and Their Applications. *IEEE Signal Processing Magazine* **13**(6), 22–37.

Tanrikulu O, Baykal B, Constantinides AG and Chambers J 1997 Residual Echo Signal in Critically Sampled Subband Acoustic Echo Cancellers Based on IIR and FIR Filter Banks. *IEEE Transactions on Signal Processing* **45**(4), 901–912.

Tarokh V, Seshadri N and Calderbank AR 1998 Space-time codes for high data rate wireless communication: Performance criterion and code construction. *IEEE Transactions on Information Theory* **44**(2), 744–765.

Telatar IE 1999 European transactions on telecommunications. *European Transactions on Telecommunications* **10**(6), 585–595.

Thng I, Cantoni A and Leung Y 1993 Derivative constrained optimum broad-band antenna arrays. *IEEE Transactions on Signal Processing* **41**(7), 2376–2388.

Thng I, Cantoni A and Leung YH 1995 Constraints for maximally flat optimum broadband antenna arrays. *IEEE Transactions on Signal Processing* **43**(6), 1334–1347.

Tkacenko A, Vaidyanathan PP and Nguyen TQ 2003 On the eigenfilter design method and its applications: a tutorial. *IEEE Transactions on Circuits and Systems – II: Analog and Digital Signal Processing* **50**(9), 497–517.

Tong L, Sadler BM and Dong M 2004 Pilot-assisted wireless transmissions: general model, design criteria and signal processing. *Signal Processing Magazine, IEEE* **21**(6), 12–25.

Trucco A, Crocco M and Repetto S 2006 A stochastic approach to the synthesis of a robust frequency-invariant filter-and-sum beamformer. *IEEE Transactions on Instrumentation and Measurement* **55**(4), 1407–1415.

Unnikrishna Pillai S and Kwon BH 1989 Forward/backward spatial smoothing techniques for coherent signal identification. *IEEE Transactions on Acoustics, Speech and Signal Processing* **37**(1), 8–15.

Vaidyanathan P and Nguyen TQ 1987 Eigenfilters: A new approach to least-squares FIR filter design and applications including Nyquist filters. *IEEE Transactions on Circuits and Systems* **34**(1), 11–23.

Vaidyanathan PP 1993 *Multirate Systems and Filter Banks*. Prentice-Hall, Englewood Cliffs, NJ, USA.

Van Trees HL 2002 *Optimum Array Processing, Part IV of Detection, Estimation and Modulation Theory*. John Wiley & Sons, Inc., New York, NY, USA.

Van Veen BD 1991 Minimum variance beamforming with soft response constraints. *IEEE Transactions on Signal Processing* **39**(9), 1964–1972.

Van Veen BD and Buckley KM 1988 Beamforming: a versatile approach to spatial filtering. *IEEE Acoustics, Speech and Signal Processing Magazine* **5**(2), 4–24.

Van Veen BD and Roberts RA 1987 Partially adaptive beamformer design via output power minimization. *IEEE Transactions on Acoustics, Speech and Signal Processing* **35**(11), 1524–1532.

Vary P 1979 On the Design of Digital Filter Banks Based on a Modified Principle of Polyphase. *Archiv Elektrische Übertragung* **33**(7/8), 293–300.

Vaughan RG, Scott NL and White DR 1991 The Theory of Bandpass Sampling. *IEEE Transactions on Signal Processing* **39**(9), 1973–1984.

Vis 2002 *Visual Numerics IMSL Fortran Numerical Libraries*, Houston, TX, USA.

Vook EW and Compton, Jr RT 1992 Bandwidth performance of linear adaptive arrays with tapped delay-line processing. *IEEE Transactions on Aerospace and Electronic Systems* **28**(3), 901–908.

Vorobyov SA, Gershman AB and Luo ZQ 2003 Robust adaptive beamforming using worst-case performance optimization: A solution to the signal mismatch problem. *IEEE Transactions on Signal Processing* **51**(2), 313–324.

Walach E 1984 On superresolution effects in maximum likelihood adaptive antenna arrays. *IEEE Transactions on Antennas and Propagation* **AP-32**(3), 259–263.

Wang H and Kaveh M 1985 Coherent signal-subspace processing for the detection and estimation of angles of arrival of multiple wide-band sources. *IEEE Transactions on Acoustics, Speech and Signal Processing* **33**(4), 823–831.

Wang YY and Fang WH 2000 Wavelet-based broadband beamformers with dynamic subband selection. *IEICE Transactions and Communications* **E83B**(4), 819–826.

Wang YY, Fang WH and Chen JT 1999 Improved wavelet-based beamformers with dynamic subband selection *Proceedings of the International Symposium on Antennas and Propagation*, Vol. 2, pp. 1464–1467, Orlando, FL, USA.

Ward DB, Ding Z and Kennedy RA 1998 Broadband DOA estimation using frequency invariant beamforming. *IEEE Transactions on Signal Processing* **46**(5), 1463–1469.

Ward DB, Kennedy RA and Williamson RC 1995 Theory and design of broadband sensor arrays with frequency invariant far-field beam patterns. *Journal of the Acoustic Society of America* **97**(2), 1023–1034.

Wax M, Shan TJ and Kailath T 1984 Spatio-temporal spectral analysis by eigenstructure methods. *IEEE Transactions on Acoustics, Speech and Signal Processing* **ASSP-32**(4), 817–827.

Weiss S and Proudler IK 2002 Comparing efficient broadband beamforming architectures and their performance trade-offs *Proceedings of the International Conference on Digital Signal Processing*, Vol. 1, pp. 409–415, Hellas, Greece.

Weiss S and Stewart RW 1998 *On Adaptive Filtering in Oversampled Subbands*. Shaker Verlag, Aachen, Germany.

Weiss S and Stewart RW 1999 On the Optimality of Subband Adaptive Systems *Proceedings of the IEEE Workshop on Applications of Signal Processing to Audio and Acoustics*, pp. 59–62, New Paltz, NY, USA.

Weiss S and Stewart RW 2000 Fast implementation of oversampled modulated filter banks. *IEE Electronics Letters* **36**(17), 1502–1503.

Weiss S, Dooley SR, Stewart RW and Nandi AK 1998a Adaptive Equalization in Oversampled Subbands. *IEE Electronics Letters* **34**(15), 1452–1453.

Weiss S, Dooley SR, Stewart RW and Nandi AK 1998b Adaptive Equalization in Oversampled Subbands *Proceedings of the Asilomar Conference on Signals, Systems and Computers*, Monterey, CA, USA.

Weiss S, Lampe L and Stewart RW 1998c Efficient Implementations of Complex and Real Valued Filter Banks for Comparative Subband Processing with an Application to Adaptive Filtering. *Proceedings of the International Symposium on Communication Systems and Digital Signal Processing*, pp. 32–35, Sheffield, UK.

Weiss S, Rice GW and Stewart RW 1999a Multichannel Equalization in Subbands *Proceedings of the IEEE Workshop on Applications of Signal Processing to Audio and Acoustics*, pp. 203–206, New Paltz, NY, USA.

Weiss S, Stenger A, Stewart RW and Rabenstein R 2001 Steady-state performance limitations of subband adaptive filters. *IEEE Transactions on Signal Processing* **49**(9), 1982–1991.

Weiss S, Stewart R and Liu W 2002 A broadband adaptive beamformer in subbands with scaled aperture *Proceedings of the Asilomar Conference on Signals, Systems and Computers*, pp. 1298–1302, Monterey, CA, USA.

Weiss S, Stewart RW and Schabert M 1999b Subband based structures and algorithms for adaptive beamforming. Project Report, University of Southampton, Southampton, UK.

Weiss S, Stewart RW, Schabert M, Proudler IK and Hoffman MW 1999c An efficient scheme for broadband adaptive beamforming *Proceedings of the Asilomar Conference on Signals, Systems and Computers*, Vol. I, pp. 496–500, Monterey, CA, USA.

Werner S, Apolinrio Jr JA and De Campos MLR 2003 On the equivalence of RLS implementations of LCMV and GSC processors. *IEEE Signal Processing Letters* **10**(12), 356–359.

Widrow B and Stearns SD 1985 *Adaptive Signal Processing*. Prentice-Hall, Englewood Cliffs, NJ, USA.

Woods JW 1990 *Subband Image Coding*. Kluwer Academic Publishers, Inc., Norwell, MA, USA.

Woods JW and O'Neil SD 1986 Subband coding of images. *IEEE Transactions on Acoustics, Speech and Signal Processing* **34**(5), 1278–1288.

Yamada Y, Ochi H and Kiya H 1994 A Subband Adaptive Filter Allowing Maximally Decimation. *IEEE Journal on Selected Areas in Communications* **12**(9), 1548–1552.

Yan S and Ma YL 2005 Design of FIR beamformer with frequency invariant patterns via jointly optimizing spatial and frequency responses *Proceedings of the IEEE International Conference on Acoustics, Speech and Signal Processing*, pp. 789–792, Philadelphia, PA, USA.

Yan SF 2006 Optimal design of FIR beamformer with frequency invariant patterns. *Applied Acoustics* **67**(6), 511–528.

Yan SF, Ma YL and Hou CH 2007 Optimal array pattern synthesis for broadband arrays. *Journal of the Acoustic Society of America* **162**(11), 2686–2696.

Yang GY, Cho NI and Lee SU 1995 On the performance analysis and applications of the subband adaptive digital filter. *Signal Processing* **41**(3), 295–307.

Yang H and Ingram MA 1997 Design of partially adaptive arrays using the singular-value decompostion. *IEEE Transactions on Antennas and Propagation* **45**(5), 843–850.

Yang K, Ohira T, Zhang Y and Chi CY 2004 Super-exponential blind adaptive beamforming. *IEEE Transactions on Signal Processing* **52**(6), 1549–1563.

Yu L, Lin N, Liu W and Langley R 2007 Bandwidth performance of linearly constrained minimum variance beamformers *Proceedings of the IEEE International Workshop on Antenna Technology*, pp. 327–330, Cambridge, UK.

Yu L, Liu W and Langley R 2008a Performance comparison of two sets of constraints in multi-path environment *Proceedings of the International ITG/IEEE Workshop on Smart Antennas*, pp. 370–374, Darmstadt, Germany.

Yu L, Liu W and Langley R 2008b A robust adaptive beamformer for multi-path signal reception *Proceedings of the International Congress on Image and Signal Processing*, Vol. 5, pp. 34–38, Hainan, China.

Yu L, Liu W and Langley RJ 2009a A novel robust beamformer based on sub-array data subtraction for coherent interference suppression *Proceedings of the IEEE Workshop on Statistical Signal Processing*, pp. 525–528, Cardiff, UK.

Yu L, Liu W and Langley RJ 2009b Robust beamforming with combined worst-case performance optimization and soft constraints in a multipath environment *Proceedings of the European Signal Processing Conference*, pp. 583–586, Glasgow, UK.

Yu L, Liu W and Langley RJ 2010 Robust beamforming methods for multipath signal reception. *Digital Signal Processing*, doi:10.1016/j.dsp.2009.08.010

Zhang M and Er MH 1997 Robust adaptive beamforming for broadband arrays. *Circuits Systems Signal Processing* **16**(2), 207–216.

Zhang ST and Thng ILJ 2002 Robust presteering derivative constraints for broadband antenna arrays. *IEEE Transactions on Signal Processing* **50**(1), 1–10.

Zhang YM, Yang KH and Amin MG 2001 Adaptive array processing for multipath fading mitigation via exploitation of filter banks. *IEEE Transactions on Antennas and Propagation* **49**(4), 505–516.

Zhang YM, Yang KH and Amin MG 2005 Subband array implementations for space-time adaptive processing. *EURASIP Journal on Applied Signal Processing* **2005**(4), 99–111.

Zhao Y, Liu W and Langley RJ 2008 Efficient design of frequency invariant beamformers with sensor delay-lines *Proceedings of the IEEE Workshop on Sensor Array and Multichannel Signal Processing*, pp. 335–339, Darmstadt, Germany.

Zhao Y, Liu W and Langley RJ 2009a An application of the least squares approach to fixed beamformer design with frequency invariant constraints. *IET Signal Processing*, under review.

Zhao Y, Liu W and Langley RJ 2009b Design of frequency invariant beamformers in subbands *Proceedings of the IEEE Workshop on Statistical Signal Processing*, pp. 201–204, Cardiff, UK.

Zhao Y, Liu W and Langley RJ 2009c An eigenfilter approach to the design of frequency invariant beamformers *Proceedings of the International ITG Workshop on Smart Antennas*, pp. 341–346, Berlin, Germany.

Zhao Y, Liu W and Langley RJ 2009d A least squares approach to the design of frequency invariant beamformers *Proceedings of the European Signal Processing Conference*, pp. 844–848, Glasgow, UK.

Zhou YP and Ingram MA 1999 Pattern synthesis for arbitrary arrays using an adaptive array method. *IEEE Transactions on Antennas and Propagation* **47**(5), 862–869.

Index

Wideband Beamforming Wei Liu and Stephan Weiss